U0318438

病虫害

对马铃薯生长机理影响的研究

STUDY ON EFFECT OF PEST AND DISEASE ON POTATO GROWTH MECHANISM

◎ [荷] A.J.Haverkort [英] D. K. L. MacKerron 编著

◎ 何英彬 罗其友 王卓卓 译

中国农业科学技术出版社

图书在版编目（CIP）数据

病虫害对马铃薯生长机理影响的研究 /（荷）哈文考特（A. J. Haverkort），（英）麦克凯文（D. K. L. MacKerron）编著；何英彬，罗其友，王卓卓译. —北京：中国农业科学技术出版社，2016. 12

ISBN 978 -7 -5116 -2800 -8

Ⅰ. ①病…　Ⅱ. ①哈…②麦…③何…④罗…⑤王…　Ⅲ. ①马铃薯 - 病虫害防治 - 研究　Ⅳ. ①S435. 32

中国版本图书馆 CIP 数据核字（2016）第 253572 号

All rights reserved
© 1995 Springer Science + Business Media Dordrecht
Originally published by Kluwer Academic Publishers in 1995
Softcover reprint of the hardcover 1 st edition 1995
No part of the material protected by this copyright notice may be reproduced or utilized in any form or by any means, electronic or mechanical, including photocopying, recording or by any information storage and retrieval system, without written permission from the copyright owners.

著作权合同登记号：图字 01 -2016 -5877

责任编辑　李冠桥
责任校对　杨丁庆

出 版 者　中国农业科学技术出版社
北京市中关村南大街 12 号　邮编：100081
电　　话　(010)82109705(编辑室)　(010)82109704(发行部)
(010)82109703(读者服务部)
传　　真　(010) 82106625
网　　址　http://www. castp. cn
经 销 者　各地新华书店
印 刷 者　北京科信印刷有限公司
开　　本　787mm ×1 092mm　1/16
印　　张　17. 5
字　　数　314 千字
版　　次　2016 年 12 月第 1 版　2016 年 12 月第 1 次印刷
定　　价　120. 00 元

版权所有 · 翻印必究

作者简介

何英彬，中国农业科学院农业资源与农业区划研究所副研究员。2004—2005 年赴意大利海外农业研究所（IAO）学习，获得“3S 技术”与自然资源评价专业硕士学位；2006 年 9 月至 12 月赴日本国际农林水产业研究中心（JIRCAS）作访问学者；2008 年 12 月赴澳大利亚墨尔本作访问学者，从事土地适宜性评价研究；2014 年 5 月正式受聘于澳大利亚昆士兰大学，成为其农业遥感监测与预测领域的客座教授；2015 年 11 月，正式受聘于天津工业大学，成为其作物种植适宜性等研究领域的客座教授。曾多次出访澳大利亚、美国、日本、新加坡、加拿大、马来西亚等国家从事学术交流活动。此外，近年专注于马铃薯领域研究，与国际马铃薯中心（CIP）开展了较为深入的互动与合作。先后主持国家自然科学基金青年基金、APEC 组织项目、中澳政府间合作项目、财政部专项、农业部专项等 10 余项；参加国家自然科学基金重点项目、科技部“973”课题、科技部国际合作重大项目、科技部国家科技支撑计划项目、科技部公益项目及平台项目 10 余项。以第一作者发表论文近 40 篇，SCI/EI 论文 10 篇；发表专著 2 部，以主编身份编著论著 3 部，参与编著论著 6 部；获得软件著作权 1 份；得奖 5 项。

罗其友，中国农业科学院农业资源与农业区划研究所研究员，博士研究生导师，中国农业科学院农业区域发展岗位杰出人才，农业布局与区域发展创新团队首席，国家马铃薯产业技术体系产业经济岗位专家，长期从事农业发展区域问题、马铃薯产业经济研究。

王卓卓，天津工业大学硕士生，宁夏回族自治区固原市人。2016 年 7 月毕业于天津工业大学管理学院土地资源管理专业，2016 年 9 月就读天津工业大学管理学院公共管理系继续攻读土地资源管理方向的硕士学位，师从中国农业科学院农业资源与农业区划研究所的何英彬副研究员，主攻马铃薯种植研究。学习期间两次获得国家励志奖学金，获得全国大学生英语竞赛三等奖，校级三好学生、优秀团干等多个荣誉称号。

《病虫害对马铃薯生长机理影响的研究》译著提纲

本译著重点阐述了马铃薯感染胞囊线虫生理反应机理及马铃薯与胞囊线虫相互影响和作用，介绍了模拟胞囊线虫影响马铃薯生长的模型及管理方法；分析了马铃薯晚疫病的成因，列举了马铃薯晚疫病模拟模型。本译著还介绍了马铃薯黄萎病菌的生命周期和生态学原理，描述了黄萎病菌动态模型建立过程。本著作列举基于作物生长的传染病模型预测种薯单产及病毒感染影响的相关研究及预定模型与虫害管理软件在马铃薯种植中的应用。

前　言

马铃薯是位列水稻、小麦和玉米之后的世界第四大作物，而我国是世界马铃薯第一大生产国，马铃薯因其耐旱耐贫瘠的特点，在我国，尤其是欠发达的内陆省份种植较多；它也是农业扶贫项目中重要的种植作物。2016年，农业部正式发布了《关于推进马铃薯产业开发的指导意见》（后简称意见），将马铃薯作为主粮产品进行产业化开发，马铃薯将成为水稻、小麦、玉米三大主粮品种之后的我国第四大主粮品种。该《意见》提出，到2020年，马铃薯种植面积扩大到1亿亩（1亩≈667m^2，全书同）以上，适宜主食加工的品种种植比例达到30%，主食消费占马铃薯总消费量的30%。马铃薯将对保障我国粮食安全起到举足轻重的作用。

马铃薯的产量波动受多种因素影响，如环境因素、品种因素等，但病虫害是马铃薯产量减少绝对不可忽视的因素；在很多情况下，病虫害的影响往往占据主导地位。随着栽培技术的普及和推广，马铃薯的种植面积和产量都有很大提升。但是马铃薯病虫害也随即蔓延，其病虫害种类多，发生的态势比较复杂，如何有效防控马铃薯病虫害，提高马铃薯的质量和产量是非常重要的课题。

国际上，马铃薯种植培育强国如荷兰、美国等相关研究成果极为丰硕，较有借鉴意义。因此，我们选择了由荷兰瓦赫宁根大学 A. J. Haverkort 教授和英国邓迪大学 D. K. L. MacKerron 教授联合主编的《Potato Ecology and Modelling of Crops under Conditions Limiting Growth》论文集进行编译。该论文集是1994年参加第二次国际马铃薯大会的80余位国际顶级马铃薯研究专家研究成果的集合，较有权威性。基于此，我们将其译成中文，供国内学者及大专院校师生进行相关研究参考，为马铃薯研究添砖加瓦。

本译著重点阐述了马铃薯感染胞囊线虫生理反应机理、马铃薯和胞囊线虫相互作用研究，介绍了模拟胞囊线虫影响马铃薯生长模型和管理模型；分析了马铃薯晚疫病的成因分析，阐述了马铃薯晚疫病模拟模型；本译著还简介了马铃薯黄萎病菌的生命周期和生态学；分析了耦合作物生长及传染病模型预测种薯单产及病毒感染影响研究，预测了预定模型和虫害管理软件在马铃薯种植中的应用。在本著的翻译过程中，分别就相关专业内容请教了中国农业科学院生物研究所的白净博士、张陈博士、刘盈盈博士，植物保护所的安兴奎博士及农业资源与农业区划研究所的张文博博士；在翻译表述中请教了农业部国际交流中心张秀玲、谭茜园及北京语言大学的姚博和刘海艳。本著由天津工业大学王卓卓协助统稿。在此向上述人员一并致谢！

由于译者英文及马铃薯业务知识水平有限，本著可能尚存不完善、不准确的文字表述，因此，我们真诚地希望广大学者、专家及博硕士能与我们多多交流，对本书的缺漏不足之处提出宝贵的建议和意见。愿大家携手为我国马铃薯研究事业更上一层楼尽份微薄之力。

译　者

2016 年 11 月 25 日

目　　录

第一章　马铃薯感染胞囊线虫生理反应机理研究 ……………………… (1)
第一节　简介 ……………………………………………………… (1)
第二节　胞囊线虫对马铃薯生长影响的机理 ………………………… (2)
一、胞囊线虫对马铃薯生长的短期影响研究 ………………………… (2)
二、马铃薯胞囊线虫对马铃薯生长的长期影响研究 ………………… (5)
第三节　胞囊线虫影响马铃薯产量形成的研究 ……………………… (8)
第四节　马铃薯对胞囊线虫耐受抗性的研究 ……………………… (10)
第五节　关于胞囊线虫对马铃薯生长影响的总结和讨论 ………… (12)
参考文献 …………………………………………………………… (14)
第二章　马铃薯和胞囊线虫相互影响研究 ……………………… (18)
第一节　简介 …………………………………………………… (18)
第二节　马铃薯胞囊线虫模型生理效应模拟 ……………………… (20)
第三节　马铃薯胞囊线虫的简易模型 ……………………………… (21)
一、模型的初步实验 ………………………………………………… (23)
二、模型耐受性评估 ………………………………………………… (24)
第四节　研究内容的总结与讨论 …………………………………… (25)
参考文献 …………………………………………………………… (26)
第三章　模拟胞囊线虫影响马铃薯生长模型 ……………………… (30)
第一节　模型简介 ………………………………………………… (30)
第二节　胞囊线虫导致马铃薯减产原因分析 ……………………… (32)
一、线虫导致马铃薯产量减少的原因 ……………………………… (32)
二、线虫对马铃薯生长的影响 ……………………………………… (33)
第三节　基于胞囊线虫密度和相对马铃薯质量统计关系的
简单模型 ………………………………………………………… (35)
一、模型的含义 ……………………………………………………… (36)
二、胞囊线虫的攻击对马铃薯块茎生长的影响 ……………………… (38)
三、多重线虫导致抑制生长的机制 ………………………………… (40)

四、参数估计 t 和 m 的分析 …………………………………………（40）
第四节　相关总结和讨论 ……………………………………………（41）
参考文献 ………………………………………………………………（42）
第四章　马铃薯胞囊线虫管理模型（线虫品种） ………………（46）
第一节　简介 …………………………………………………………（46）
第二节　马铃薯胞囊线虫种群动态研究 ……………………………（47）
第三节　不易感染病毒马铃薯与胞囊线虫间的相互作用机制 ……（50）
第四节　马铃薯胞囊线虫密度和产量的数量关系研究 ……………（52）
第五节　宿主缺失后的种群数量减少原因探索 ……………………（54）
第六节　马铃薯胞囊线虫管理模拟模型 ……………………………（55）
一、模型简介 …………………………………………………………（55）
二、影响因子敏感性分析 ……………………………………………（58）
三、相对易感马铃薯轮作计划 ………………………………………（59）
第七节　相关的结论与讨论 …………………………………………（61）
参考文献 ………………………………………………………………（62）
第五章　马铃薯晚疫病的成因分析 ………………………………（66）
第一节　简介 …………………………………………………………（66）
第二节　马铃薯晚疫病菌的生命周期 ………………………………（67）
第三节　马铃薯晚疫病发展进展介绍 ………………………………（68）
第四节　马铃薯晚疫病对马铃薯生长和单产的影响 ………………（71）
第五节　影响马铃薯晚疫病菌无性繁殖周期的因素分析 …………（71）
一、马铃薯晚疫病自然环境因素 ……………………………………（71）
二、马铃薯晚疫病生理影响因素 ……………………………………（76）
三、表面微生物菌群 …………………………………………………（78）
四、生物因素——马铃薯晚疫病基因型 ……………………………（78）
五、杀菌剂 ……………………………………………………………（80）
第六节　影响马铃薯晚疫病菌有性繁殖周期的因素分析 …………（80）
参考文献 ………………………………………………………………（81）
第六章　马铃薯晚疫病模拟模型 …………………………………（91）
第一节　简介 …………………………………………………………（91）
第二节　SEIR 模型的发展历程 ……………………………………（92）
第三节　SEIR 模型植物感染机理 …………………………………（93）
第四节　潜在及感染期可替换分布 …………………………………（94）
第五节　马铃薯器官损伤生长模拟 …………………………………（95）

一、病变生长建模 …………………………………………… (95)
二、空间异质性 ……………………………………………… (96)
三、寄生主体生长 …………………………………………… (96)
第六节　马铃薯晚疫病模拟模型在杀菌剂使用中的应用 ………… (97)
第七节　马铃薯晚疫病模拟模型在抗性育种过程中的应用 ……… (98)
第八节　马铃薯晚疫病模型初始化和本地化中存在的问题 ……… (98)
第九节　有关问题的结论与讨论……………………………… (100)
参考文献……………………………………………………… (101)
第七章　马铃薯黄萎病菌的生命周期和生态学……………… (107)
第一节　马铃薯黄萎病菌简介………………………………… (107)
第二节　马铃薯黄萎病菌土壤种群动态研究………………… (107)
一、马铃薯黄萎病微菌核存活研究………………………… (107)
二、马铃薯黄萎病微菌核萌发研究………………………… (109)
三、马铃薯黄萎病菌在植物根部定殖的研究……………… (110)
第三节　宿主—病原体相互作用机理………………………… (110)
一、马铃薯根部的系统性感染研究………………………… (110)
二、马铃薯植株定殖研究…………………………………… (112)
三、马铃薯黄萎病菌的繁殖研究…………………………… (113)
第四节　马铃薯黄萎病菌宿主特异性研究…………………… (114)
参考文献……………………………………………………… (115)
第八章　综合作物毛长模型与流行病模型预测种薯单产和病毒感染…………………………………………………… (122)
第一节　简介………………………………………………… (122)
第二节　马铃薯流行病毒模型“EPOVIR”概述 ……………… (123)
第三节　马铃薯作物生长模型的应用………………………… (125)
一、马铃薯病毒感染对生长状况的影响…………………… (125)
二、马铃薯载体相互作用机制……………………………… (125)
第四节　有关问题的总结与讨论……………………………… (128)
参考文献……………………………………………………… (130)
第九章　预定模型和虫害管理软件在马铃薯种植中的应用………… (134)
第一节　简介………………………………………………… (135)
第二节　计算机应用的初步开发……………………………… (136)
第三节　软件使用带来的价值………………………………… (136)
第四节　规范虫害管理软件介绍……………………………… (139)

一、疾病管理…………………………………………………………………（139）
二、种薯块处理……………………………………………………………（140）
三、昆虫管理…………………………………………………………………（141）
四、杂草管理…………………………………………………………………（141）
五、灌溉管理…………………………………………………………………（141）
六、施氮管理…………………………………………………………………（142）
七、昆虫和作物相互作用机制……………………………………………（142）
第五节 计算机模型的应用前景……………………………………………（143）
一、增强计算机应用………………………………………………………（143）
二、计算机模型的农业应用领域…………………………………………（144）
第六节 有关管理软件的总结………………………………………………（145）
参考文献………………………………………………………………………（145）
附录 部分原文摘录……………………………………………………（147）
Crop physiological responses to infection by potato cyst nematode (*Globodera* spp.) ……………………………………………………………（1）
Modelling the interaction between potato crops and cyst nematodes ……………………………………………………………………（17）
A growth model for plants attacked by nematodes ……………………（26）
An advisory system for the management of potato cyst nematodes (*Globodera* spp) ……………………………………………………………（41）
Factors involved in the development of potato late blight disease (*Phytophthora infestans*) ……………………………………………………（57）
Simulation models of potato late blight ……………………………………（74）
Life cycle and ecology of Verticillium dahliae in potato ………………（85）
Use of a crop - growth model coupled to an epidemic model to forecast yield and virus infection in seed potatoes ………………………………（95）
Prescriptive crop and pest management software for farming systems involving potatoes …………………………………………………………（104）

图目录

图 1 – 1　线虫密度与光合利用率和水分利用效率的关系 ……………（3）
图 1 – 2　受干旱和胞囊线虫影响的马铃薯作物的红外线反射比例 …（6）
图 1 – 3　根数量与种植天数关系 ……………………………………（7）
图 1 – 4　瓦赫宁根根系试验 0 ~ 30cm 和 30 ~ 100cm 中的土壤氮素总量 ……………………………………………………………（7）
图 1 – 5　轻微和受感染较严重的 Desirtee 和 Elles 品种地面覆盖比例比较 …………………………………………………………（10）
图 2 – 1　早熟和晚熟品种在不同土壤线虫密度下的最终块茎产量模拟 ……………………………………………………………（20）
图 2 – 2　受不同数量损伤机制影响的马铃薯块茎干物质模拟 ………（21）
图 2 – 3　受感染土壤和控制处理下 Darwina 品种生长变化模拟 ……（24）
图 2 – 4　不同处理方式下马铃薯块茎重量随时间变化曲线 …………（25）
图 3 – 1　茎线虫的初始种群密度 P 和未受侵袭的洋葱马铃薯 y 间的关系 $y = Z^P$ ……………………………………………………（34）
图 3 – 2　无线虫和线虫密度为 P 的马铃薯的生长曲线………………（34）
图 3 – 3　表达总质量 Y 和线虫密度 P 之间的关系、直角横截面的三维模型曲面 …………………………………………………（37）
图 3 – 4　13 个试验中播种或种植时的线虫密度 P/T 和种植后一定时间 y' 的关系 …………………………………………………（38）
图 3 – 5　有无线虫影响马铃薯的总茎和块茎质量的增长曲线 ………（39）
图 4 – 1　根据公式（1）得出初始虫卵密度与最终虫卵密度之间的关系 ……………………………………………………………（48）
图 4 – 2　Pi 与最大密度值数量关系……………………………………（50）
图 4 – 3　Pf 与 Pi 数量关系 …………………………………………（51）
图 4 – 4　初始种群密度（P）与相对产量 y 之间的关系 …………（54）
图 4 – 5　马铃薯线虫密度与减产量之间的关系 ………………………（57）
图 4 – 6　马铃薯相对易感性与平均减产量之间的关系 ………………（59）

图5－1 致病疫霉的生命周期 ……………………………………………（68）
图5－2 马铃薯晚疫病的病害周期 …………………………………………（69）
图5－3 恒定条件下马铃薯叶上单病斑面积的增加 ……………………（69）
图5－4 整株马铃薯总叶片病斑面积的假设增加曲线 …………………（70）
图5－5 有利和不利条件下马铃薯晚疫病的进展曲线 …………………（70）
图5－6 （a）马铃薯晚疫病发病时间和（b）发病速率对病害发展影响的曲线………………………………………………………（70）
图6－1 SEIR 模型的结构 …………………………………………………（93）
图6－2 马铃薯晚疫病各种宿主品种的模拟发生曲线 …………………（99）
图6－3 马铃薯晚疫病的5个不同模型的模拟发生曲线 …………（100）
图7－1 大丽轮枝菌生命周期示意 ………………………………………（108）
图8－1 EPOVIR 模型结构（马铃薯病毒流行病学） ……………（124）
图8－2 植物对病毒的易感性与龄期的相关性方程式 …………（126）
图8－3 蚜虫相对栖息率 …………………………………………………（127）
图8－4 概率与生长日期的关系 …………………………………………（128）
图8－5 种植密度对病毒传播的影响 ……………………………………（129）
图9－1 1983年以来马铃薯产业中“马铃薯病害防治/马铃薯作物管理”软件使用情况 ……………………………………（137）
图9－2 早疫病防控采用300个生理天数的施用阈值可延迟杀真菌剂的施用时间 …………………………………………………（138）
图9－3 因应用“马铃薯病害防治/马铃薯作物管理”软件减少杀真菌剂用量而减少的费用 ………………………………（138）
图9－4 1994年使用“马铃薯作物管理”软件和相关技术在 28 300hm^2 土地上减少投入品而节省的费用预估…………（139）
图9－5 增强抗病力和施用杀真菌剂改变了早疫病的发展情况 …（140）

表目录

表 1－1　马铃薯胞囊线虫侵染和干旱对作物水关系的影响 …………（4）
表 1－2　在干湿条件下胞囊线虫对马铃薯器官的影响 ………………（5）
表 1－3　不同管理条件下马铃薯肥料使用情况 ………………………（8）
表 1－4　四种环控情况下胞囊线虫对马铃薯单产影响 ………………（9）
表 1－5　土壤 pH、块茎产量、干物质和胞囊线虫初始密度间的关系 ………………………………………………………………（11）
表 1－6　不同品种不同处理情况下马铃薯生长状况 …………………（12）
表 2－1　受胞囊线虫影响的马铃薯生长简单模拟模型的基本公式 ………………………………………………………………（22）
表 4－1　不同品种相对感染度 …………………………………………（53）
表 4－2　不同感染度与平均减产量之间的关系 ………………………（57）
表 4－3　不同马铃薯品种对致病型 Pa 3 两个种群的相对易感性 …（60）
表 6－1　致病疫霉主要模拟模型 ………………………………………（92）
表 6－2　植物传染病抗性五个组成部分的定义 ………………………（94）
表 9－1　使用“马铃薯作物管理”软件和相关技术生产晚熟 Russet Burbank 马铃薯可减少的投入品分类 ………………（138）
表 9－2　以马铃薯为主要作物的 2～3 年轮作的六种轮作安排 ……（144）

第一章 马铃薯感染胞囊线虫生理反应机理研究

本章主要介绍了胞囊线虫感染对马铃薯生长的短期和长期影响，并讨论了耐受型马铃薯的育种和栽培方法。

文中阐述了新鲜马铃薯块茎产量由作物绿叶截获的光合有效辐射总量及其转换为营养物质的转化效率、分配至块茎的营养物质比例及块茎干物质含量四方面决定。生长季节开始和结束时光截获量减少能较好地解释马铃薯胞囊线虫感染所造成的产量损失。冠层形成时期，冠层郁闭度推迟是由于光合速率降低（由于养分吸收率降低、激素信号和作物水分关系混乱造成）、分配至根部的干物质增加、形成的茎数量减少和叶片变小变厚造成。生长季节后期，同化率降低导致新叶片形成数量减少，根系深度变浅致使水分和养分吸收减少，而根部区域水分和养分减少会使叶片脱落，感染胞囊线虫的马铃薯提前衰老。此外，因为马铃薯胞囊线虫感染导致块茎数量减少，块茎干物质含量降低，所以新鲜块茎产量减少。

第一节 简介

作物特性由以下几个方面决定：产量形成因素（如温度和太阳辐射）、产量限制因素（如缺乏水分和养分）和减产因素（如病虫害）。限制和减少产量的因素可能是生物或非生物因素，马铃薯生长过程中这些因素可能对其有影响，例如不利的土壤条件（冷或潮湿的土壤）、土壤中传播的病原体（如立枯丝核菌和线虫）或种子传播的病毒都可能会影响马铃薯发芽。发芽至林冠发育期间，相同的减产因素会影响马铃薯生长。冠层逐渐会受到空气传播的疾病影响（如由致病疫霉、蚜传病毒引起的晚疫病）。马铃薯生长周期后期，水和矿物质消耗及由黄萎病菌引起的马铃薯萎蔫对产量影响日渐突出。作物管理和疾病控制的目标是减少不利条件对生长的限制。

马铃薯对线虫攻击的反应可被划分为短期和长期响应两类。短期响应包括对植物和细胞水分、气孔导度、光合作用、呼吸作用、营养吸收和化学成分的影响，监测方法包括压力室、测孔术、红外气体分析仪和同位素标记等技术；长期影响往往表现为生长发育迟缓、提前衰老、枯萎、发黄和叶片脱落现象等。可通过分析作物生长、测量截获辐射和辐射利用效率、对干物质进行分区和计算比叶面积来考量这些影响，其他长期影响可通过特殊测量如作物反射率、蒸腾效率、稳定的同位素分馏和作物矿物质含量进行检测。

本章的研究目的是回顾马铃薯胞囊线虫对作物生长、发育和生产力的短期和长期影响，将讨论马铃薯胞囊线虫损伤机制与作物生长和发育模拟、育种策略和栽培方法的相关性。本章主要研究线虫感染与非生物因子，特别是水分、养分和土壤 pH 之间的相互作用。

第二节　胞囊线虫对马铃薯生长影响的机理

一、胞囊线虫对马铃薯生长的短期影响研究

已有研究结果清晰表明（Fatemy 等，1985；Haverkort 等，1991a）胞囊线虫影响马铃薯植株水分关系。植物细胞中的水分条件可以用水势（ψ）表示，水势由渗透组分（ψ_o）和压力组分（ψ_{pi}）组成，$\psi = \psi_o + \psi_{pi}$，单位为 MPa。由于蒸腾作用丧失的水分量多于传输到根部的水分量，作物和细胞含水量减少（由于蒸发强烈、土壤干旱和根系损伤），ψ 和 ψ_{pi}、细胞或组织的相对含水量都会降低（Turner，1988）。使用压力室或湿度计（Brown 和 Van Haveren，1972）可以观察叶片水势，借助渗压计可以观察渗透趋势。Vos 和 Oyarzun（1987）观察生长在潮湿土壤中的马铃薯发现，ψ 和 ψ_o 的典型值分别为 -0.5MPa 和 -0.8MPa。干旱与根部感染胞囊线虫使叶水势从 -0.6MPa 降到 -1.1MPa（Haverkort 等，1991a）。Fatemy（1985）发现线虫感染能增加气孔阻力，降低光合速率和减少植物用水量。植物体外施用脱落酸可以短暂增强这些效果（Pentland Dell 品种 2d，耐受性 Cara 品种 6d），显然当马铃薯生长适应了升高的脱落酸水平，这种作用就会消失。水分胁迫也可以增加脱落酸水平，因此干旱对气孔调节的初期影响是通过激素水平影响发挥作用。

Schans 和 Arntzen（1991）试验证明，感染胞囊线虫降低了马铃薯叶片的光合速率。这种影响会随着线虫密度的增加变得更加严重，但在每单

位线虫密度，但在某一线虫密度下，马铃薯生长时间越长，影响越小（图1－1）。作者认为光合作用逐渐增加来自于两方面因素。

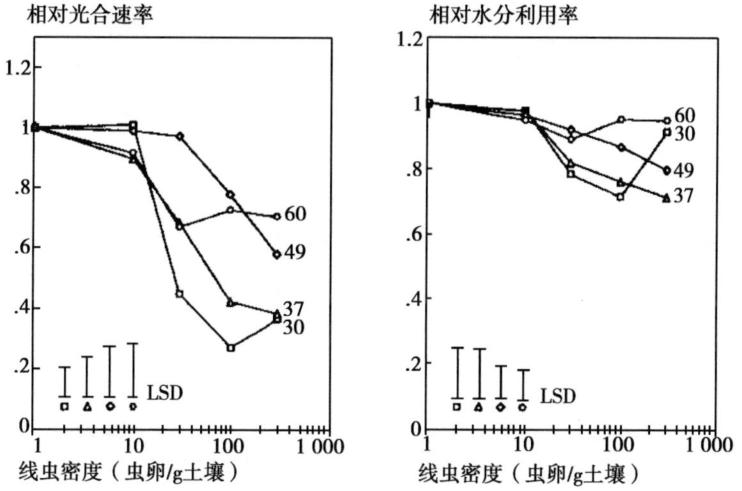

图1－1　线虫密度与光合利用率和水分利用效率的关系

（引自 Schans，Arntzen，1991）

注：不同图形表示种植不同天数后的测量值，分别为30d（□）、37d（△）、49d（◇）和60d（○）

首先，随着线虫的消耗，攻击减弱，感染的根尖减少，使激素信号减弱；其次，随着叶面积增长，叶片中信号相对减弱，从而导致气孔响应能力降低。作者没有记录ABA或其他激素的水平。马铃薯胞囊线虫感染同时对光合作用和水分利用效率的影响图（图1－1）显示并非只影响了气孔进程（因为接下来水分利用效率提高），同时也影响了光化学和生物化学进程。营养吸收减少作为一个主要的组成部分将在稍后讨论。

Schans 和 Arntzen（1991）认为马铃薯胞囊线虫引起单个叶片水分利用效率下降，赞同1982年 Evans 的试验结果，即整体植株水分利用效率从栽培到第32天呈递减趋势，32d后开始上升。但到目前为止报道的气体交换和水分利用效率关系的文献仍未能明确证明两种机制（信号和根损伤导致水分和营养吸收受阻）。试验方法和获得的结果仍存在矛盾，其中的过程需进一步阐明。

线虫侵染，如同水分和热胁迫，通过气孔关闭（瞬时效应）间接作用，或通过降低叶片光合能力（几天后）直接降低光合速率。气孔关闭的副作用同化/蒸腾率升高，像植物体内稳定同位素^{13}C比例提高一样（Farquhar

等，1982）。线虫对光合作用的主要影响以及类囊体膜光反应与卡尔文循环的生化反应是否受影响最严重目前仍不能达成共识。可以通过体内荧光信号检测光利用效率和内囊体内电子流速率。结合气体交换检测方法，分析荧光信号可以分析气相中流动阻力以及内部光合作用相关过程中的速率限制，但是这些方法在马铃薯胞囊线虫对马铃薯影响的研究中还未应用到。

表 1 – 1 显示的是马铃薯胞囊线虫（G. pallida）侵染以及干旱对马铃薯作物水关系的影响。如表 1 – 1 显示，压力室测量结果表明：线虫和干旱均会导致叶片水势降低，提高植株的干物质含量；但不同的是，线虫对气孔扩散阻力和蒸腾几乎没有影响，干旱会大幅提高气孔阻力，从而大幅降低蒸腾速率（Haverkort 等，1991a）；干旱增加了水分利用效率，但是如图 1 – 1 所示，马铃薯胞囊线虫却降低了水分利用效率，与干旱相比，线虫主要影响光合作用而不是蒸腾作用。

表 1 – 1　马铃薯胞囊线虫侵染和干旱对作物水关系的影响

观察报告	处理		
	对照	感染	干旱
叶水势	6.9a	10.8b	11.8b
扩散阻力（s/cm）	1.36a	1.51a	5.11b
蒸腾作用（$\mu g H_2O\ cm^{-2} s^{-1}$）	11.0a	10.21a	4.09b
水分利用效率（g/kg）	7.34a	6.43b	9.19c
叶片干物质浓度（%）	8.5a	10.5b	12.3c
茎干物质浓度（%）	6.5a	8.0b	8.0b
块茎干物质浓度（%）	15.0a	15.2a	21.7b

注：胞囊线虫侵染试验每克土壤含 15 个活的虫卵，干旱处理的马铃薯自出苗后不再浇水，试验为盆栽实验，栽培后 42d 收集数据，一列数据中不同字母表示差异在 1% 置信区间内（引自 Haverkort 等，1991a）

在马铃薯生长阶段后期，感染植株水分利用效率大幅增加，甚至比健康植株更高。Evans 等人（1975）也获得了相似的实验结果，生长后期植株吸收水分器官受损，水分吸收效率减弱，易感品种会受到水分胁迫并过早衰老。Haverkort 等（1991a）得出结论称：至少存在两种产量减少的机制，并且在不同生长时期其相对重要性有所变化；首先，是明显的同化率降低，这部分是由胞囊线虫初始侵染导致水平衡变化；其次，是由水分和/或营养吸收降低导致的干物质积累降低。干旱和胞囊线虫的影响经常叠加（由大田实验观察数据得出，稍后将具体讨论）。但感染植株比健康植株利用的水分少，因而受到的水分胁迫也相对较小，只有这种情况下其影响不叠加。

二、马铃薯胞囊线虫对马铃薯生长的长期影响研究

1. 胞囊线虫对马铃薯形态的影响

观察由环境因素造成的作物发育差异的便捷方法是对冠层覆盖地面比例进行比较评价，因为其与光截获的比例、叶面积指数、由冠层反射的红外线辐射比例密切相关，即叶片反射率比土壤反射率更高（Haverkort 等，1991b）。受干旱（未灌溉）、胞囊线虫（未受烟熏）和夜间霜（种植后 7 周）影响的马铃薯作物红外线反射的比例如图 1－2 所示。与干旱比，受线虫影响地面覆盖和红外线反射率减少幅度更为剧烈。夜间霜并未减少绿叶覆盖的地面比例，但却能降低冠层反射率。

线虫入侵和涝灾提前发生对雨遮容器中马铃薯植株某些形态特征的影响如表 1－2 所示。干旱期结束时，受感染马铃薯的叶面积还不到环控植株 3 181cm^2 的一半，而非植株 8 253cm^2 的一半，因为叶片变厚（SLA 从 301cm^2/g 减少到 263cm^2/g），叶面积减少至每片 72cm^2 而非 150cm^2。干旱对这些形态特征的影响相同。与环控马铃薯相比，受一种或两种压力因素影响的马铃薯寿命更短，块茎产量更低，根比例更低。再次给干旱环境影响马铃薯浇水 4 周后（尽管马铃薯之前曾遭受干旱，叶片肥厚且叶面积比率较低），干旱和非干旱马铃薯叶面积相同。未受影响的马铃薯在种植 70d 后开始衰老，而早期遭受干旱影响的马铃薯继续生长，可能是因为其未耗尽土壤养分。

表 1－2　在干湿条件下胞囊线虫对马铃薯器官的影响

植株特性	马铃薯胞囊线虫（每克土壤的卵数）			
	0 潮湿	18.5 潮湿	0 干旱	18.5 干旱
叶片，干物质（g/株）	27.4	12.1	20.2	10.4
叶面积（cm^2/株）	8 253	3 181	4 661	2 258
叶片大小（cm^2/片）	150	72	108	58
SLA（cm^2/g 叶干重）	301	263	228	218
LAR（cm^2/g 植株干重）	93.1	89.0	74.2	73.1
茎的数量	5.4	4.5	5.2	4.5
茎长（cm）	52	30	34	24
特殊茎重（mg/cm）	62	46	62	52
块茎干重（g/株）	42	15	27	13
每株块茎数量	30	16	28	16
茎根比	17.3	8.6	15.1	7.9

注：种植 43d 后，马铃薯胞囊线虫和干旱对马铃薯 cv. Mentor 作物特点的主要影响。$P < 0.05$，控制和线虫或干旱处理间的作物特征差异较大（Haverkort，1991a）

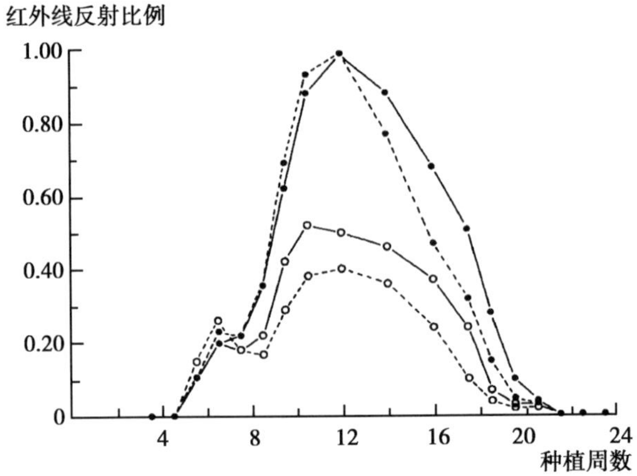

图1-2 受干旱和胞囊线虫影响的马铃薯作物的红外线反射比例

注：——灌溉；---未灌溉；· 线虫受控制；o 线虫未受控制（Haverkort 和 Schapendonk，1994）

表1-2结果表明，干旱影响和受胞囊线虫感染的马铃薯水分情况相似。这两个压力因素对植物形态的影响也存在相似之处。然而，与干旱相比，胞囊线虫（从30块和28块减少至16块）能使块茎数量大量减少。胞囊线虫感染使茎数量减少只能部分解释其不同（每株从5.4减少至4.5）。试验中的根冠比略受干旱影响，而由于胞囊线虫感染明显下降，这种情况表明干物质分配的重大转变发生在根部。

根据瓦赫宁根马铃薯根系实验室（Smit 等人，1994）经常观察受各种土壤因素如线虫和干旱影响的马铃薯根部生长情况。研究使用瓦赫宁根马铃薯根系实验室设备（Haverkort 等人，1994）比较未受感染和受 Globodera pallida 感染的马铃薯（每克土壤40个线虫）生长，证明在土壤表层30cm，根部生长较快，30cm 土壤以下，根部生长较慢。马铃薯胞囊线虫导致表层土壤形成的根部减少，但在种植120d 后，根系形成仍在继续，而在环控的条件下，种植50d 后，表层土壤根部形成停止，开始开发底土。

2. 胞囊线虫对马铃薯养分吸收的影响

每两周测定表层土壤0~30cm 和底土30~100cm 土壤深度（与图1-3同试验）可溶性氮的总量（硝酸盐和氨）。无胞囊线虫情况下，至马铃薯衰老，深度至少1m 的土壤剖面矿质氮已经耗尽，然而土壤线虫密度较高情况下，底土30~100cm 氮浓度没有变化。由胞囊线虫侵害造成的土壤中根部（图1-3）和氮（图1-4）的空间异质性（因为底土中几乎没有

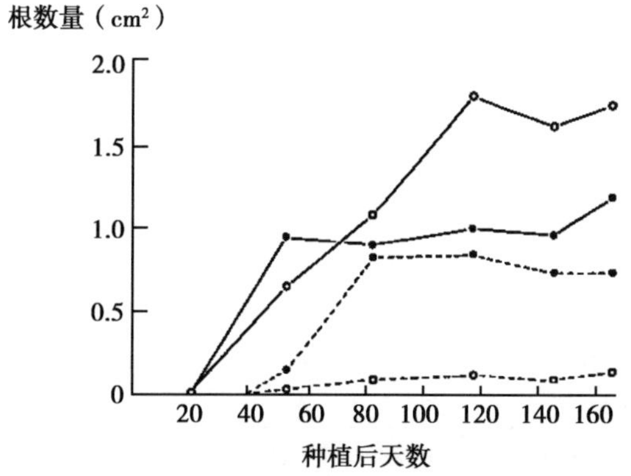

图 1－3　根数量与种植天数关系

注：土壤深度在 0～30cm（－）和 30～100cm（---）、每克土壤存活的 Globodera pallida 幼虫数量在 0d（·）和 40d（o）时观察到的每平方厘米表面根系的平均根数量（Haverkort，1994）

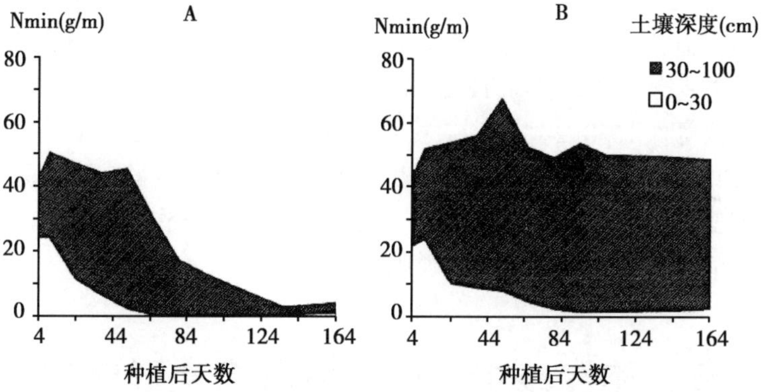

图 1－4　瓦赫宁根根系试验 0～30cm 和 30～100cm 中的土壤氮素总量

（A）没有马铃薯胞囊线虫；（B）每克土壤最初 40 只幼虫

根部存在，因此作物不能吸收土壤层的氮素）显然是一种重要的损伤机制，尤其是在生长季节的第二阶段，可能会导致作物提前衰老，这是在受虫害感染地块中观察得到的结论。

在使用灌溉、烟熏和抗病品种的一项长期试验中，作者分析了马铃薯胞囊线虫对马铃薯生长的长期影响（Trudgill，1975）。基于观察，作者假设：由于胞囊线虫造成的土壤表层马铃薯生长速率的降低是根部损伤的一个重要原因，根部损伤使根部营养吸收的比例降低，导致最低有效性养分的长期不足。图 1－3 和图 1－4 所示结果与这一假设相符。

表 1－3　不同管理条件下马铃薯肥料使用情况

观察	杀线虫剂肥料增加标准	N	P	K	无杀线虫剂肥料增加标准	N	P	K
茎（g）	298	448 *	306	251	98	265 *	208 *	160 *
根（g）	60	84 *	77	34	11	64 *	56 *	22
块茎（g）	363	372	363	357	100	158	168	142
N（%）	2.5	3.4 *	2.5	2.4	3.3	3.5	3.3	3.3
P（%）	0.16	0.22 *	0.26 *	0.15	0.18	0.20	0.20	0.20
K（%）	3.5	3.3	3.4	5.2 *	2.6	3.1	3.1	4.58

注：杀线虫剂和追肥的影响（盆栽试验种植后的 14 周，每株作物的重量（用 g 表示）、茎部干物质的营养含量、追肥标准为 2 倍氮（N）标准剂量或 2 倍钾（K）标准剂量或 3 倍磷（P）标准剂量。同一线虫处理中，每株作物的重要差异显著（$P < 0.01$）

在两个严重感染胞囊线虫的地块及一个盆栽试验中，通过观察施肥频率及非熏杀线虫剂的相互作用，验证了这一假设。在一项实地试验中，耐受性较强的 Cara 品种比耐受性较差的 Pentland Dell 品种对杀线虫剂或肥料响应更弱。施肥率从每公顷 0.5t 增加到 1.5t，可使未使杀虫剂的 Pentland Dell 品种块茎产量从每株 0.12kg 增加到 0.82kg，受杀线虫剂保护的每株作物产量从每株 1.30kg 增加到 1.60kg。在盆栽试验中为独立区分 P、K 和 N，Trudgill（1980）实验证明：线虫破坏与 N 和 P 的数量有很强的相互作用。钾肥虽量小但却非常显著（表 1－3），盆栽试验中其影响比田间试验影响大，可能是该地加重黏土对磷肥的吸收。

第三节　胞囊线虫影响马铃薯产量形成的研究

前文已提到，作物产量（Y）由其截获的进行光合作用的有效太阳辐射（R）、转换成干物质的效率（E）、干物质总量与收获的比率（H）及其干物质含量（D）决定。公式为：$Y = R\ E\ H\ /\ D$。

这一简单作物生长模型的参数数值由一系列基础流程决定。R 依赖于叶膨胀率和冠层的消光特性，E 取决于光合作用和呼吸率，H 取决于与马铃薯生长相关的干物质分配过程。在荷兰东北部的一系列田间环控试验中有 4 个品种，环控分别为有灌溉和无灌溉，熏蒸和未熏蒸（平均每克土壤 5 ~44 只线虫），定期收获以确定干旱和胞囊线虫对马铃薯产量的影响。马铃薯胞囊线虫和干旱加重对辐射截获和辐射利用效率会产生负面影响。

表 1-4　四种环控情况下胞囊线虫对马铃薯单产影响

年份	品种	灌溉	无灌溉	产量 =	*R* ×	*E* ×	*H*/	*D*
1988—1990	Darwina	+	-	48	61	90	94	105
	Desiree	+	-	52	71	86	93	102
	Elles	+	-	73	86	95	102	107
	Mentor	+	-	49	57	92	100	101
1989—1990	Darwina	-	+	55	62	99	94	105
	Desiree	-	+	77	88	99	94	105
	Elles	-	+	80	93	90	95	101
	Mentor	-	+	73	87	97	97	111

注：非灌溉和非熏蒸地块产量构成因素的相对值（1989 年和 1990 年的平均数据，灌溉和熏蒸地块的数值 = 100）（Haverkort 等人，1992）

表 1-4 显示无灌溉和无熏蒸地块相对于有灌溉和有熏蒸地块，4 个品种（包括 Mentor）的辐射截获、辐射利用效率、收获指数和干物质产量的数值。

显然，胞囊线虫和干旱对产量因素的影响效果相似：由于作物拦截的太阳辐射较少导致产量减少，低辐射利用效率、低收获指数和低干物质产量也是使新鲜块茎产量减少的重要原因。对线虫破坏抵抗力最差的品种（Darwina）对干旱抵抗力也最差，Elies 品种对这两个因素抵抗力最强。Trudgill 等人（1990）发现，有抵抗力的 Cara 品种辐射利用效率数值为 1.05g/MJ，没有抵抗力的 Pentland Dell 的品种数值为 1.25g/MJ，这些品种覆盖地面时间较长，因此能拦截更多的太阳辐射。较低的辐射利用效率与较高的截获辐射值有关，如表 1-4 所示，无熏蒸耐旱 Elies 品种 *E* 值较低，为 5%，*R* 值较低，为 14%，熏蒸处理中 Mentor 品种（对马铃薯胞囊线虫的耐受力较弱）在未熏蒸时 *R* 值较低，为 43%。随着干旱发生，这一现象会更加显著：无熏蒸条件下，Darwina 品种只损失 1% 的效率，在无灌溉时，却损失 38% 的截获辐射。难以保持冠层生长状态良好的品种受到压力时也难以保持其正常生长，光合速率下降，而呼吸作用并没有降低。除了应激因素对作物形态的长期显著影响及对截获辐射和辐射利用效率可测的影响外，应激因素同样会影响蒸腾效率、^{13}C 的标识、矿物质吸收。Haverkort 等人（1991a；表 1-1）报告，受应激因素影响的作物蒸腾效率为 7.34g/kg，受胞囊线虫影响的马铃薯的蒸腾效率为 6.43 g/ kg，受干旱影响的马铃薯的蒸腾效率为 9.19g/kg 。

Mulder（1994）证明：马铃薯生长过程中，不同品种对胞囊线虫感染的反应也不同。通过比较健康作物和受感染马铃薯的地面覆盖值（图 1-

5）可以发现，与马铃薯生长后期阶段相比，Desiree 品种生长早期地面覆盖发展受影响较小，而 Elies 品种（耐受品种），观察到的现象则正好相反。

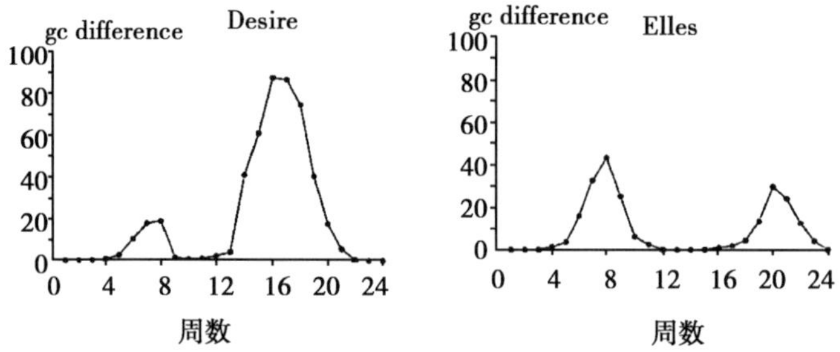

图 1－5 轻微和受感染较严重的 Desirtee 和 Elles 品种地面覆盖比例比较

马铃薯胞囊线虫和干旱都能减少^{13}C 的标识，这表明线虫能够破坏马铃薯水分关系（Haverkort 和 Valkenburg，1992）。干旱和胞囊线虫对马铃薯的影响不同，感染线虫后水分利用效率降低，遭受干旱后水分利用率提高。马铃薯胞囊线虫感染对植物生长过程如光合作用和根部的干物质分配的负面影响比对水分利用率影响显著。干旱和胞囊线虫对马铃薯影响之间的另一个差异性是干旱降低 Ca 的吸收，而胞囊线虫感染却恰恰相反（Fatemy 和 Evans，1986）。根尖吸收钙，Ca 通过蒸腾作用留在植物体内运输；当多核体或合胞体突破内皮层，使得质外体产生移动，Ca 吸收和运输的过程会被加速。如图 1－3 所示，胞囊线虫导致马铃薯根部分支减少，表土层根尖增多，因此有助于增加 Ca 的吸收。

第四节 马铃薯对胞囊线虫耐受抗性的研究

马铃薯生长后期与前期线虫种群密度的比率可以表明马铃薯对胞囊线虫的耐受性。耐受性指标对降低线虫密度非常重要，但并不一定能反映预测的马铃薯损害程度。专家发现线虫密度较高时，不同品种受影响程度不同。这是一个相对量度。Evans 和 Haydock（1990）回顾了由 Brown、Oostenbrink 和 Seinhorst 等分析的初始线虫密度和相应的块茎产量损失之间的线性关系。尽管当预测马铃薯产量损失时这些数量关系才有效，但其并未显示相关发生的损伤机制。Evans 和 Haydock（1990）在其评论中将品种间不同的耐受性归因于线虫孵化、线虫入侵、根的定殖、根系活力、根系统的形式、根部对线虫攻击根部的局部反应、植物对线虫攻击的生理反应和其他生物的相互作用等因素。他们并未发现任何证据能证明作物对 lobodera rostochiensis 胞囊线虫的抵抗力要比 Globodera pallida 胞囊线虫更强。已有研究证明，生物土壤传播疾病如水稻纹枯病菌、黄萎病菌及黄萎病能叠加马铃薯胞囊线虫造成的危害。非生物的土壤相关因素与生物因素同等重要，有时比生物因素更加重要。Mulder（1994）将马铃薯胞囊线虫造成的损害与土壤类型、土壤肥力、土壤水分供应和土壤 pH 造成的损害相比较，发现较高的土壤 pH 很重要。Haverkort（1993）等人发现，在荷兰东北部的沙质土壤，产量损失和 pH 之间有很强的相互作用。尽管每年都会浸灰，pH 值在 4.5 ~6.5 的泥炭土地块仍能获得 22% 的有机质。这些地块的马铃薯产量（表 1 –5）从 pH 值为 4.5 时的每公顷 45t 下降到 pH 值为 6.5 时的每公顷 33t，线虫密度从每克土壤 18 只减少到 9 只。在配置相同土壤类型的一项容器试验中观察到，土壤 pH 和胞囊线虫之间具有较强的相互作用。在无线虫的情况下，pH 值为 6.5 要比 pH 值为 4.5 块茎产量低，幅度为 11‰，但当每克土壤最初有 27 个幼虫时，幅度为 44%。pH 值为 4.5 时作物生长更加茂盛，即使初始线虫密度比 pH 值为 6.5 时更高。据观察，pH 值较低时，马铃薯生长旺盛，可能会导致较高的增值速度。

表 1 –5　土壤 pH、块茎产量、干物质和胞囊线虫初始密度间的关系

pH_{KCl}	块茎干物质量（kg/m^2）	初始密度（卵/克土壤）
4.5	1.09	17.5
5.0	1.04	13.8
5.5	1.03	14.4
6.0	0.92	11.3
6.5	0.75	8.8
LSD_{005}	0.11	8.3

注：引自（Haverkort 等，1993）

为了准确估计未感染和感染胞囊线虫的模型参数值，在模拟胞囊线虫对马铃薯生长、发展和块茎产量的影响时，应考虑这些相互作用。培育对马铃薯胞囊线虫有耐受性且在受感染地块仍具有高产量的新品种，这需要有耐受性筛选技术。Evans 和 Hay dock（1990）考察了几种实地和盆栽实验，以及直接和间接的耐受性筛选方法。Trudgill 和 Cotes（1983）证明：与未受感染的不耐受品种相比，盆栽中受感染的耐受品种的根部总长度要更长。

表 1-6　不同品种不同处理情况下马铃薯生长状况

品种	线虫	根部长度		
		主茎	侧根	总长
Maris Piper	未接种	138	384	522
（耐受的）	接种	275	490	766
Epicura	未接种	118	298	416
（中间的）	接种	87	307	397
Maris Anchor	未接种	179	973	1 153
（不耐受的）	接种	219	260	480

注：每个小坑有 1 万个卵的接种效果 品种根部长度的耐受性不同（Trudgill 和 Cotes，1983），不同品种和根类具有显著差异（$P<0.05$）

Arntzen（1993）发现，《荷兰品种推荐表》中推荐的马铃薯耐受品种和成熟品种之间没有任何联系；然而，Haverkort 等人（1992）和 Trudgill 等人（1990）得出结论：枝叶茂盛且能拦截到更多光能的品种（一般为晚熟品种）要比那些枝叶较少且覆盖地面时间较短的品种（早熟品种）受胞囊线虫感染影响的损失更小。Arntzen 还发现：在温室中，胞囊线虫感染导致叶面积减少的主要原因是由于植株形成的叶片较小，而不是由于他们形成的叶片较少。此外，温室盆栽马铃薯总重量的减少与实地产量的减少有一定的相关性。盆栽作物生长衰退主要与线虫的孵化率和根部生长差异相关，体外试验几天后会变得非常明显。

盆栽筛选技术对确定马铃薯基因类型对胞囊线虫的耐受程度非常有效，但其对实现马铃薯的生态种植这一目的没有任何作用。盆栽马铃薯的接种物和营养物比田间马铃薯损耗更快，水供应量的波动可能会在较短的时间跨度内发生。作者经常观察到盆栽马铃薯（Fasan 和 Haverkort，1991）受感染与未受感染马铃薯相比，生长初期未受感染的马铃薯植株会继续生长更长一段时间，而受感染的马铃薯，接种物耗尽后会继续吸收土壤中存在的养分。

第五节　关于胞囊线虫对马铃薯生长影响的总结和讨论

三个主要生理过程体现了线虫对马铃薯生长和发育的影响。在马铃薯生长期的两个关键阶段，胞囊线虫感染将导致马铃薯产量降低。而胞囊线虫对马铃薯的侵害主要在以下几个方面：①降低同化率；②影响叶片干物质分配；③影响叶片扩展的形态变化。考虑到光能拦截及其作用，对产量形成最重要的生长阶段是：①从出苗至冠层生长停止的冠层形成时期；②生长季节后期光截获从100%减少至0%的作物衰老时期。对马铃薯生理过程的三大主要影响过程在这些阶段都会发挥作用。胞囊线虫对马铃薯的损伤机制可解释如下：为获得最大单产，播种和出苗阶段应尽可能缩短，出苗至光截获率为100%阶段应尽可能缩短，光截获率为100%生长阶段应尽可能拉长。播种至出苗期间不受胞囊线虫的影响（Haverkort等人，1992），但出苗光截获率为100%生长初期及地面覆盖从100%下降为0%（冠层衰老）的最后阶段都会受到影响。

马铃薯胞囊线虫能直接导致光合作用功能趋弱。这种趋弱最初可通过脱落酸等激素信号跟踪。受感染马铃薯水分吸收减少导致作物遭受干旱（气孔关闭、叶水势降低和蒸腾减少）；然而，光合速率比蒸腾更易受到虫害影响。矿物质（氮磷）吸收减少也可能导致光合效率降低，即导致光合机构效率降低，进而使辐射利用效率降低。感染胞囊线虫后，马铃薯冠层生长停止时间推迟，这可能主要是由于水分关系受到破坏；而在生长季的第二阶段，由于感染马铃薯根部深入土壤较浅且表层土壤水分和养分已快耗尽，因此同化作用减少。由于辐射截获减少及辐射利用效率降低，同化率降低导致新叶形成减少（叶片脱落导致辐射截获减少），块茎膨大率降低。感染胞囊线虫后，马铃薯块茎形成数量减少，可能会导致同化率降低，因为形成的营养物质减少，导致同化率降低。

对感染胞囊线虫马铃薯的损害与分配至根部的干物质的相对增加有关，瓦赫宁根根系实验证明，茎根比降低，根数量增加，有利于根部生长的干物质分配能减少分配至叶片的干物质，从而减少截获光辐射总量，减少马铃薯生长率。块茎形成时间几乎不受感染胞囊线虫影响，这意味着与健康马铃薯相比，受感染的马铃薯块茎形成开始于较小的植株，将导致不利于叶片生长的干物质分配模式形成，造成冠层生长停止推迟，衰老提前开始。

分配至叶片的干物质应尽可能广泛覆盖土壤表面，才能对地面覆盖及辐射截获产生最大影响，这意味着，较大的比叶面积（薄叶的形成）、较高的叶水含量（导致叶片变大）和更多、更长的茎（允许叶片更好地覆盖土壤表面，重叠较少）是导致光截获量、决定分配至马铃薯顶部干物质总量的主要因素。然而，胞囊线虫对这四个因素会产生不利影响。如之前所述，受胞囊线虫影响的马铃薯叶面积变小，叶片水分含量降低。由于水分关系受破坏，马铃薯的茎变少变短，受损根部氮、磷、钾的吸收率降低。

市场销路好的产品和马铃薯盈利能力不仅取决于分配到块茎的干物质总量，也依赖于块茎大小分布。由于作物生长速率降低，块茎含水量较低，受感染马铃薯产生的块茎较小。

上述3个过程对块茎产量影响的相对重要性各不相同。哪一个基础过程影响同化物生产（发送信号、缺水或营养）、营养分配（有利于根或生殖器官）和形态（叶厚度和茎长）取决于多个因素。出苗至冠层衰老期间，缺氮很可能会限制最佳的光合作用，并可能逐步取代直接信号效应，这在马铃薯生长的早期阶段对减少光合作用发挥着关键作用。有利于根部生长的干物质分配初期会阻碍同化作用，在未受感染的条件下有助于叶片生长。生长季中期，分配至块茎的干物质可能发挥着更加显著的作用：较小马铃薯的块茎形成导致几乎所有块茎的同化物提前分配，因此导致提前衰老，因为没有剩余的同化物供新叶片形成、或维持旧的叶片。厚的叶片将导致冠层生长停止推迟，受马铃薯胞囊线虫感染光截获减少可能是由于形成的叶片数量减少。

马铃薯胞囊线虫感染后，3个主要过程对产量减少的相对重要性根据品种（耐受程度、耐受性所基于的生理过程）、栽培方法（土壤类型、pH、整个生长季节土壤水分和养分有效性）、种植者无法改变的环境因素（如温度和辐射）而不同。植物的生长过程、环境和遗传因素之间的相互作用不可能用单一的马铃薯生长和发育综合模拟模型进行量化，因此，建模主要关注相互作用的过程，例如，受品种、土壤类型及水分可获取性影响的根的生长及氮素吸收等方面。在非常高的集成条件下，如基于辐射截获、辐射利用效率和收获指数的方程式，简单的作物引导系统中可以使用这种模型。

参考文献

Arntzen F K. 1993. Some Aspects of Resistance to and Tolerance of Potato Cyst Nematodes in Potato. Ph. D. Thesis, Wageningen, The Netherlands.

Brown R W, Haveren B. V. van. 1972. Psychrometry in water relations research. Utah Experimental Station, Logan, 45 pp.

Evans K. 1982. Water use，calcium uptake and tolerance of cyst-nematode attackin potatoes. Potato Research 25：71 –88.

Evans K，Haydock P P J. 1990. A review of tolerance by potato plants of cystnematode attack with consideration of what factors may confer tolerance and methods of assaying and improving it in crops. Annals of Applied Biology117：703 –740.

Evans K，Parkinson P L，Trudgill D L. 1975. Effects of potato cyst nematodes onpotato plants III. Effects on the water relations and growth of a resistant and a susceptible variety. Nematologica 21：273 –280.

Farquhar G D，O'Leary M H，Berry J A. 1982. On the relationship between carbon isotope discrimination and intercellular carbondioxide concentration in leaves. Australian Journal of Plant Physiology 9：121 –137.

Fasan T，Haverkort A J. 1991. The influence of cyst nematodes and drought on potato growth. 1. Effects on plant growth under semi-controlled conditions. Netherlands Journal of Plant Pathology 97：151 –161.

Fatemy F，Evans K. 1986. Effects of globodera Rostochiensis and water stress on shot and root growth and nutrient uptake of potatoes. Revue de Nematologie 9：181 –184.

Fatemy F，Trindler P K E，Wingfield J N，Evans K. 1985. Effects of Globodera rostochiensis，water stress and exogenous abscissic acid on stomatal function and water use of Cara and Pentland dell potato plants. Revue de Nematologie 8：249 –255.

Haverkort A J，Boerma M，Velema R，Van de Waart M. 1992. The influence of drought and cyst nematodes on potato growth. 4. Effects on crop growth under field conditions of four cultivars differing in tolerance. Netherlands Journal of Plant Pathology 98：179 –191.

Haverkort A J，Fasan T，Van de Waart M. 1991a. The influence of cyst nematodes and drought on potato growth. 2. Effects on plant water relations under semi-controlled conditions. Netherlands Journal of Plant Pathology 97：162 –170.

Haverkort A J，Groenwold J，Waart M van de. 1994. The influence of cyst nematodes and drought on potato growth. 5. Effects on root distribution and nitrogen depletion in the soil profile. European Journal of Plant Pathology 100: 381 -394.

Haverkort A J, Mulder A, Waart, M van de. 1993. The effect of soil pH on yield losses caused by the potato cyst nematode Globodera pallida. Potato research 36: 219 -226.

Haverkort A J, Schapendonk A H C M. 1994. Crop reactions to environmental stress reactions. In Struik P C (ed) Plant production on the threshold of a new century. Kluwer Academic Publishers: 339 -347.

Haverkort A J, Uenk D, Veroude H, Vande Waart M. 1991b. Relationships between ground cover, intercepted solar radiation, leaf area index and infrared reflectance of potato crops. Potato Research 34: 113 -121.

Haverkort A J, Valkenburg G W. 1992. The influence of cyst nematodes and drought on potato growth. 3. Effects on carbon isotope fractionation. Netherlands Journal of Plant Pathology 98: 12 -20.

Mulder A. 1994. Tolerance of the Potato to Stress Associated with Potato Cyst Nematodes, Drought, and pH. An Ecophysiological Approach. Thesis, Wageningen, The Netherlands.

Schans J. 1993. Population Dynamics of Potato Cyst Nematodes and Associated Damage to Potato. Ph. D. Thesis, Wageningen, The Netherlands.

Schans J, Arntzen F K. 1991. Photosynthesis, transpiration, and plant growth characters of different cultivars at various densities of Globodera pallida. Netherlands Journal of Plant Pathology 97: 297 -310.

Smit A L, Groenwold J, Vos J. 1994. The Wageningen Rhizolab - a facility to study soil-rootatmosphere interactions in crops. II. Methods of root observations. Plant and Soil 161: 289 -298.

Trudgill D L. 1987. Effects of rates of nematicide and of fertilizers on the growth and yields of cultivars of potat which differe in their tolerance of damage by potato cyst nematode (Globodera rostochiensis and G. pallida. Plant and Soil 104: 235 -243.

Trugill D L. 1980. Effects of Globodera rostochiensis and fertilizers on the mineral nutrient content and yield of potato plants. Nematologica 26: 234 -254.

Trudgill D L, Cotes L M. 1983. Tolerance of potato cyst nematodes (Globodera rostochiensis and G. pallida) in relation to the growth and efficiency of the root system. Annals of Applied Biology 102: 385 -397.

Trudgill D L, Evans K, Parrot D M. 1975. Effects of potato cyst nematodes on potato plants. Nematologica 21: 183 - 191.

Trudgill D L, Marshall B, Phillips M. 1990. A field study of the relationship between pre-planting density of Globodera pallida and the growth and yield of two potato cultivars differing in tolerance. Annals of Applied Biology 117: 107 - 118.

Turner N C. 1988. Measurement of the plant water status by the pressure chamber technique. Irrigation Research 9: 289 - 308.

Vos J, Oyarzun P J. 1987. Photosynthesis and stomatal conductance of potato leaves. - Effects of leaf age, irradiance and leaf water potential. Photosynthesis Research 11: 253 - 264.

第二章 马铃薯和胞囊线虫相互影响研究

胞囊线虫对马铃薯生长影响的模拟研究仍处于起步阶段。20 世纪 70 年代，首次尝试模拟线虫群体动态及寄生生长。多年来，模型始终关注线虫群体的增长率、并使用过于简单的描述函数来观察寄生生长及其对线虫的反应。此外，缺乏所涉及的生理损伤机制方面的信息也进一步阻碍了模型的发展。

然而，近年来，复杂的组合模型得以迅速发展，这种模型同时研究害虫和寄生生长，方式更加实用。胞囊线虫对寄主和营养物质代谢影响的研究越来越清楚。作者将探讨将不同的损伤机制纳入模拟模型的可能途径。

第一节 简介

Elston 等人对使用统计回归模型结合马铃薯产量和种植前线虫密度进行了综述（Elston 等人，1991）。在本章中，作者将只讨论模拟模型。在此类模型中，对马铃薯与土壤和根部线虫数量互作中涉及的生理过程进行了动态模拟。作者将重点关注模型如何解释线虫对寄生马铃薯生长的影响。

最早的模拟模型开发于 20 世纪 70 年代。Ferris（1976；1978）在对葡萄根结线虫的研究中表明，马铃薯生长速度受温度控制，但会随着线虫对根部系统的感染而降低。与马铃薯生长的简单模型相比，该模型比较复杂，包括线虫生命周期的众多信息及其对各种环境条件的依赖性。这种马铃薯线虫模拟模型方法，涉及线虫与马铃薯之间复杂的互相影响机理，但对马铃薯生长描述较为简单。在接下来的几年里，这种方法仍是常用方法（McSorley 等，1982）。Jones（Jones 和 Perry，1978；Jones 等，1978；Jones 和 Kempton，1980）提出了马铃薯金线虫模型。Jones 模型研究的新特征之一是考虑了线虫的基因因素，该模型可用于毒性筛选。本研究还解决了如何利用模型研究线虫、不同轮作方式的抗病品种和线虫种群间的竞争对线虫数量的影响。

这些年来，线虫群体动态建模和马铃薯生长建模之间仍存在差距。1982年的一次会议上，线虫学家表示希望能够缩小这种差距。如果建立有效的马铃薯模型，需要农学家参与进来，并与他们密切合作，因为农学家建立了营养、光合作用、生长和碳分配模型，这些模型具有很好的解释功能。但是在病虫害研究中，没有学者在病虫害研究中接触这些模型（Freckman，1982）。Boote 等人（1983）再次强调将害虫与马铃薯生长模型整合的必要性，并展示了如何对大豆生长模型进行简单调整来实现这一目的。

然而，将一个非常复杂的线虫子模型与一个过于简单化的马铃薯生长模型结合起来仍然是一种趋势（Bird 等，1985；Caswell 等，1986；Schmidt，1992；Van Haren 等人，1994）。Ward 等（1985）的模型是一个例外，其将马铃薯胞囊线虫的动态纳入复杂的马铃薯生长模型 SUCROS（Van Keulen 等，1982）。遗憾的是，他们没有充分利用马铃薯生长模型，因为他们限制了线虫及改变后的同化物分配的影响，而忽视了对光合作用和叶片衰老的影响。Den Toom（1990）后来使用 SUCROS 模型来验证外寄生性线虫 Tylenchorynchus dubius 群体对多年生黑麦草 Lolium perenne 影响的假设。Den Toom 的模型研究表明，T Dubius 能够通过阻碍水分吸收影响寄生马铃薯生长。

Wallace（1987）综述了线虫对作物光合作用的影响，大力提倡将线虫影响纳入到现有的马铃薯生长模型中。Schans（1993）最后在其模型中详细模拟了马铃薯胞囊线虫的群体动态。SUCROS 被再次用来模拟寄生马铃薯生长。Schans（1991）早前从实验中得出结论，第二阶段马铃薯胞囊线虫幼虫入侵根部能触发从根部至茎部的激素信号，引起气孔关闭，并降低光合作用。通过按比例降低光合速率、根系渗透速度和叶面积指数，将此损伤机制应用到模型中，并应用该比例解释渗透根激素生成的模拟和如何分散叶组织的激素信号。Schans 利用其模型来解释线虫对马铃薯损伤的几个现象。他的模型解释了晚熟品种（图 2-1）较高水平的耐受性。该模型能够解释晚熟品种耐受性的生理原因，晚熟品种与早熟品种相比，叶面积更大，能稀释更多激素，且当大部分线虫对根部入侵停止后，块茎才开始膨大。

Schans 的模型非常复杂，但缺少了 Haverkort 和 Trudgill（1995）确认的一些损伤机制。在本章的其余部分，作者将展示如何开发一个简单却能综合考虑损伤机制的模型。

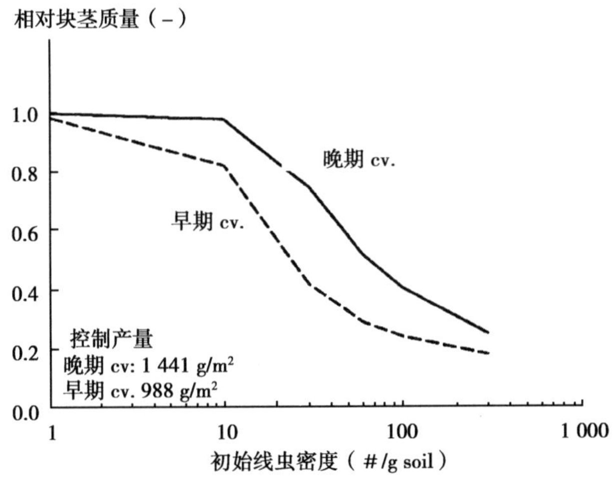

图 2-1　早熟和晚熟品种在不同土壤线虫密度下的最终块茎产量模拟

注：引自（来源：Schans，1993）

第二节　马铃薯胞囊线虫模型生理效应模拟

马铃薯生长速率可以理解为马铃薯光截获的产物（利用该效率，截获光被用来生产生物质）（Monteith，1977）。已有研究表明此类分析对研究马铃薯胞囊线虫对马铃薯生长的影响非常有用（Trudgill 等，1990；Haverkort 和 Trudgill，1995）。线虫主要影响叶面积动态特征，如光截获量，同时降低了光利用效率。Haverkort 和 Trudgill（1995）的综述中，将胞囊线虫对马铃薯叶面积特征和光能利用效率（LUE）的影响划分为 4 个方面：①线虫感染降低了比叶面积（SLA）；②叶片和茎部同化物分配减少，促进了根系统的分配；③叶片衰老加快；④光合作用受影响导致 LUE 降低。

这 4 个方面对产量损失的影响程度各不相同，因此模拟模型不一定全部包含上述 4 种。作者将展示如何使用一个简单的模型来衡量这 4 个方面的重要性。因此，作者采用了马铃薯生长模拟模型 LINTUL（Spitters 和 Schapendonk，1990，可以找到一个程序列表）。在 LINTUL 模型中，光截获和 LUE 决定马铃薯生长速率，同化物分配和叶片衰老都取决于有效积温，使用恒定的 SLA，根据叶重计算叶面积。为衡量 4 个损伤的影响程度，作者将 SLA 和 LUE 降低 10%，叶片衰老速度和至根部的分配加倍。

这些变化对生长在严重感染土壤中的马铃薯非常典型（Haverkort 和 Trudgill，1995；Van Oijen 等人，1995a）。因为线虫影响被当作驱动函数，因此作者并没有运用线虫子模型或计算根长密度扩展该模型。Boote 等人（1993）也使用类似的方法，即使用标准的马铃薯生长模型识别马铃薯线虫系统的主要损伤机制，来监测大豆根结线虫。

模拟结果如图 2－2 所示。叶片衰老加速和较小程度的 LUE 减少造成大部分的最终产量损失。因为作者认为不必将线虫感染对 SLA 的影响纳入到马铃薯胞囊线虫模型中。根部同化物的刺激分配对马铃薯的生长潜力没有太大直接影响（图 2－2）。但是，如果作者打算将马铃薯生长模型和线虫数量动态模拟模型结合起来，就应该准确模拟根系生长，还应考虑线虫对分配至根部同化物的影响。同时，还应考虑线虫能加速根部衰老（Haverkort 和 Trudgill，1995）这一影响。

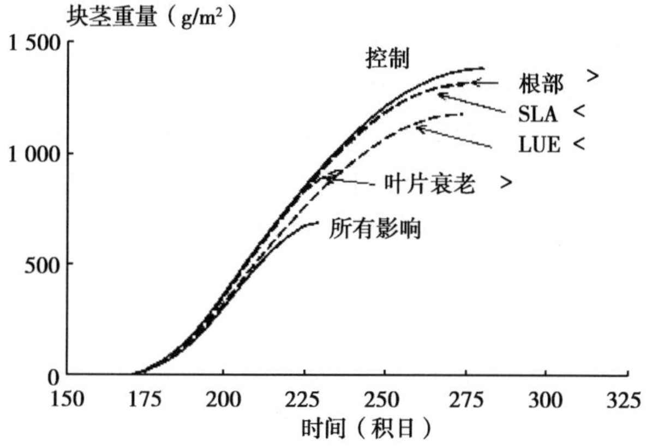

图 2－2　受不同数量损伤机制影响的马铃薯块茎干物质模拟

注：零（“控制”），一个（增加至根部的分配、减少 SLA 或 LUE、或加速叶片衰老，分别表示为：“根部 >”、“SLA <”、“LUE <”、“叶片衰老 >”）或四种损伤机制同时存在（“所有影响”）

第三节　马铃薯胞囊线虫的简易模型

本节中作者将介绍如何将上文提到的损伤机制纳入到简单的马铃薯生长模型中。之前作者已经展示过一个初始模型（Van Oijen 等人，1993）。上文已经提到，作者选择的马铃薯生长模型是 LINTUL。Spitters 和 Schapendonk（1990）总结了该模型的所有公式，但为了方便，作者只列了 LIN-

TUL 的 6 个主要方程式（表 2 - 1：公式 1 ~ 6）。

为计算二龄幼虫的根系渗透速度，作者增加了 3 个方程（表 2 - 1：公式 7 ~ 9），这实际上简单代表了 Schans（1993）的群体动态模型。这些方程基于以下 3 个假设：①比根长是恒定的；②根长随机分布在土壤中；③二龄幼虫对根系统的渗透速率与土壤中的线虫密度和根部土壤体积的增加率呈正比。

最后，作者需要 4 个或 4 个以上的公式来评估上述段落中（表 2 - 1：公式 10 ~ 13）的损伤机制（表 2 - 1：公式 10 ~ 13）。作者需要详细地说明这些方法。首先，假设叶片衰老加速与马铃薯生长速率和根长密度（公式 10）呈正比。这反映了一个事实，线虫引起的根长变短可能导致养分供给不足，难以维持生长、保持叶片面积（Haverkort 和 Trudgill，1995）。

其次，作者假设根死亡与二龄幼虫在根系统中的最近渗透密度呈正比（表 2 - 1：公式 11）。生长后期根部的线虫很大程度上导致了根部损伤。作者并没有验证该模型的这种可能性，因为在未对模型行为产生重大影响的条件下，可能会增加模型的复杂性。

纳入模型中的第 2 个和第 4 个损伤机制分别代表了分配至根部的同化物增加（表 2 - 1：公式 12）和 LUE 下降（表 2 - 1：公式 13）。作者假设这两个机制都依赖于根渗透与叶面积指数的比率。可以证明 Schans（1993）模型中的假设，这两种变化都是受从根部至茎部的激素信号影响，激素信号可能与根渗透率呈正比。

表 2 - 1 受胞囊线虫影响的马铃薯生长简单模拟模型的基本公式

序号	参数	公式
(1)	LAI	$= W_l \times SLA$
(2)	$I_{intercepted}$	$= I_0 \times (1-e^{-k \times LAI})$
(3)	CGR	$= I_{intercepted} \times LUE$
(4)	$d\ W_l/dt$	$= f_l \times CGR\text{-}Senescence_l$
(5)	$d\ W_r/dt$	$= f_l \times CGR\text{-}Senescence_r$
(6)	$d\ W_t/dt$	$= f_l \times CGR$
(7)	L_r	$= Wr \times SRL$
(8)	$V_{rhizosphere}$	$= V_{total} \times (1 - e^{-Lr \times c/Vtotal})$
(9)	Penetration	$= k_1 \times P_i \times d\ V_{rhizosphere}/dt$
(10)	Senescencel	$= k_2 \times CGR/Lr$
(11)	$Senescence_r$	$= k_3 \times Penetration/Wr$
(12)	fr	$= k_4 \times Penetration/LAI$
(13)	LUE	$= LUE_{max} - k_5 \times Penetration/LAI$

（续表）

序号	参数	公式
(14)	LAI	=叶面积指数（Leaf Area Index）（m^2/m^2）
(15)	W_l, W_r, W_t	=叶重、根重、块茎重量（Weight of/eaves，roots，and tubers）（g/m^2）
(16)	SLA	=比叶面积（Specific Leaf Area）（m^2/g）
(17)	I_0，$I_{intercepted}$	=入射和截获光［MJ／（$m^2 \cdot d$）］
(18)	k	=消光系数（-）
(19)	CGR	=马铃薯生长率［g／（$m^2 \cdot d$）］
(20)	LUE，LUE_{max}	=实际和最佳条件下的光利用效率（g/MJ）
(21)	f_l，f_r，f_t	依赖于热时间的分配系数（f_r 可参考公式 12）
(22)	Lr	根长（m/m^2）
(23)	SRL	=比根长（m/g）
(24)	$V_{rhizosphere}, V_{total}$	=根际体积和总根际面积（m^3/m^2）
(25)	c	=根部周围根际的剖面面积
(26)	$Senescence_l$，$Senescence_r$	=叶片和根部的枯死率［g／（$m^2 \cdot d$）］
(27)	Penetration	=二龄幼虫根部渗透率（m^3）
(28)	P_i	=土壤线虫密度（m^3）
(29)	k_1，k_2，k_3，k_4，k_5	=正比常数（－，m/m^2，g^2/m^2，m^2d，gm^2d/MJ）

显然，通过增加马铃薯营养和激素状态变量，可以更加全面模拟损伤机制。但是，作者没有如此，主要是因为缺少这些过程的相关数据，有关模型参数的相关问题，增加模型复杂性并不一定能提高模型性能。

一、模型的初步实验

该模型的第 1 个试验通过实地观察，检查其模拟马铃薯线虫互作关系的能力。为此，作者收集了 1991 年荷兰北部，一块感染 G pallida 沙质土的实验数据（Van Oijen 等，1995a，b）。在本次实验中，使用高密度线虫侵染 Darwina 品种（每立方厘米有 15 只线虫），对照处理中，大多数线虫已被杀线虫剂杀死。模型使用本地天气作为输入数据，马铃薯生长参数保持在荷兰生长条件下的标准值。量化马铃薯线虫关系（表 2－1 中 9～13 方程的 $k_1 \sim k_5$）的参数值应用到了测试中。因此，在本次模拟中，只有对照处理代表了模型的一次独立检查，然而，用几个简单的公式模拟与线虫的相互作用只能展示模型捕捉马铃薯对线虫复杂反应的能力。模拟的结果如图 2－3 所示。作者认为，就对照处理和线虫感染处理而言，模型能很好模拟根长的时间进程、根部渗透、地面覆盖和块茎生长。这些结果表明无需更复杂的模型对其机理进行模拟。

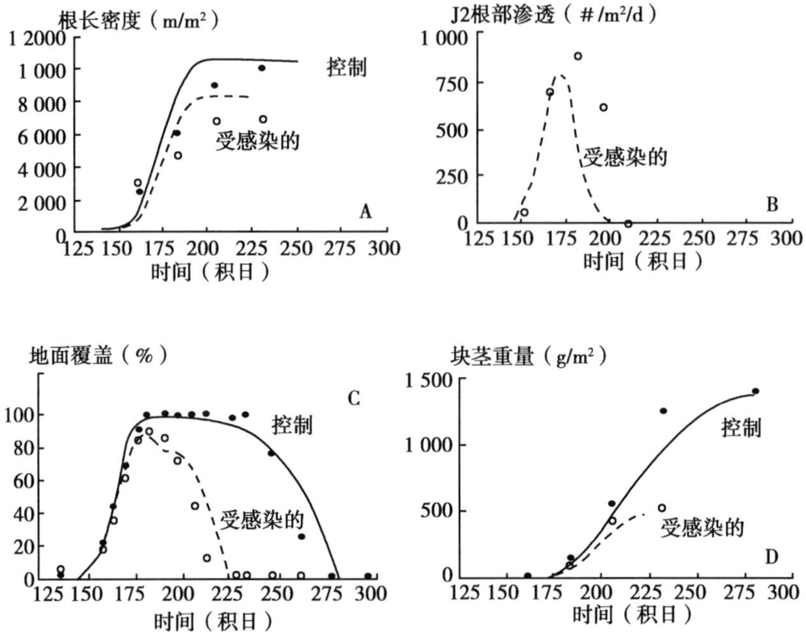

图 2-3 受感染土壤和控制处理下 Darwina 品种生长变化模拟

注：（点：实验观察；线：模拟）A. 根长；B. 二龄幼虫根系渗透；C. 绿叶覆盖地面；D. 块茎干物质增长

二、模型耐受性评估

在另一项该模型的实验中，作者运用马铃薯特性的知识分析了 Elies 品种的高线虫耐受性。Elies 和 Darwina 不同，Elies 晚熟，根冠较大，比根长较短。并且，在之前描述的 Darwina 品种实验条件下（Van Oijen 等，1995a），Elies 几乎不会遭受任何产量损失。作者使用模拟模型来证明 Elies 品种的典型特性是否能充分说明其维持产量的原因。模拟结果如图 2-4 所示。由于马铃薯晚熟，分配至根部的生物量增加似乎确实有利。考虑晚熟的影响，作者的简单模型因而能证实 Schans 复杂模型的结果（图 2-1）。另一方面，Elies 栽培变种的比根长较低只能增加产量损失。在一个模型中结合 3 个特性，作者可以看到与 Darwina 品种相比，产量损失大大减少。然而，这样模拟线虫的耐受性机理还不全面，而 Elies 栽培变种的测量值表示其完全耐受。测量和模拟之间的不同可以用不同方式进行解释。应对线虫渗透，Elies 品种减少了刺激激素生成，且对信号不太敏感。没有数据证明或反驳这一点（Van Oijen 等，1995b）。或者，与Darwina 相比，对 Elies 品种来说，根死亡和组织营养浓度相对减少可能不太明显，无论哪种解释正确，不管作者的模型是否有效，模型已经考虑了调查中需要假设条件。

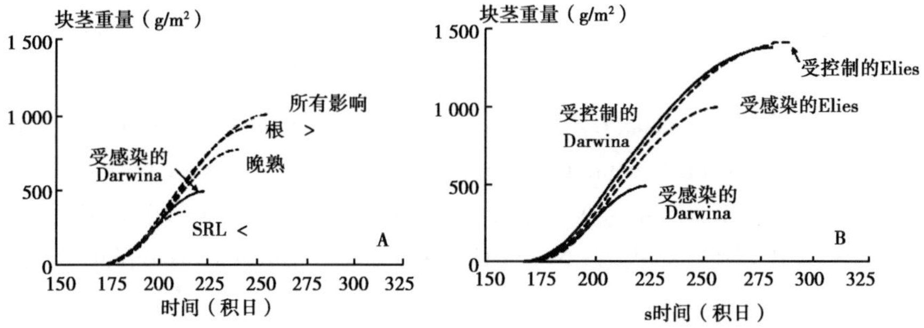

图 2－4　不同处理方式下马铃薯块茎重量随时间变化曲线

注：模拟易受线虫感染的品种 Darwina 块茎生长的时间进程，假设无马铃薯特性变化。每一变化对应一个线虫耐受品种 cv. Elies 的观测特征。A. 改变一种马铃薯特征（增加对根部的分配、延迟成熟、减少比根长），或同时改变三种，对受感染土壤中马铃薯生长的影响；B. 同时改变三个特性对马铃薯在受感染土壤和对照处理生长的影响

第四节　研究内容的总结与讨论

在本章中，作者重点关注使用模型模拟线虫破坏的生理过程，以及不同品种间的差异。已被作者证明一个简单模型在识别主要的损伤机制、评估不同寄生马铃薯对线虫耐受性的重要性方面是有用的。当然，模拟模型也用于研究环境与损伤关系的影响。胞囊线虫在种植前的密度和最终产量之间的关系受许多环境因素相互作用的影响，因此也是多变的（Trudgill，1986）。Barker 和 Noe（1987）建议使用马铃薯线虫系统的模拟模型来解释这一变化。

作者已经证明，没必要在马铃薯胞囊线虫模型中明确考虑营养或激素关系，尽管潜在的损伤机制会影响这些关系。如果想要使用作者的模型来解释观察到的相互作用，如线虫密度的影响和施肥量（Trudgill，1980；1986），与增加模型的复杂性相比，重新参数化根长和叶片衰老速率间的建模关系似乎更有远见。作者打算以此种方式运用该模型。

参考文献

Barker K R, Noe J P. 1987. Establishing and using threshold population levels. Pages 75 - 81 in Veech J A, Dickson D W (Eds.) Vistas on Nematology. Society of Nematologists, Hyattsville, USA.

Bird G W, Tummala R L, Gage S H. 1985. The role of systems science and data management in nematology. Pages 160 - 173 in Barker K R, Carter C C, Sasser J N (Eds.) An Advanced Treatise on Meloidogyne, Vol. II: Methodology. North Carolina State University Graphics, Raleigh, NC, USA.

Boote K J, Batchelor W D, Jones J W, et al. 1993. Pest damage relations at the field level. Pages 277 - 296 in Penning de Vries F W T, Teng P, Metselaar K (Eds.) Systems Approaches for Agricultural Development. Kluwer Academic Publishers, Dordrecht, The Netherlands.

Boote K J, Jones J W, Mishoe J W, et al. 1983. Coupling pests to crop growth simulators to predict yield reductions. Phytopathology 73: 1 581 - 1 587.

Caswell E P, MacGuidwin A E, Milne K, et al. 1986. A simulation model of Heterodera schachtii infecting Beta vulgaris. Journal of Nematology 18: 512 - 519.

Den Toom A L. 1990. Simulation of the host-parasite system Lolium perenne-Tylenchorynchus dubius. 2. The effect of T. dubius on L. perenne. Netherlands Journal of Plant Pathology 96: 211 - 225.

Elston D A, Phillips M S, Trudgill D L. 1991. The relationship between initial population density of potato cyst nematode Globodera pallida and the yield of partially resistant potatoes. Revue de Nematologie 14: 213 - 219.

Ferris H. 1976. Development of a computer simulation model for a plant-nematode system. Journal of Nematology 8: 255 - 263.

Ferris H. 1978. Modification of a computer simulation model for a plant-nematode system. Journal of Nematology 10: 198 - 201.

Freckman D W (Ed.). 1982. Nematodes in Soil Ecosystems. University of

Texas Press, Austin, TX, The Netherlands.

Haverkort A J, Trudgill D L. 1995. Crop physiological responses to infection by potato cyst nematodes (Globodera spp.). Pages 167 – 184 in Haverkort A J, MacKerron D K L (Eds.) Potato Ecology and Modelling of Crops under Conditions Limiting Growth. Kluwer Academic Publishers, Dordrecht, The Netherlands.

Jones F G W, Kempton R A, Perry J N. 1978. Computer simulation and population models for cyst-nematodes (Heteroderidae: Nematoda). Nematropica 8: 36 – 55.

Jones F G W, Perry J N. 1978. Modelling populations of cyst-nematodes (Heteroderidae: Nematoda). Journal of Applied Ecology 15: 349 – 371.

Jones F G W, Kempton R A. 1980. Forecasting crop damage by nematodes: Nematode population dynamics. EPPO Bulletin 10: 169 – 180.

McSorley R, Ferris J M, Ferris V R. 1982. Model synthesis and validation: primary consumers. Pages 141-156 in Freckman D W (Ed.) Nematodes in Soil Ecosystems. University of Texas Press, Austin, TX, USA.

Monteith J L. 1977. Climate and the efficiency of crop production in Britain. Philosophical Transactions of the Royal Society of London, Series B, 281: 277 – 294.

Schans J. 1991. Reduction of leaf photosynthesis and transpiration rates of potato plants by second-stage juveniles of Globodera pallida. Plant, Cell and Environment 14: 707 – 712.

Schans J. 1993. Population Dynamics of Potato Cyst Nematodes and Associated Damage to Potato. Ph. D. Thesis, Agricultural University Wageningen, The Netherlands.

Schmidt K. 1992. Zeitdiskrete Modelle zur Vorhersage der Populationsdynamik des Rubenzystennematoden Heterodera schachtii (Schmidt) in Abhangigkeit von der Fruchtfolge und des Temperaturmusters. Ph. D. Thesis, University of Bonn, Germany.

Spitters C J T, Schapendonk A H C M. 1990. Evaluation of breeding strategies for drought tolerance in potato by means of crop growth simulation. Plant and Soil 123: 193 – 203.

Trudgill D L. 1980. Effect of Globodera rostochiensis and fertilisers on the mineral nutrient content and yield of potato plants. Nematologica 26: 243 – 254.

Trudgill D L. 1986. Yield losses caused by potato cyst nematodes: a review of the current position in Britain and prospects for improvements. Annals of Applied Biology 108: 181 – 198.

Trudgill D L, Marshall B, Phillips M. 1990. A field study of the relationship between pre-planting density of Globodera pallida and the growth and yield of two potato cultivars of differing tolerance. Annals of Applied Biology 117: 107 – 118.

Van Haren R J F, Hendrikx E M L, Atkinson H J. 1994. Growth curve analysis of sedentary plant parasitic nematodes in relation to plant resistance and tolerance. Pages 172 – 184 in Grasman J, Van Straten G (Eds.) Predictability and Nonlinear Modelling in Natural Sciences and Economics. Kluwer Academic Publishers, Dordrecht, The Netherlands.

Van Keulen H, Penning de Vries F W T, Drees E M. 1982. A summary model for crop growth. Pages 87 – 97 in Penning de Vries F W T, Van Laar H H (Eds.) Simulation of Plant Growth and Crop Production. Simulation Monograph. Pudoc, Wageningen, The Netherlands.

Van Oijen M, De Ruijter F J, Ammerlaan F H M. 1993. Simulation of root growth reduction following compaction of nematode infested soil. Pages 135 – 136 in European Association for Potato Research, Abstracts 12th Triennial Conference, Paris, France.

Van Oijen M, De Ruijter F J, Van Haren R J F. 1995a. Responses of field-grown potato cultivars to cyst nematodes at three levels of soil compaction. I. Leaf area dynamics, photosynthesis and crop growth (Submitted).

Van Oijen M, De Ruijter F J, Van Haren R J F. 1995b. Responses of field-grown potato cultivars to cyst nematodes at three levels of soil compaction. II. Root length dynamics and nutrient uptake (Submitted).

Wallace H R. 1987. Effects of nematode parasites on photosynthesis. Pages 253 – 259 in Veech J A, Dickson D W (Eds.) Vistas on Nematology. Society of Nematologists, Hyattsville, USA.

Ward S A, Rabbinge R, Den Ouden H. 1985. Construction and preliminary evaluation of a simulation model of the population dynamics of the potato cyst-nematode Globodera pallida. Netherlands Journal of Plant Pathology 91: 27 -44.

第三章 模拟胞囊线虫影响马铃薯生长模型

根据三维度生长模型的直角横截面和时间轴可以得出中小初始种群密度和相对马铃薯总质量之间的关系，这三个维度包括种植时间 t，相对马铃薯总质量 Y 和相对生长速率 r_p/r_0。相对生长速率是指在一定线虫密度 P 条件下，一定质量的马铃薯生长速率 r_p 和无线虫侵袭相同质量的马铃薯生长速率 r_o 之间的比值（常数），无线虫侵袭条件下马铃薯需要达到一定质量的时间和线虫密度 P 的比值为：$t_0/t_p = r_p/r_0$（2）。

相对生长速率 r_p/r_0 在 $P > T$ 时为，$r_p/r_0 = k + (1 - k)^{0.95P/T-1}$，在 $P \leq T$ 时，$r_p/r_0 = 1$。通常，k 是最小的相对生长速率，因为 $P \to \infty$。因此，对于方程而言，$P > T$ 时，$y = m + (1 - m)^{0.95P/T-1}$，$P \leq T$ 时，$y = 1$（6）同样也适用于中、小初始种群密度和相对马铃薯总质量间的关系。T 指耐受限度，低于这一限度，生长和产量将不会受到线虫影响而减少，m 指相对最小产量。

可以通过类似的方式从生长模型中推断马铃薯胞囊线虫中、小初始种群密度和马铃薯相对块茎质量间的关系。然而，块茎形成伊始，有无线虫侵袭马铃薯的茎质量并不相同，且开始于线虫密度大茎质量较小时。在线虫密度为 P 时，有正常块茎质量和那些受线虫侵袭相同总质量马铃薯的块茎质量并不相同，但是 $r_p \Delta t$ 将更大，Δt 指块茎形成的真正时间和受线虫侵袭马铃薯的总质量与块茎形成开始时，不受线虫侵袭马铃薯的总质量相同的时间之差。

线虫密度较大的“早衰”马铃薯的相对总块茎质量要比模型和公式 2 预测的小。这表明生长减少机制在线虫密度较大时变得活跃，其在线虫密度较小时并不发挥作用。

第一节 模型简介

在许多生物和线虫研究中，研究人员不会过多强调需求。一些线虫学家或生物学家制定了种植前线虫密度和马铃薯产量之间的数学关系，这些关系确是基于一些观察结果而得到，但大部分具有一定的局限性，且缺乏理论基础和预测价值。如线性和对数函数（Lownsberry 和 Peters，1956；Hesling，1957；Seinhorst，1960；Hoestra 和 Oostenbrink，1962；Brown，1969）；逆线性函数（Elston 等，1991）和指数函数（Van Haren 等，1993）。这些函数并没有代表线虫-马铃薯相互作用机制内容，计算一些函数的变量或参数会产生偏离实际的结果。

其他学者（例如，Evans 和 Haydock，1990；Trudgill，1992）试图在不使用模型研究线虫攻击对马铃薯的影响基础之上建立一些理论，这一方法体系很难有条理地阐述分析生物科学问题，或检验出推理链条的缺陷，也没有将这些理论与线虫学家交流，供进一步改进或批评。

部分满足霍金要求之一的方法是“综合模拟模型”。这种类型的模型试图使用大量的子模型和参数等详细信息描述一种经常受限的研究领域。Ward 等人（1985）首次尝试将这种模型应用于线虫学关于马铃薯胞囊线虫种群动态的初步模型中，随后 Schans（1993）扩展了该模型。其尝试描述马铃薯胞囊线虫种群动态，并利用一些马铃薯生理过程和与环境条件改变的相关信息描述胞囊线虫对马铃薯带来的相关损害。由于子模型和大量参数的任意性特征，有些参数通常被视为常数，但一般情况下是变量。这些模型无法推动理论构建，且预测价值有限。在 Schans 的模型中，对植物生理学的描述仅限于对一片或几片叶子光合作用、呼吸作用和蒸腾作用的短期测量，这并不能解释在一定线虫密度范围内产量的不同，此外，马铃薯生长期间内线虫密度并不一定相同。在这些模型中，外部条件是马铃薯生长过程中不可或缺的因素，制定线虫病害控制措施不适合考虑这些条件。

在线虫学家 Seinhorst 的模型中，马铃薯胞囊线虫模型（Seinhorst 1986a，b；1993）与 Hawking 推荐的经典方法最接近。这些模型基于描述线虫—线虫和线虫—马铃薯相互作用主要机制的理论，这些机制与国际上很长一段时间内对不同线虫种类所做的大量研究结果并不冲突。由较复杂模型推导出的综合、简单和随机模型主要用于科学目的（Been 等，1994），只包含几个参数和已知的分布函数，允许在较少限制内做出预测，尽管有一些约束条件。Been 等人（1994）证明这些模型的预测价值，其预测价值适用于作为控制马铃薯胞囊线虫对农民田地影响的咨询系统；线虫对马铃薯生长和产量的影响是 Seinhorst 模型的一个重要原理，将在本章中予以说明。

第二节 胞囊线虫导致马铃薯减产原因分析

一、线虫导致马铃薯产量减少的原因

线虫攻击导致马铃薯产量减少，单位面积产量减少，例如，受根结线虫攻击的胡萝卜可能毫无价值，因为其主根出现分支和变形，尽管这些胡萝卜每单位面积的质量和不受线虫攻击的胡萝卜相同。洋葱质量正常，但感染少数茎干线虫，在收获后储藏时可能果实会损失。受胞囊线虫攻击的马铃薯不仅块茎质量减少，体积也会变小；然而，攻击马铃薯的胞囊线虫和所有攻击马铃薯根部的感染线虫几乎不会影响收获产物每单位质量的价值，除非较大的初始种群密度使马铃薯质量大大减少，如此从开始种植马铃薯时经济效益就不高。因此，基于播种或种植时的线虫密度和收获时单株马铃薯平均质量关系间的模型，可以预测受线虫影响导致马铃薯产量减少的结果。下文中将避免使用“单产”这一术语，考虑到上述限制，农艺学意义上的单产必须从单株马铃薯质量获得。

为建立种植前种群密度（P）与受感染马铃薯（洋葱、球根花卉）的比例（y）或作物生长一段时间后的相对马铃薯质量（y）间的数学关系模型（相同生长条件下，线虫密度为 P 时作物产量和无线虫影响作物产量间的比值），可通过与播种或种植一段时间后观察到的线虫密度、作物生长和作物重量间的数学关系模式进行分析，验证该理论。能估计不同实验条件下系统参数数值。

Seinhorsfs（1965）的模型研究了种植时茎线虫（Ditylenchus dipsaci）的种群密度和受感染洋葱比例间的关系，其模型与 Nicholson（1933）的感染寄生蜂幼虫的模型相同。推理过程如下：只有受感染和未受感染的洋葱有区别，单株作物受感染的程度没有任何讨论价值，因为只有未受感染的洋葱才可以出售。建立模型只需要考虑以下三个假设（其公式虽略有不同但同样适用于根部感染物种）：①不同密度下线虫平均数量相同。这意味着种群密度并不影响平均大小或线虫活动；②线虫不会影响彼此的行为。不直接或间接吸引或排斥对方；③线虫随机分布在某一地区的作物上。

如果在单位面积或单位质量的土壤里，一定密度的某种线虫攻击了某一区域比例为 d 的洋葱作物（未受感染的比例则为 $1-d$,），那么单位面积或单位质量的土壤里，一定密度的两种线虫将同时攻击已遭受损失的作物，使作物已损失比例 d，扩大了 d 倍（对产量减少没有额外影响），还剩一部分未受感染的作物的比例 $1-d-d\ (1-d)\ =\ (1-d)^2$。

广义而言：线虫密度 P，未受感染的洋葱为 $y=\ (l-d)\ p=z^p$（1）。图 3-1 中绘制了 z^p 的值，不使用 P，而使用 $\log P$。这种处理具有一定的优势，针对所有 z，曲线的形状都相同［d（log$P$）/dy 只由 y 决定］，而且如果通过计算一个土壤样品中的线虫数量估计 P，$\log P$ 的方差较少取决于真正的 P 的数值。z 的数值由影响线虫发现和侵袭作物效率的条件决定。对于茎线虫的斑状感染，侵袭斑状的中心似乎比侵袭边界更加有利，因此离中心越远 z 值越大。这导致斑状性的持久性。该模型还适用于线虫从随机分布的受感染作物蔓延至相邻的作物，使受感染的作物具有重叠的圆形斑状。

线虫的斑状侵袭并不一直如此。例如，感染马铃薯胞囊线虫的感染源往往在田地里引进囊胞的时候出现：少量囊胞由种薯传播，而大部分的囊胞则是由农业机械传播。通过马铃薯繁殖后，这些胞囊将利用耕作和收割机传播。导致在荷兰的所有马铃薯种植区，斑状都具有相同的形状（Seinhorst，1982；1988；Schomaker 和 Been，1992）。如果优质马铃薯实行短期轮作，斑状侵袭将变得越来越严重，直到整块土地受到感染。在格罗宁根东南部和德伦特省东部就发现了这样的虫害感染案例，当地即实行每隔一年种一季马铃薯的轮作制度。

二、线虫对马铃薯生长的影响

与茎线虫相比，侵袭根部的线虫（如胞囊线虫、根结线虫和短体线虫）减少马铃薯产量的方式并不直接。受侵袭马铃薯的生长速率将降低且生长受到抑制，导致与播种或种植一段时间后未受线虫影响的马铃薯相比，其质量变小。通常情况下，马铃薯生长季节结束时的外部条件能阻止马铃薯成熟的推迟。

Seinhorst（1979b；1986b）的生长模型基于两个简单的概念：一是马铃薯的性质，随时间的推移马铃薯的质量将会增加；一是马铃薯寄生线虫的性质，这一因素能减少马铃薯质量的增长速率，且能使线虫的种群密度变大。除了上述提到的因素，以下原则同样适用于茎线虫密度和受攻击马铃薯比例间的关系模型：①根部感染线虫随机分布在土壤中；②线虫进入马铃薯根部的空间和时间比较随机，因此，在单位时间进入单位数量根的平均线虫数量是恒定的。这一数字和线虫密度 P 呈正比（每单位质量或土壤体积的线虫数量)；③如果在播种或种植后的给定时间 t，马铃薯的生长速率等于每单位时间（dY/dt）总质量的增长，r_o 表示不受线虫侵袭的马铃薯，r_p 表示线虫密度为 P 的马铃薯，则无线虫和线虫密度为 P 时，相同总质量马铃薯（因此，不同年龄）的比值 r_p/r_o 在整个生长季节恒定。根据图 3－2，$r_0 = \tan\alpha = \triangle Y/\triangle t_0$，$r_p = \tan\beta = \triangle Y/\triangle t_p$。因此，$r_p/r_0 = t_0/t_p$ (2)。

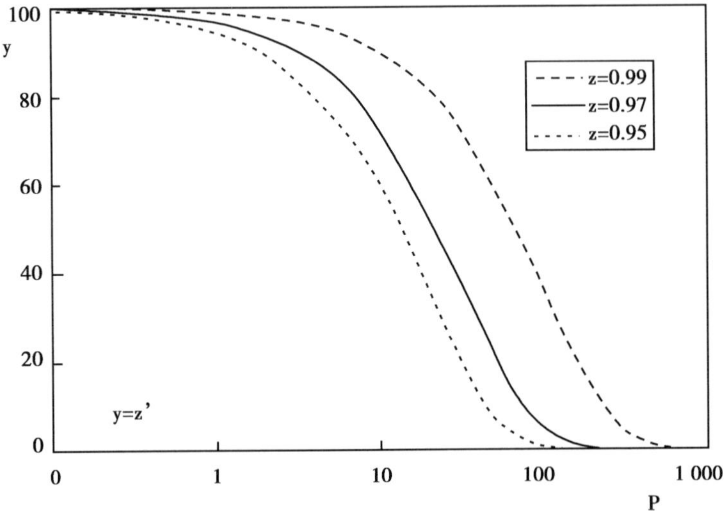

图 3－1　茎线虫的初始种群密度 P 和未受侵袭的洋葱马铃薯 y 间的关系 $y = Z^P$

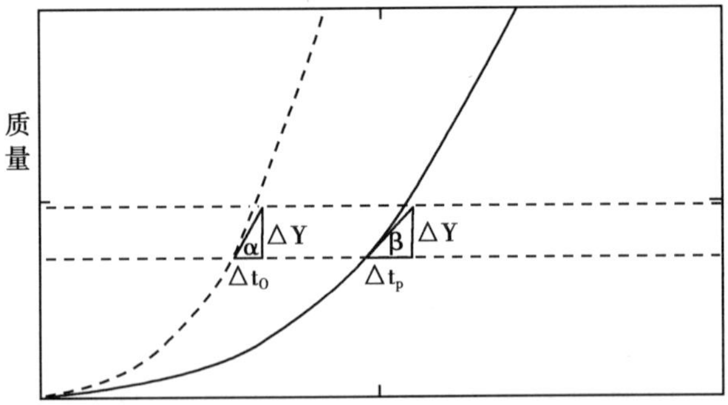

图 3－2　无线虫和线虫密度为 P 的马铃薯的生长曲线

注：Y 指总马铃薯质量；t 指播种或种植后的时间；t_0 和 t_p 指有无线虫影响马铃薯达到相同质量的时间；r_0 和 r_p 分别是无线虫和线虫密度为 P 时马铃薯的生长速率

根据 Nicholson 的竞争模型（公式 1 和图 3－1）可分析线虫种群密度和其对马铃薯生长速率影响间的数量关系。公式 1 是连续函数，$0 \leqslant P \leqslant \infty$。关于不同线虫种类（包括马铃薯胞囊线虫在内）的种群密度 P 对不同马铃薯质量的影响不同；有充分的观察结果表明，存在最低的密度 T，在此密度下线虫不会减少马铃薯质量。因此，可以得出结论，线虫密度 < T 时，马铃薯的生长速率不会减小。几项实验的结果证实了这一结论，在这几项实验中，生长速率明确。此外，在线虫密度较大时，马铃薯质量在少数这样的实验中几乎为 0，而生长速率则并非如此。用 $P-T$ 取代 P，引入最小的相对生长速率 $r_p/r_{0=}$ k 来改变公式 1，因为 $P \rightarrow \infty$。则构成模型的第二个方程为：

$P > T$ 时，$r_p/r_{0=}k + (1-k)\ z^{P-T}$；$P \leqslant T$ 时，$r_p/r_{0=}1$（3），z 是小于 1 的常数。k 的数值与播种或种植后的线虫密度和时间无关，但对于不同的实验，k 的数值可能有所不同。应用公式 2 和公式 3，从一个无线虫侵袭的马铃薯生长曲线可以得出不同线虫密度下马铃薯生长曲线。如果这些曲线连续，且播种或种植后不久后生长速率呈单调形式降低，则这些曲线的形状可能不同。图 3－3 是一个三维模型，对于一个给定的数值 k，总马铃薯质量为 Y，相对线虫密度为 P/T，种植后的时间为 t。

第三节 基于胞囊线虫密度和相对马铃薯质量统计关系的简单模型

试验的主要结果通常指播种或种植后在给定时间受已知线虫密度攻击的马铃薯质量。检验这些关系是否符合该模型，必须将它们与垂直于模型时间轴的截面图作比较，即通过与马铃薯生长曲线的不同密度范围 P/T 和不同 k 的数值作比较。这些截面图与以下公式非常一致：P > T 时，$y = m + (1-m)\ z^{P-T}$（4），m 是最小的相对马铃薯质量，z 是一个小于 1 且与公式 3 的数值相等或略小的一个常数，T 是耐受限度，且与公式 3 的数值相同。

Seinhorst（1986b）在盆栽试验中，说明了不同垫刃亚科线虫包括燕麦胞囊线虫（Seinhorst，1981）、马铃薯金线虫和马铃薯白线虫的 P/T，播种或种植数月后的相对植株干重（y）之间的数量关系。13 个盆栽试验中，$y' = (y-m) / (1-m) = z^{P-T}$（5）的数值相对于 P/T 的数值绘制在图 3－4 中。平均 y' 和 P/T 的关系与 $y' = z^{P-T}$ 非常一致，这表明不同试验中依据这种关系得出的单个数据偏差是由于试验误差造成。种植后不同时间不同抗感品种的平均相对植株和块茎质量与根据公式 5（T = 1.8 卵/克土壤）得出的结果非常吻合。因为公式 4 和 5 相同，可以得出结论，这一模型较好地描述了图 3 - 4 和其他试验中的结果，这显然适用于与胞囊线虫、穿刺短体线虫和矮化线虫饲养和繁殖方式不同的垫刃亚科线虫；这也表明播种或种植一段时间后，垫刃亚科线虫的密度和受侵袭马铃薯的相对植株质量间的一般关系与无线虫影响马铃薯的生长曲线及外部条件无关，尽管相对于不同线虫和马铃薯种类，z 和 T 的数值会有不同。对于相同线虫和马铃薯种类组合的试验，m 的数值也不相同，不同试验 z^T 的数值与 0.95 相比变化较小，难以区别由试验误差造成的变化。因此，根据公式 3 将曲线与试验数据相拟合，通常假定 z^T 的数值为 0.95，将公式 4 变换成：$y = m + (l - m)\ 0.95^{P-T}$（6）。

T 的数值取决于线虫种类、致病型和马铃薯种类，但似乎一般不受外部条件的影响。对于在春季播种的感染胞囊线虫的马铃薯，T 约为 1.8 卵/克。一个例外是马铃薯胞囊线虫对昼长的敏感度（Been 和 Schomaker，1986；Greco 和 Moreno，1992）。与此相反，对于一个给定的线虫种类，m 的数值随实验马铃薯品种的不同而不同，且对于同一品种，不同试验在未知外部条件影响下的 m 数值也不相同。不同马铃薯品种间是否存在一致的差异性尚未成立。因此，T 是一种马铃薯品种对特定线虫种类或致病性的耐受程度的主要测量方法，而 m 的重要性还有待进一步展开调查。大多数关于不同马铃薯品种耐受性差异讨论的问题是没有同时考虑这两个参数。

一、模型的含义

不同线虫种类对线虫攻击反应的相似性并不限于其对马铃薯大小的影响。线虫密度至少达到 $16T$，但一般都超过 $32T$，不会影响单位马铃薯质量和单位持续时间的用水量，也不会影响干物质的含量（Seinhorst，1981）。也有证据表明，这也适用于冠和根重的比值（Seinhorst，1979a；Been and Schomaker，1986）。相同质量未感染线虫的马铃薯和那些轻度和中度感染线虫的马铃薯相比，唯一的区别可能是后者稍高些。

这一模型也对受线虫攻击、相同质量和不同年龄的马铃薯生长速率进行了比较。从禾谷胞囊线虫 H avena 对青涩马铃薯（Seinhorst，1979）茎根比影响的调查及对有无 G. Pallida 影响的马铃薯马铃薯的质量、硝酸盐和钾含量（Been 和 Schomaker，1986）数量关系的分析可以明显体现出这一模型的重要性。

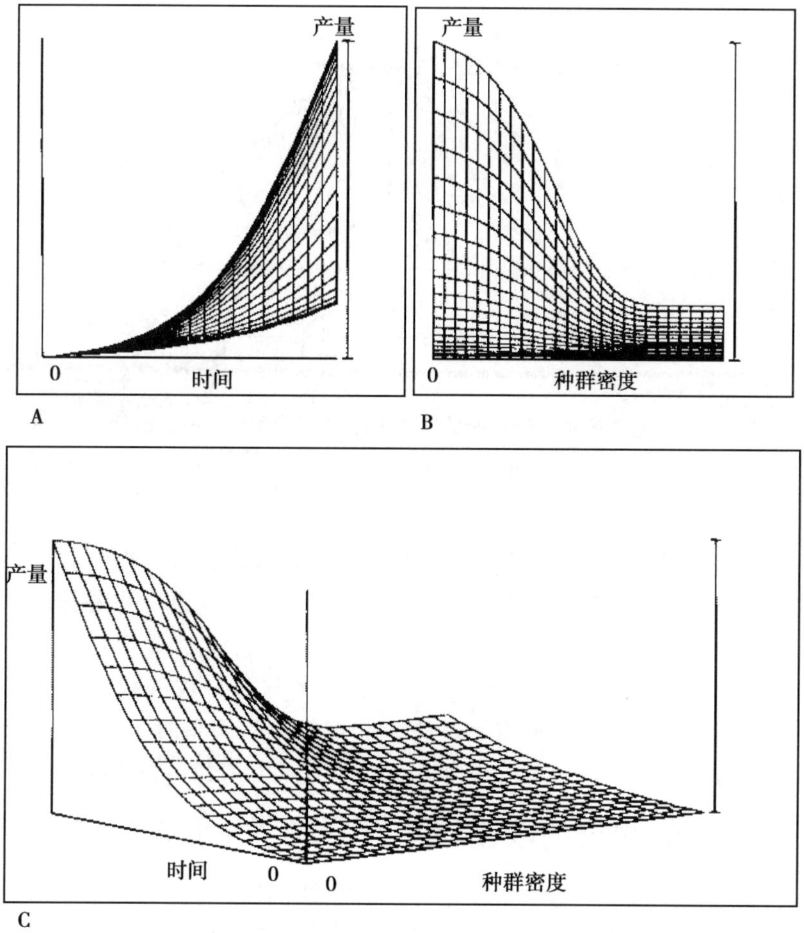

图 3－3　表达总质量 *Y* 和线虫密度 *P* 之间的关系、直角横截面的三维模型曲面

注：不同线虫密度的马铃薯生长函数的时间轴为 t，Y（r_p，t）；无线虫密度 Y（r_0，t）；A. 展示了旋转为 0°时，Y 和不同线虫密度 t 间的关系；B. 展示了旋转 270°时，Y 和线虫密度为 P 时的关系；C. 展示了旋转 230°时，（Y，P）和（Y，t）间的关系。所有旋转都是顺时针。无线虫 r_0 和线虫密度为 P（r_p）相同马铃薯的生长率都是根据方程 1 和 2 得来的

Seinhorst（1986b）的评价指出，线虫降低马铃薯生长速率的机制（影响营养素的提取、对根组织的机械损伤阻碍了水和矿物质的吸收、植物导管阻塞导致萎蔫及食物的提取）在一些文献中被认为是无法实现。对只在根部感染线虫期间才会生成生长抑制物质机制的结论来说，更是如此。由于单位根数量和单位持续时间，线虫侵染的数量恒定，因此，对于单位质量的马铃薯，生长减少刺激物也恒定。有待检验随着线虫密度增加

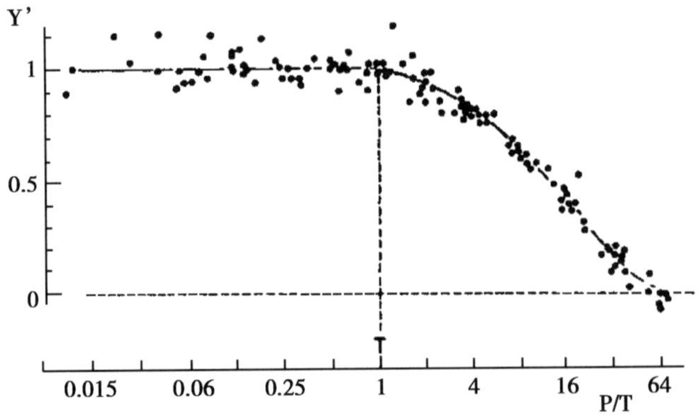

图 3－4　13 个试验中播种或种植时的线虫密度 *P/T* 和种植后一定时间 *y*' 的关系

注：$y' = (y - m_x) / (1 - m_x) = 0.95^{P/T-1}$。线虫和马铃薯种类结合时，$Y = m_x + (1 - m) 0.95^{P/Tx-1}$，所有的 Tx 的数值都一致

每只线虫效率降低的原因。其他问题如对于一定线虫密度，控制其生长减少的方式，什么因素决定了 k 的比值使其不依赖于 T，也需要展开进一步的研究。

严格而言，该模型只适用于每个生长季节只有一代的线虫种类的情况，如马铃薯胞囊线虫、燕麦胞囊线虫和根结线虫；然而，图 3－3 表明其也适用于繁殖速率较快的线虫（如一个生长季节增长 10～20 倍）。

二、胞囊线虫的攻击对马铃薯块茎生长的影响

块茎形成开始后，茎质量变化的速率急剧下降，直至达到最小值。在此模型中，块茎生长开始不久后，茎叶重量增加就会停止，总质量增加完全是块茎增重，而茎叶重量增长停止。

如果线虫攻击对马铃薯的影响仅仅是发育延缓，则对于有无线虫影响的环境，块茎形成开始时，马铃薯的茎重相同。如果 t_{s0} 指无线虫条件下马铃薯块茎形成的开始时间，则线虫密度为 P 时，块茎形成的开始时间为：$t'_{s}p = r_0 t_{s0}/r_p - \Delta t$（7）。$\Delta t$ 随 r_p/r_0 增加，$t'_{s}p$ 将在两周内接近最大值 t_{s0}，r_p/r_0 和 $t'_{s}p$ 间的关系根据图 3－5 中的直线 ab）。因此，线虫密度为 P（Wp'）时马铃薯的块茎质量为 $r_p \Delta t$，要比无线虫影响的相同重量的马铃薯茎重大。

在上述构建的模型中，给定的时间 t，线虫密度和相对块茎质量间的关系也是根据公式 6 计算，$P/T = 50$ 卵/克土壤，$k > 0.4$（$m > 0.05$），从块茎和总干重比大于 0.5 开始，T 值相同，m 小于总马铃薯质量。在块

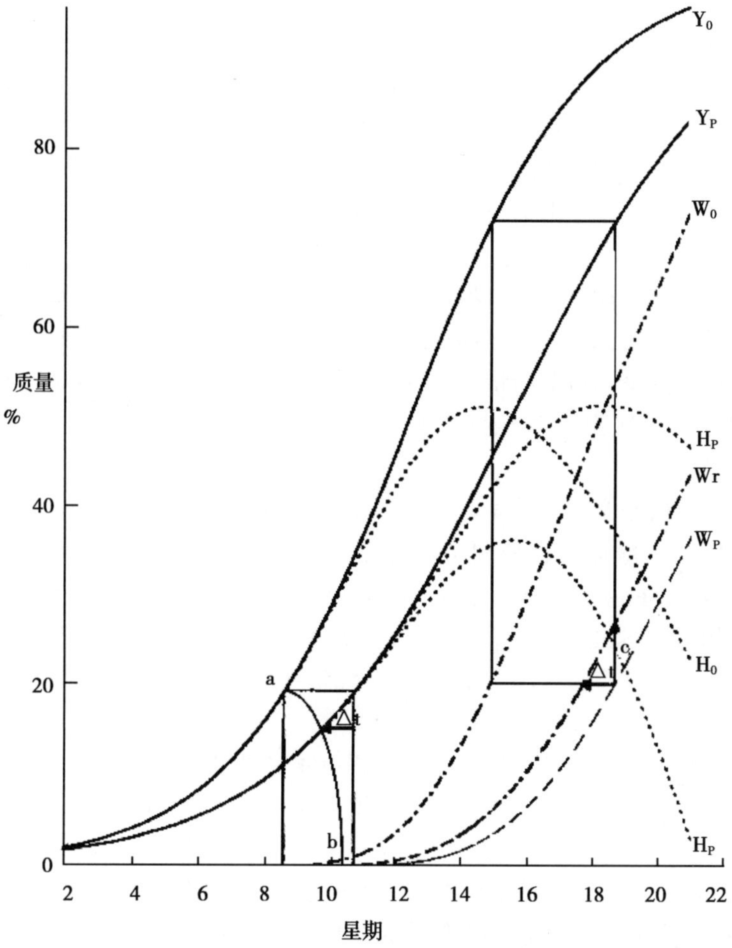

图 3-5 有无线虫影响马铃薯的总茎和块茎质量的增长曲线

注：有无线虫影响条件下，相同总质量或相同块茎质量的马铃薯增长率与公式 2 一致，$r_p/r_0 = t_o/t_p$；有无线虫影响相同总质量的马铃薯块茎质量曲线与方程 $W'_p = W_0 + r_p\Delta t$ 一致。在图中，曲线 ab 表示茎质量和块茎生长开始的时间之间的关系 $c = r_p\Delta t$；Y_0 和 Y_p 分别指无线虫和线虫密度为 P 时的马铃薯总质量；W_0 和 W_p 指如果受线虫影响和不受线虫影响的马铃薯块茎形成开始时茎质量相同，无线虫和线虫密度为 P 时马铃薯的块茎质量；H_0 和 H_p 指如果受线虫影响和不受线虫影响的马铃薯块茎形成开始时茎质量相同，无线虫和线虫密度为 P 时马铃薯的茎质量

茎和总干重的比值范围内，如果模型不受其他抑制生长的机制影响，P/T 的数值、线虫密度和相对块茎质量间的实际关系 k 都是根据模型中的公式进行计算。

三、多重线虫导致抑制生长的机制

根据 Seinhorst（1981），研究结果线虫导致两种抑制生长的机制产生：第一种抑制生长机制对所有的线虫种群密度都发挥了作用，第二种抑制生长的机制是第一种机制的补充，仅在中等和较大线虫密度条件下有显著影响。该模型仅适用于由第一种机制导致的生长抑制，与不受线虫影响的具有相同质量的马铃薯相比。“第一种机制”会阻碍马铃薯生长，有时也会增加茎的长度。如果仅第一种机制发挥作用，在短期时间内耗水量与马铃薯的质量呈正比，因此，在播种或种植后，不同线虫密度和不同时间的相对耗水量是测量马铃薯质量的一种计算方法。实际马铃薯质量是这些相对质量乘以不受线虫影响相同年龄马铃薯的实际质量的结果（Seinhorst，1981）。

第二种机制会减少单位马铃薯质量的耗水量，减少 K^+ 和 Na^+ 的主动吸收和排出，增加马铃薯对 Ca^{2+} 的被动吸收和减少干物质含量（Seinhorst，1981；Been 和 Schomaker，1986）。马铃薯年龄和线虫密度之间可能存在负相关，“第二种机制”对线虫密度的影响变得显著。对于马铃薯胞囊线虫，这一密度很少小至 16T。与“第一种机制”相反，“第二种机制”倾向于促进块茎开始生长。

第三种抑制生长的机制是受马铃薯白线虫攻击导致的马铃薯“早衰”，指茎长度和质量增加的突然停止。种植后这一现象发生的时间可能与线虫的密度呈负相关。对早期品种 Ehud 来说，最早出现在种植后的 9 周内，对 Darwina 品种来说，最早出现在线虫密度最小值即为 25T 时。并非所有的品种都同样敏感（Seinhorst 等人）。“早衰”的原因尚不可知。

四、参数估计 t 和 m 的分析

有充分的证据表明，春季种植马铃薯的 T 值不受外部条件差异的影响，因此，可以在温室和田间的盆栽实验中确定 T 的数值。温室实验的唯一要求是使用足够大的花盆来保证土壤中马铃薯的根部密度和田间种植马铃薯根密度大致相同，防止根占满花盆，如果根占满花盆将影响线虫密度和块茎重量之间的关系数量，获取并掩盖 T 的真值（Seinhorst 和 Kozlowska，1977）。根据模型和盆栽实验的结果，T 为种植后短时间的马铃薯总质量，足以作为最终块茎质量 T 的估值，可以允许使用较小的花盆。应在不同线虫密度（连续线虫的密度比值在 1～2 之间）和密度 $< T$ 的一定数量条件下开展实验，为 $P < T$ 时提供一个准确的马铃薯质量估计，该范围

内的最大密度应约为 30T。估值的准确性主要在于马铃薯物质和生长条件（光、水含量）的一致性，仔细填充和处理花盆。

m 是线虫和马铃薯种类结合的特征，不同实验间有很大的差异性。因此，温室中的少量实验足以获得可靠 T 的估值，不会产生马铃薯品种和胞囊线虫病原型组合的估值。然而，从温室实验的结果中，可以得出不同马铃薯品种间的差异。此外，对于一些马铃薯品种，必须在田间实验中估计足够数量的 m 值，才能建立分布函数。

由于在胞囊线虫密度较大时，种植马铃薯经济效益低，因此用于咨询目的的模型，可以忽略受“第二种机制”影响的线虫密度和生长、产量减少之间的数量关系。

在田间实验中估计 T 和 m 要比在盆栽实验中更加烦琐。需要相同的密度范围且线虫密度是唯一的变量。如果与难处理的空间分布、块茎产量变化的原因无关，那么马铃薯胞囊线虫侵袭的斑块中线虫密度范围最接近这一要求。不能通过施用不同剂量的杀线虫剂或其他生物剂来控制线虫密度的范围，因为这对马铃薯产量的影响不可预测。线虫密度的范围（对正在讨论的估计参数有用）必须根据来自每个田间地块足够的样品来确定，以保证线虫数量的变异系数不大于 15%（密度为 1：2，这是可区分的）。根据 Seinhorst（1988），考虑到胞囊线虫数量的变异系数为 16%，在较小地块（每千克土壤的集中系数为 50）样品的线虫密度呈负二项分布，需要每个地块 4kg 的土壤样品来估计每克土壤放 1 个卵的种群密度。由于单位土壤质量的变异系数与线虫密度呈负相关，所需的样本量也是如此，例如，以相同精度估计每 0. 5 卵/克土壤或 0. 25/卵/克土壤的线虫密度，必须分别计算 10kg 和 20kg 土壤样品中的卵数量；另一个要求是地块要小（如 1m^2），以便减少中等范围密度变化的影响，且每个小密度间隔必须有足够大的面积来保证每单位面积块茎质量的较小变化；同样，必须有足够数量的地块的线虫密度小于 T，才能准确估计最大产量。

因此，为了避免处理较大的土壤样品，应尽可能从温室实验中估计 T 和 m。田间试验能用在更加贴近自然环境的外部条件下，有效检验不同病原体和品种组合的估值或估值比值。

第四节　相关总结和讨论

由于研究的所有根部寄生线虫对受侵袭马铃薯生长影响的惊人一致

性，无论马铃薯的宿主状态如何，线虫以何种方式攻击，受感染组织的反应如何，上述讨论的生长模型都适应于所有垫刃亚科线虫中低程度的种群密度。基于在播种或种植时线虫密度和相对马铃薯总质量间相同的数量关系，只要后代有较大的繁殖率（例如根结线虫），都不会导致其他抑制生长的现象出现；大量适用于这一目的的盆栽实验和少数实地实验可以证明这一现象。用两个参数值描述一个特定线虫种类在一种马铃薯（一年生和多年生）播种或种植时，中低种群密度对马铃薯生长和产量减少的敏感性，要优于不参考线虫密度的单一表征模拟。研究两个参数最重要的是耐受极限 T 的确定，极可能对通常出现在马铃薯生长时期的外部条件不敏感。胞囊线虫对马铃薯攻击的影响取决于其生长时期的昼长变化，这些参数值，尤其是耐受限度，在计算一个给定线虫种群密度而采取控制措施的成本减少时是关键因素，由存活线虫引起的马铃薯损失的成本能带来最大的纯收益。这些分析结果对所有垫刃亚科线虫、年生马铃薯和控制措施来说效果相同。

另一方面，该模型能够表明线虫攻击马铃薯后抑制生长机制的性质，因此，可作为研究生物化学的基础，这一抑制影响的反作用导致耐受限度的存在，大量线虫对受攻击马铃薯生长速率最大程度的降低起到促进作用，通常比无线虫影响马铃薯的实际生长速率低。到目前为止，线虫攻击抑制生长的生理机制的研究还不深入。根据这一生长模型，这一类型的研究可以开辟管理这些线虫的新方式，在未来将取得更好的进展。

参考文献

Been T H，Schomaker C H. 1986. Quantitative analyses of growth，mineral composition and ion balance of the potato cultivar Irene infested with Globodera pallida（Stone）. Nematologica 32：339 – 355.

Been T H，Schomaker C H，Seinhorst J W. 1994. An advisory system for the management of potato cyst nematodes（Globodera spp.）. Pages xxx in Haverkort A J，McKerron D K L（Eds.）Ecology and Modelling of Potato Crops under Conditions Limiting Growth. Kluwer Academic Publishers，Dordrecht，The Netherlands.

Brown E B. 1969. Assessment of the damage caused to potatoes by the potato cyst eelworm，Heterodera rostochiensis Woll. Annals of Applied Biology 103：471 –476.

Elston D A，Phillips M S，Trudgill D L. 1991. The relationship between initial population density of potato cyst nematode Globodera pallida and the yield of partially resistant potatoes. Revue de Nematologie 14：213 – 219.

Evans K，Haydock P P J. 1990. A review of tolerance by potato plants of cyst nematode attack，with consideration of what factors may confer tolerance and methods of assaying and improving it in crops. Annals of Applied Biology 117：703 –740.

Greco N，Di Vito M，Brandonisio A，et al. 1988. The effect of Globodera rostochiensis and G. pallida on potato yield. Nematologica 28：379 – 386.

Greco N，Moreno L J. 1992. Influence of Globodera rostochiensis on yield of summer，winter and spring sown potato in Chile. Nematropica 22：165 –173.

Hawking S. 1988. A Brief History of Time. Space Time Publications. Bantam Press，Toronto，Canada.

Hesling J J. 1957. Heterodera major O. Schmidt 1930 on cereals - a population study. Nematologica 2：285 –299.

Hoestra H，Oostenbrink M. 1962. Nematodes in relation to plant growth. IV Pratylenchuspenetrans（Cobb）on orchard trees. Netherlands Journal of Agricultural Science 10：286 –296.

Lownsberry B F，Peters B G. 1956. The relation of the tobacco cyst nematode to tobacco growth. Phytopathology 45：163 –167.

Nicholson A J. 1933. The balance of animal populations. Journal of Animal Ecology 2：132 –178.

Peters B G. 1961. Heterodera rostochiensis population density in relation to potato growth. Journal of Helminthology. R. T. Leiper Supplement：141 – 150.

Schans J. 1993. Population Dynamics of Potato Cyst Nematodes and Associated Damage to Potato. Ph. D. Thesis，Wageningen Agricultural University，Wageningen，The Netherlands.

Schomaker C H，Been T H. 1992. Sampling strategies for the detection of potato cyst nematodes：Developing and evaluating a model. Pages 182-

194 in Gommers F J, Maas P W Th (Eds.) Nematology from Molecule to Ecosystem. Dekker and Huisman, Wildervank, The Netherlands.

Seinhorst J W. 1960. Over het bepalen van door aaltjes veroorzaakte opbrengst-vermindering bij cultuurgewassen. Mededelingen Landbouw Hogeschool Gent 25: 1025 - 1040.

Seinhorst J W. 1965. The relation between nematode density and damage to plants. Nematologica 11: 137 - 154.

Seinhorst J W. 1979a. Effect of Heterodera avenae on the shoot: root ratio of oats. Nematologica 25: 495 - 497.

Seinhorst J W. 1979b. Nematodes and growth of plants: Formalization of the nematode-plant system. Pages 231-256 in Lamberti F, Taylor C E (Eds.) Root-Knot Nematodes Species Systematics, Biology and Control. Academic Press, London, UK/New York/San Francisco, USA.

Seinhorst J W. 1981. Water consumption of plants attacked by nematodes and mechanisms of growth reduction. Nematologica 27: 34 - 51.

Seinhorst J W. 1982. The relationship in field experiments between population density of Globodera rostochiensis before planting potatoes and the yield of potato tubers. Nematologica 28: 277 - 284.

Seinhorst J W. 1986a. The development of individuals and populations of cyst nematodes on plants. Pages 101 - 118 in Lamberti F, Taylor C E (Eds.) Cyst Nematodes. Plenum Press, New York, USA/London, UK.

Seinhorst J W. 1986b. Effects of nematode attack on the growth and yield of crop plants. Pages 191 - 210 in Lamberti F, Taylor C E (Eds.) Cyst Nematodes. Plenum Press, New York, USA/London, UK.

Seinhorst J W. 1988. The Estimation of Densities of Nematode Populations in Soil and Plants. Vaxtskyddrapporter. Jordbruk 51. Research Information Centre of the Swedish University of Agricultural Sciences (Ed.), Uppsala, Sweden.

Seinhorst J W. 1993. The regulation of numbers of cysts and eggs per cyst produced by Globodera rostochiensis and G. pallida on potato roots at different initial egg densities. Nematologica 39: 104 - 114.

Seinhorst J W. 1995. The reduction of the growth and weight of plants added by a second and later generation to that caused by a first generation.

Nematologica (in press).

Seinhorst J W, Kozlowska J. 1977. Damage to carrots by Rotylenchus uniformis, with a discussion on the cause of increase of tolerance during the development of the plant. Nematologica 23: 1 – 23.

Seinhorst J W, Schomaker C H, Been T H. 1995. Effect of attack by Globodera rostochiensis and G. pallida on growth, weight and water consumption of various susceptible and resistant potato cultivars. II Effects on tuber and total plant weight. Journal of Fundamental and Applied Nematology (submitted).

Trudgill D. 1992. Mechanisms of damage and tolerance in nematode infested plants. Pages 182 – 194 in Gommers F J, Maas P W Th (Eds.) Nematology from Molecule to Ecosystem. Dekker and Huisman, Wildervank, The Netherlands.

Van Haren R J F, Hendrikx E M L, Atkinson H J. 1993. Growth curve analysis of sedentary plant parasitic nematodes in relation to plant resistance and tolerance. Pages 1 – 11 in Grasman J, Van Straten G (Eds.) Predictability and Nonlinear Modelling in Natural Sciences and Economics. Proceedings in Nonlinear Science. Wiley, New York, USA.

Ward S A, Rabbinge R, Den Ouden H. 1985. Construction and preliminary evaluation of a simulation model of the population dynamics of the potato cyst nematode Globodera pallida. Netherlands Journal of Plant Pathology 91: 27 – 44.

第四章 马铃薯胞囊线虫管理模型（线虫品种）

文中咨询系统是为加强对马铃薯胞囊线虫（白色胞囊线虫）的管理而提出，主要着眼于使用具有部分抗性的马铃薯栽培品种，可以在固定的短时期轮作时间内为控制胞囊线虫种群密度保持低水平提供可能。使用基于马铃薯胞囊线虫病的种群动态、种植前线虫密度与相对产量的关系作为随机模型，便可能计算出线虫病种群发展的概率以及由此类种群密度导致的单产减少量。为此，一种结合了两种模型，并应用一些与马铃薯栽培品种及线虫种群这种特定组合相关的极可变参数频率分布的仿真模型便得以应用。而且，该模型还包括非寄主生长时线虫种群密度自然递减的情况。倘若了解栽培的某一马铃薯品种、线虫种群以及轮作时期，那么计算出一定的马铃薯减产量则成为可能。因此，对农民来说，在固定的轮作时期内，评估出风险以及不同控制措施所耗成本是切实可行的。将这一模型应用在淀粉马铃薯种植区可以极大地提高收益，并且所使用的线虫杀虫剂量则会显著降低。

第一节 简介

在荷兰，马铃薯是耕作农业中最具收益的农作物之一，因此，人们经常频繁地种植马铃薯，这在农民几乎无权选择其他高经济收益作物的地区尤为明显。但是，这种种植频率还受马铃薯胞囊线虫病的限制。如果对此种病情不加控制，将会导致农作物损失惨重。除了实施作物轮作（每 5 ~ 7 年，实行易受感染的马铃薯品种与非寄主作物轮流耕作）以外，其他可行的控制措施包括种植高抗性或具备部分抗性的马铃薯变种，并进行化学防控。运用化学防控不仅会使效益成本比率降低，还会通过法律法规为其使用设置更多的限制。在荷兰政府制定的“多年作物保护计划”中，其主要目标是：到 1995 年，将所使用的所谓的化学成分熏蒸剂减少 50%；到

2000 年，将其减少 80%。对胞囊线虫致病型 Pa2 和 Pa3 具备高抗性的品种不是数量极少（由于工业生产过程导致）就是不可被利用（主要因为人类消费）。由于作物减少与种植期的线虫密度关系紧密，所以控制措施应当主要针对阻止线虫密度过大来实行。一旦种植区的线虫种群密度减小至可接受水平，那么由一种马铃薯品种自身相对易感性与作物轮作结合形成的这一适宜的组合就有能力使线虫密度保持在小范围内。将具备部分抗性的马铃薯品种与农业实践中实行的其他控制措施结合应用，即通过控制成本以及产量减少成本的总数最小化，从而达到收益最大化的方式是符合马铃薯种植户利益与环境效益的。考虑到一个注重使用部分抗性的马铃薯品种的轮作方式，这也就需要一个基于可预测线虫种群发展与减产量的咨询系统。因为使用线虫杀虫剂处理成本过高（约 1 200荷兰盾/hm^2），而且在减少种群密度方面，杀虫剂的效果经常言过其实。当将部分抗性的马铃薯品种种植在经过适当短期轮作的种植区后，由于线虫密度很小，从而导致的作物损失自然也随之降低，这样与化学防控相比，利用杀虫剂带来的纯收益就要大得多。如果可以根据咨询系统提供的建议实行，随之带来的结果则是，“多年作物保护计划”要求降低使用的化学成分大部分得以实现，但却不会附加不利的经济后果。

作者所建立的咨询系统主要依据一个随机模拟模型，这一模型可以将那些种群密度空间分布（这里需要使用新的土壤采集方法）的一系列子模型，以及对经济方面最大收益的计算包括其中。本章主要讨论在特定作物轮作中，使用具有部分抗性的马铃薯品种，从而实现减产量以及控制成本最小化。在这里，作者将会讨论一体化仿真模型中所涉及的 4 个子模型：①马铃薯胞囊线虫种群动态简化版在种植期中小型线虫密度中的应用；②相对易感性概念（“易感性”相当于“寄主状态”）；③种植前线虫密度与相对产量（相对产量即产量与种植马铃薯后，不存在线虫时产量的比例）之间的关系；④因种植无寄主作物导致的线虫种群减少。

第二节 马铃薯胞囊线虫种群动态研究

Seinhorst（1967；1970；1986a；1993）（图 4－1）主要描述了在种植或播种时期的线虫（马铃薯、燕麦、白车轴草胞囊线虫及根结线虫）种群密度（*P*i）与收割时期新生一代线虫的种群密度（*P*f）（因为线虫这一物种一季只有一代）之间的关系。他对这一关系所扩展出方程式包含 10 个变量，其中最重要的包括：繁殖的最大速率（这一过程发生在数量较少的初始种群密度期）、理论密度最高值以及每单位土壤重量繁殖出的虫卵数量等。在初始线虫密度较大，线虫会减小植物的尺寸。产区内每株马铃薯作物上线虫实际密度的最大值出现在初始密度中间，还包括亲代未孵化出的虫卵（Seinhorst，1967；1984；1986a）。

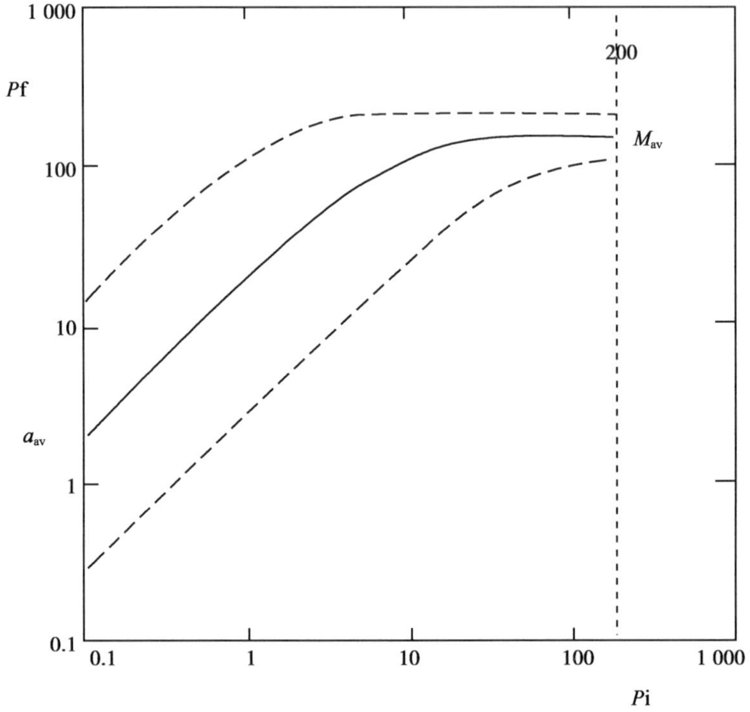

图 4－1　根据公式（1）得出初始虫卵密度与最终虫卵密度之间的关系

注：虚线 = 表示对数 a 和 M（0，145a < a_{av} < 6，9a 和 0，71M < Mg_{av} < 1，41M）正态分布置信区间的 95%。Pf（虫卵个数/每克土壤含量）的规模适用于白色胞囊线虫（易感性作物的 a_{av} = 20 个虫卵/每克土壤含量和 M_{av} = 150 个虫卵/每克土壤含量）

Seinhorst（1993）为这一关系创建的方程式如果延伸至尚未被植物根系利用的土壤里存活的亲代种群（Seinhorst，1986a）中，则包含 8 个变量；但是这些变量在具备一定程度抗性的马铃薯品种种植活动结束后，对有关种群密度及其频率分布的预测中显得多而无用。然而，这一公式可以作为基础，为较为简单的公式提供制约条件。

$$Pf = M \times (1 - e^{-aPi/M}) \tag{1}$$

“Pi”代表初始虫卵密度（种植前）（单位：虫卵个数/每克土壤含

量)；“*Pf*”代表最终虫卵密度（收割后）（单位：虫卵个数/每克土壤含量)；“*a*”代表繁殖最高速率；“*M*”代表理论密度最高值。

这一公式适用于描述初始虫卵密度与最终虫卵密度之间的关系。根据拓展方程式可知，*Pf* 能否达到最大值要取决于 *Pi* 的值。根据公式（1），将 *Pf* 的最大值作为衡量 *M* 的条件并不会导致 *Pf* 所得值有显著的差异，而且在田间试验中观察到的数据（Seinhorst，1986a）可知变量逐级变动的幅度巨大。

由于 *a* 和 *M* 的值不仅由马铃薯品种决定，而且与外部条件有很大关系。随着产区与年份的不同，某一初始种群密度的最终密度值也会产生强烈变动。不仅不同产区与不同年份会造成这两种变量产生差异，即使在同一产区，两个变量的值也是变化的（图 4－2）。因此，虽然在当前现有的咨询系统与立法规定内这是习惯性的做法，但是作者不能只用个别产区内 *a* 和 *M* 的平均值来预测种群密度的发展趋势，更不必说仅用 *a* 的值，这是根本不可能实现的。相比较而言，最好建立出 *a* 和 *M* 的频率分布，从而穷尽 *a* 和 *M* 所有可能的组合，并计算出这些组合以及因而产生的种群密度 *Pf* 与它所对应的概率。作者用了几年的时间，从 17 位农民所拥有的不同类型土壤的种植区对马铃薯金线虫的繁殖最高速率 *a* 和理论密度最高值 *M* 进行观测，根据观察结果显示，*a* 的值一直在 3～157 之间变动，其几何平均数为 25；*M* 的值则处于 200～400 个虫卵/每克土壤之间，平均数为 300 个虫卵/每克土壤（Seinhorst，1986c）。有关这些数值的具体分析将在其他处刊登的著作中显示。倘若进行进一步计算，这两种变量的频率分布应当呈对数正态分布。但是这些分布并不表示具备部分抗性的马铃薯品种就应当另当别论。因此，对于极易受感染的品种来说，其计算主要基于具有相同可变性品种的较小平均值（取决于品种所具有的抗性程度）。

对估计白色胞囊线虫的繁殖最高速率 *a* 和理论密度最高值 *M* 及两个变量的变动情况，目前田间试验所收集的观测数据并不充分。从盆栽试验中种植的 Irene 易病品种监测所得值可以推断出，马铃薯金线虫致病型 Ro 1 与白色胞囊线虫致病型 Pa 3 的比值为 0.8；而这两种致病型 *M* 值之间的比值则为 0.5（Den Ouden，1974a；Seinhorst 和 Oostrom，1984；Seinhorst，1986b）。

因此，根据推测，被测量的种植区内白色胞囊线虫易感品种的 aav 和 Mav 分别为 20 个虫卵/每克土壤和 150 个虫卵/每克土壤。一旦这些田间试验可实行时，上述推测必须要经过观测来加以验证。

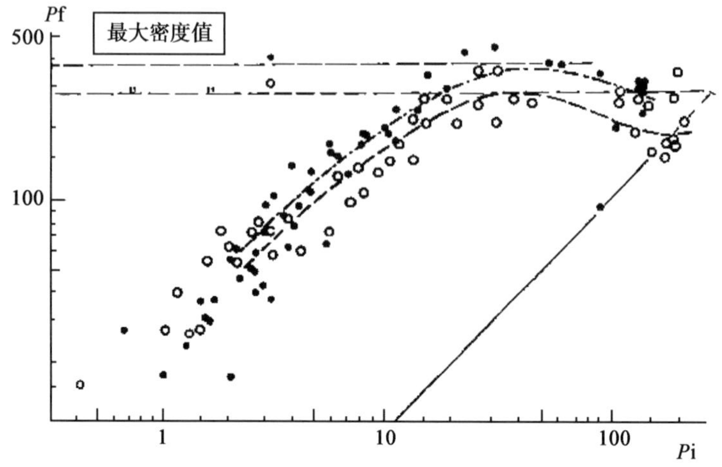

图 4 -2 *Pi* 与最大密度值数量关系

注：表示一位农民同一种植区内两部分所种 Bintje 品种感染马铃薯金线虫致病型 Ro 1 后 *Pi* 和 *Pf* 之间的关系（Seinhorst 1986a）。*Pi* 和 *Pf* 的单位为虫卵个数/每克土壤量。实线表示 *Pi*/*Pf* = 1

第三节　不易感染病毒马铃薯与胞囊线虫间的相互作用机制

当具有部分抗性的马铃薯品种长成时，这时，雌性线虫很少会在除易感品种以外的地方发育成熟；而且，虫卵或胞囊数量可能很小。因此，线虫在这些品种上的繁殖远不如在易受感染品种上那样迅速，而且仅维持较小范围内的最大种群密度。然而，a 和 M 的值太易发生改变，所以不太适合作为部分抗性的测量参照。因此，基于马铃薯胞囊线虫种群动态建立的相对易感性概念就得以引进。相对易感性即被测试品种线虫种群最大增殖速率 a 与该品种易感性参照或相当于这些品种上最大种群密度 M 值的比率。如果被测试品种与该品种易感性参照都在同一实验中、同一条件下生长，那么就有两种针对部分抗性或相对易感性的测量方法。外部环境会改变 a 和 M 的值，造成二者发生显著变化，但是却对相对易感性影响甚小。图4 -3 表明，根据公式（1），具有部分抗性的 Darwina 品种与易感品种 Irene 上白色胞囊线虫致病型 Pa 3 的初始虫卵密度 *Pi* 和最终虫卵密度 *Pf* 之间的关系。

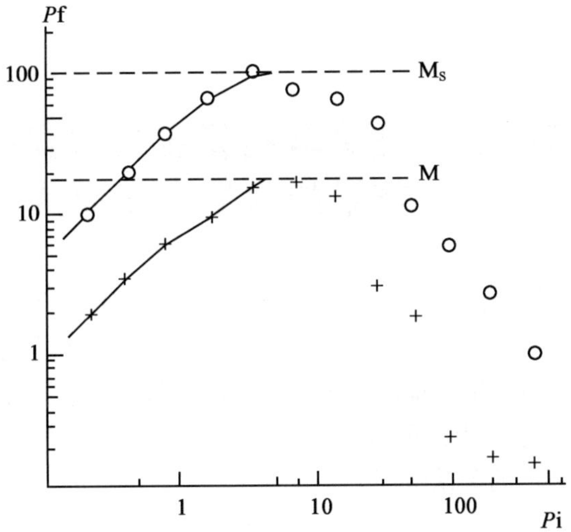

图 4-3　*Pf* 与 *Pi* 数量关系

注：图中线条根据公式（1）绘制，表示具有部分抗性的品种 Irene（o）和 Darwina（+），（Seinhorst 和 Oostrom，1984）上白色胞囊线虫致病型 Pa 3 的初始虫卵密度 *Pi* 和最终虫卵密度 *Pf* 之间的关系

据 Jones 等人（1981）和 Phillips（1984）推测，（$a_{partially\ resistant}/a_{susceptible}$）× 100% 和（$M_{partially\ resistant}/M_{susceptible}$）× 100% 这两个比率在数值上相等。根据 Seinhorst（1984）、Seinhors、Oostrom（1984）和 Seinhorst 等人的出版物，这一推测应当也适用于被测试的 11 种具备致病型 Pa 3 品种中的 9 种。但是，Activa 品种、育成品系 Karna77/281（Seinhorst 等人出版物）和 Ehud 品种（Seinhorst 1984）的 rs_M 要比 rs_a 小。Seinhorst 和 Oostrom 在 1984 年的数据也表明，尽管在不同试验中同一品种相同致病型的初始低虫卵密度的繁殖率差异显著；相对感染度的波动不仅因实验误差，且由于有限的盆栽数量及虫卵重复统计过程中数据不全也会导致其变化。

Forrest and Holliday（1979）、Phillips、Trudgill（1983；1985）和 Phillips 等人（1987）通过观测，根据盆栽与田间试验中品种的易感性程度总结出了品种排名之间的很大相似性，这一相似性肯定了相对易感性并非依赖于外部条件的结论。运用更高的精确度判断马铃薯品种相对易感性的各种方法在 1993 年 Seinhorst 等人出版的著作中有详细描述。

表 4-1 中呈现出在荷兰植物保护研究所测量出的 101 种盆栽中 16 种马铃薯品种对白色胞囊线虫不同致病型的相对感染性。在这些致病型中，包括目前所发现最具毒性的种群之一 Pa 3（1）（Rookmaker 种群）。

当然，被测量品种的相对易感性可以在持续不断、更广阔的范围内获得。

第四节　马铃薯胞囊线虫密度和产量的数量关系研究

一体化仿真模型的另一个重要组成部分是种植时期的虫卵种群密度与收获时期的块茎产量之间的关系，这里用不存在线虫干扰情况下的产量比例表示。为了避免产量大幅度减少，必须使用额外控制措施减少 *Pi* 的值，在作物长成前使之保持在可接受的取值范围内。因此，与中小线虫密度（Seinhorst 1986b）的关系应当满足：

$$y = m + (1 - m) \times 0.95^{(P-T)/T} \quad 当 P > T 时 \qquad (2)$$
$$y = 1 \quad 当 P \leqslant T 时$$

式中：y 表示种植时期受虫卵密度 P 的影响，并作为 $P \leqslant T$ 条件下的一部分时的产量；m 表示相对产量最小值（因此应为一个小于 1 的常数）；T 表示公差极限，当密度 P 达到公差极限时，不会造成减产。

在对致病型 Ro 1 所做的平均数为 2 个虫卵/每克土壤的田间试验（Seinhorst 1982a；1986b）、对致病型 Ro 1、Pa 3（Greco 等人，1982）所做的微区试验和对致病型 Ro 1，Ro 3 和 Pa 3（Seinhorst，1982b）所做的盆栽试验中，易感品种的公差极限 T 几乎无变化。而 m 值的变化程度很大；根据 Seinhorst（1982a）的研究表明，m 值应在 0.2 ~ 0.6 之间。但是运用 m 的平均数即 $m = 0.4$ 进行预测的密度为≤15 个虫卵/每克土壤，并不会导致实际与计算中的误差偏失达到不可接受的程度。过高或过低估计的范围也不会超过 5%（图 4－4）。对更大范围内在线虫密度的影响下造成的相对产量损失进行预测并无实际价值，因为此种情况下这类密度造成的减产量过大，超出了可承受的范围。如若没有首先采取控制措施，农民将不会在这样的线虫密度条件下种植马铃薯作物。在荷兰，这种情况就意味着需要使用线虫杀虫剂来减少 *Pi* 的值，而且由于化学熏蒸法在绝大部分情况下并不能将线虫密度降至可接受程度，因此，应当早在种植以前就先进行系统地杀虫工作。后一方法可以通过延缓线虫入侵作物的时间，从而达到暂时性保护作物的效果，进而提高最低产量值 m。

表 4－1　不同品种相对感染度

品种	Pa 1	Pa 2	Pa 3（1）	Pa 3（2）
Irene	100	100	100	100
Amalfi	—	3	—	39
Amera	76	63	—	98
Producent	15	3	40	34
Multa	—	12	43	40
Pansta	—	6	32	36
Promesse	—	4.5	35	21
Proton	—	1	31	18
Darwina	4.7	0.3	12	5
Santé	—	1	18	8
Atrela	—	0.7	20	9
Karna 77/270	—	—	28	—
Karna 77/281	—	—	12	—
Activa	—	—	25	—
Ellen	—	—	17	—
Seresta	—	—	2	—

注：表示 16 种马铃薯品种对白色胞囊线虫致病型 Pa 1，Pa 2 和 Pa 3 的相对感染度（a/a_s）用易感性品种 Irene 对这类致病型的易感性百分比表示。Pa 3（1）又称之为“Rookmaker”种群，表示这一致病型具有很高的毒性，Pa 3b 则表示具有一般毒性的种群。符号“-”和“=”代表未测量出的相对易感性；荷兰还未曾发现致病型 Pa 1

除了对品种 Darwina 易感的致病型 Pa 3 的公差极限 $T=2$ 个虫卵/每克土壤时有少许了解以外，作者对其他具备部分抗性品种的 T 值知之甚少。当然，对品种 Darwina 进行田间试验的结果与其他品种的 T 值并不矛盾。作物品种 M_r/M_s 与 a_r/a_s 比值的相似性，包括对致病型 Pa 3（Seinhorst 等人出版的著作）具备部分抗性的品种 Darwina 在内，表明作用在这类品种上的致病型 Pa 3 的 T 值并非为品种 Darwina 的 T 值 $=2$ 个虫卵/每克土壤的倍数。有两个原因需要作者对公差极限做出准确的评估。在种植时期已知虫卵种群密度后，就需要对预期产量减少计算评估，因为当 Pf 的值接近 M 值时，最大种群密度就趋向于与植物本身大小成正比，这时这个最大值就将与公差极限呈正相关。然而，公差中的一些细微差异（如极限公差比率在 1～2 之间）却并不会对 M 的值造成明显的影响。

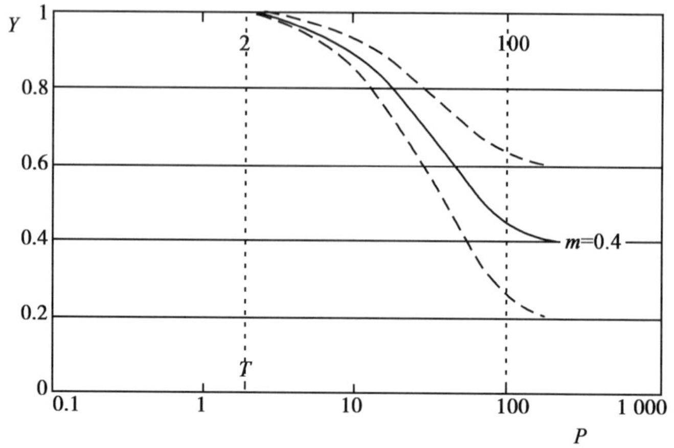

图 4-4 初始种群密度（P）与相对产量 y 之间的关系

注：根据方程式可知，初始种群密度（P）与相对产量 y 之间的关系为：$y = m + (1 - m) \times 0.95^{(P-T)/T}$。这时，$T = 2$ 个虫卵/每克土壤，$m = 0.4$（平均数），而 m 的值应当介于 0.2～0.6 之间（取置信区间的 95% 为界限）

第五节 宿主缺失后的种群数量减少原因探索

马铃薯种植时期的线虫密度是以先前该种作物在停止生长时期的繁殖存活率为基础，而留存下来的密度。这里的存活率可以视为与线虫种群密度无关。缺少寄主作物情况下的种群减少绝大部分程度上是由于在春天这一短时期内，比例较为固定的虫卵进行自然孵化导致的。根据 Huijsman（1961）的观点，在 6 年的时间节点内，存活率为每年 65%，而且这一存活率几乎与种群年龄无关，而是与存活种群的生存能力相关（Den Ouden，1963）。Cole 和 Howard（1962）发现，以 3 年为时间节点，每年的存活率可以达到 80%；Den Ouden（1970）的报告中指出，在 1969 年 14 个种植区的平均存活率为 79%，但是这一数字在 1973 年中的 6 个种植区仅有 51%（Den Ouden，1974b）。然而，作者尚未了解种植区与种植区之间、每个年份，不同因素之间的变动对存活率影响的程度不同，因为通常情况下，抽样误差抑或可以忽略不计，抑或误差过大，无法区分出差异。但是，也有证据表明，种植马铃薯作物后，第 1 年的存活率（Den Ouden，1960；Cole 和 Howard，1962；Andersson，1987；1989）会比接下来的几年更低一些。因此，仿真模型会选择使用第 1 年相对较低的存活率，而舍弃了随后几年的数据（大约从50%变到65%）。存活率一般用来为作物轮作计算因数 c，如果因数 c 的取值为0.5，则轮作的比例就是1：4，如果是 0.5×0.65^3，则轮作的比例就为1：5。但是想要获取更多关于不存在寄主马铃薯的情况下导致存活率下降的信息，就需要在必要的条件下，对模型进行调整。

第六节 马铃薯胞囊线虫管理模拟模型

一、模型简介

为了达到计算出马铃薯品种对白色胞囊线虫致病型 Pa 3 易感程度的目的，这时，a 和 m 的频率分布应当成对数正态，所以分别取平均值1.30103和1.17609，标准差分别为0.419和0.07525。两种频率分布都被分成24个等级，广度为标准差 s 的0.25倍。因此，作者使用的等级范围介于（$\log a$ 或 M）$+3s$ 和（$\log a$ 或 M）$-3s$ 之间，包括99.7的频率分布。每个由 a 和 m 构成的等级都由其等级内中间数的逆对数分别除以 a_{av} 和 M_{av} 与相对频数共同表示。超出这一范围的 $\log a$ 和 M 的值可以忽略不计。所有相对频数都要乘以 0.997^{-1} 得到该范围内累计概率的总数为1。Pf 的值是根据已给出的马铃薯品种 Pi 和 rs，并利用公式（1），外加每个等级 a 分布的中间数乘以 $rs \times a_{av}$ 与等级 M 分布的所有等级中间数乘以 $rs \times M_{av}$ 计算得出的。每个 Pf 的相对频数都是所使用的那些 a 和 M 共同作用的结果。想用如上所述运用 24^2 个值作为 Pi 值的方法，为下一次作物轮种计算出 Pf 的值则需要做更多不必要的工作。因此，作者也将 Pf 的值分为24个等级，并计算出各个新的中间数。每个等级中间数的相对频数都是所属等级中 Pf 值的相对频数。目前，作者用24个 Pi 值开始进行计算，其结果则是在第二代马铃薯作物长成后，Pf 的值为 $24^3 = 13\ 824$。这一过程不断循环，直至模拟出所需的马铃薯作物数量为止。

因为本章仅限于使用部分抗性，将受土壤样本决定的线虫密度保持在小范围内，并且实际种群密度不作为输入值使用。相反，作者会选择使用经过5年生长、几乎呈稳定状态，并且具有部分抗性、拥有相同相对感染性的马铃薯种群密度频率分布。这些数据事实上在作物生长的第1年与种群密度没有关系。对仿真模型这一部分更为细致的描述将会在另一著作中呈现。

为了评估作物的相对减产量（$1-y$），根据公式（2），并运用线虫密度与块茎相对产量之间的关系计算 Pf 值的等级里中间数的 y 值。为得知公差极限 T，作者选择使用的值为 2 个虫卵/每克土壤，最低产量 m 的值为0.4。图 4－5 将展示出在给定的年份里，对线虫密度与作物减产量超出一定范围的种植区所占百分比的估算；以及在给定的种植区内的某一特定轮作范围内，具有一定量 rs 的马铃薯品种对线虫密度与作物减产量超出一定范围的年份所占百分比的估算。作物减产量大于或等于一定值（横坐标表示）的概率用纵坐标表示。从这些数据中可以计算出平均减产量。当采用作物轮作（这一方法同样适用于其他控制措施）来减少两种马铃薯品种的种群密度时，作者会利用因数 c 而不是 Pi 作用于下一代马铃薯作物，cPf 作为最后一代马铃薯作物的种群密度。这时，种群密度的发展就依赖于 ca_{gem} 和 cM_{gem} 了。这就意味着，图 4－5 中的预期作物减产频率分布是当马铃薯作物具备一定的、连续不断的 rs 值才得以显示的，这与在相对应的轮作周期内种植一种具备 rs/c 的马铃薯作物是相同的。为了就关于相对易感性范围这一问题有一个更佳的概览，表 4－2 中展示出不同的平均减产量，即减产量大于 0%、5%、10%、12.5% 和 15% 的概率；以及在运用五种不同作物轮作、拥有不同 rs 值的 95% 的种植区内减产量的最大值。从这些图表中作者可以总结出，作物品种具备的部分抗性 $\leq 8/c\%$ 时，作物本身就可以在无须附加控制措施的情况下将马铃薯胞囊线虫有效地控制住。然而，rs 的值仍然有用，而且它的等级取决于控制成本以及因为实施此种控制带来的额外产量。此外，马铃薯种植者的财务状况也是另一大影响因素。如果种植者有稳定的财务储备，那么他就可以根据预测的平均减产量来选择具有不同相对易感性的作物品种。但是，如果没有可利用的财务储备，而且作物减产量可能并不会超出一定数量，那么就要考虑一下一定减产量所带来的风险，以及在选择具有不同相对易感性的作物品种方面受限会更多。图 4－2 和表 4－2 中的百分比也可以理解为像这种种群密度以及减产量是可能在任何一块种植区、任何一个年份发生的。

种植易感品种后，通过在所有种植区施行一定量的控制，并利用计算达到即使是在害虫聚居为患的产区，也能预防产量损失发展到不可接受的程度，但是在大批种植区内所施加的控制已经过度。倘若这种程度的控制是为达到平均收益的最大化，那么在一定比例的种植区，产量损失将会过高，然而在其他种植区，过量的控制仍然屡见不鲜。因此，种植易感马铃薯后，调整对通过土壤取样估计实际种群密度的控制量是正确的做法（Seinhorst，1982b；Schomaker 和 Been，1992）。这个时候，就需要运用抽样法这一误差很小的方法来估计虫卵以及幼虫密度。

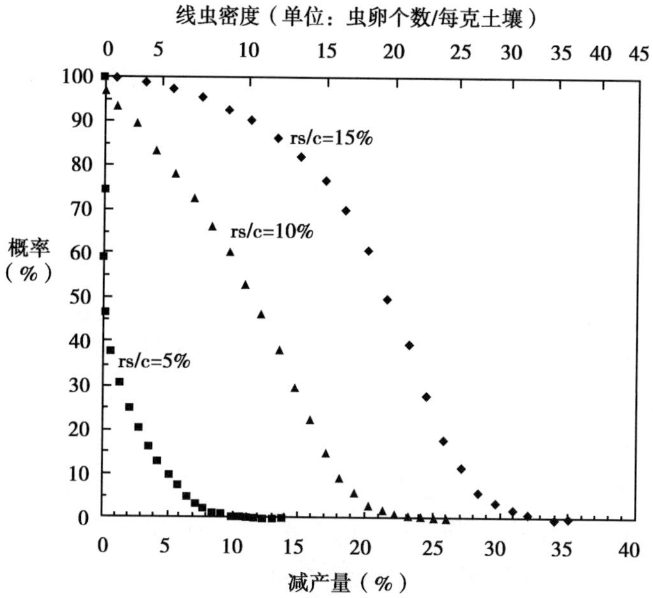

图 4－5　马铃薯线虫密度与减产量之间的关系

注：表示种植 5 种具备部分抗性、且第一年初始密度为 5 个虫卵/每克土壤（种植频率为 1∶2 时，$c=0.5$；频率为 1∶3 时，c 值为 0.32；种植频率为 1∶4 时，c 值为 0.21；种植频率为 1∶5 时，c 值为 0.14）、rs 的值分别为 $5/c$（■）；$10/c$（▲）和 $15/c$（◆）的马铃薯品种后线虫密度概率（横坐标轴顶部）与相应的减产量 $(1-y)\times 100\%$（横坐标轴底部）。根据公式（2）可以得出相对产量（y）的值、根据图 4－4 可以得知 $T=2$ 个虫卵/每克土壤和 $m=0.4$，$M_{av}=150$ 和 $a_{av}=20$ 个虫卵/每克土壤

表 4－2　不同感染度与平均减产量之间的关系

种植频率为下列情况下的相对感染性（%）					平均减产量（%）	减产量大于下列情况的概率（%）					95% 的产区中减产量百分比小于下列情况
1∶1	1∶2	1∶3	1∶4	1∶5		0%	5%	10%	12.5%	15%	
1	2	3	5	7	0.0	0.0	0.0	0.0	0.0	0.0	0.0
2	4	6	9	14	0.0	0.1	0.0	0.0	0.0	0.0	0.0
3	6	9	14	20	0.1	3.7	0.0	0.0	0.0	0.0	0.0
4	8	12	18	27	0.4	18.0	1.2	0.0	0.0	0.0	2.6
5	10	15	23	34	1.2	37.8	7.4	0.0	0.0	0.0	5.8
6	12	18	27	41	2.5	60.0	20.0	1.8	0.0	0.0	8.6
7	14	21	33	47	4.2	70.9	35.4	8.9	2.1	0.0	11.1
8	16	24	36	54	6.1	86.4	51.1	21.9	9.2	2.3	13.7
9	18	27	41	61	8.1	90.5	63.8	37.0	21.4	7.9	16.1

（续表）

种植频率为下列情况下的相对感染性（%）					平均减产量（%）	减产量大于下列情况的概率（%）					95%的产区中减产量百分比小于下列情况
1∶1	1∶2	1∶3	1∶4	1∶5		0%	5%	10%	12.5%	15%	
10	20	30	45	67	10.1	93.3	74.0	51.0	35.4	19.5	18.2
11	22	33	50	74	12.2	95.7	81.4	62.7	49.7	32.9	20.2
12	24	36	54	81	14.1	96.8	86.6	72.3	61.6	47.0	22.2
13	26	39	59	88	16.1	100	91.0	79.3	70.8	59.5	24.2
14	28	42	63	94	17.9	100	93.8	84.4	77.8	69.3	25.7
15	30	45	68	100	19.7	100	95.6	88.4	83.5	76.6	27.5

注：表示相对易感性、种植频率、平均减产量以及在某一特定产区，以第一年 *Pi* 的值为 5 个虫卵/每克土壤种植五种具备部分抗性的马铃薯品种概率之间的关系

但是，当种植了具备高抗性的马铃薯品种以后，这时在这些品种限制种群上的小范围种群密度最大值 *M* 增长十分迅速，随之而来的附加控制成本就会超出所得纯收益的增长，而所要求的马铃薯种植频率却是由其他因素决定，并非由马铃薯胞囊线虫病导致的作物减产决定，因此，对土壤取样的做法就变得陈腐老旧、不合时宜。这时候的土壤取样仅限在几个轮作周期内对是否存在毒性更强的线虫致病型进行一次检测。

二、影响因子敏感性分析

依据几个因数，在给定一个线虫致病型后，ca_{av} 与 cM_{av} 的值应当适用于某一种植区内具备部分抗性的品种。这几个因数，如品种易感性（*rs*）、易感性品种参照物中所包括的致病型的 a_{av} 和 M_{av} 值（在田间试验中，a_{av} 是估计难度最大的）以及在未种植马铃薯的几年内线虫致病型的存活率 *c* 都是试验误差中所涉及的主体。为了能够实际应用，从易感性品种致病型 Pa 3 中间接得到的 a_{av}、M_{av} 值应当尽可能快地用从不同产区、不同年份充足的数据中直接观测到 *a*、*M* 更准确的平均数以及估算出的频率分布被替换掉。

通过采用对易感性控制 a_{av} 值（20）的最佳估算，以及取值为 25 时（假设对最佳估算低估了 20%）计算出产区平均减产量（%）与减产量 > 10% 时的概率。表 4－3 中展示了几个具有不同 *rs* 值的作物品种的值。对 a_{av} 的低估只会造成计算出的作物损失平均值比使用对 a_{av} 最佳估算时略微大些。然而，对 a_{av} 的低估却会导致在单一的种植区内，对作物损失 ≥10% 时的概率低估出的误差相当之大。假设对 a_{av} 的最佳估算值高估

20%（这时 a_{av} 的值应为16），只会导致相应程度的改变。计算出的平均作物损失仅是略微小些，但是对于作物损失≥10%时情况的估算，其概率就会比最初估量的小太多。

图4-6展现的是相对产量对品种相对易感性所采用取值的敏感性。在图中，作者可以看到，与估计 a_{av} 值时产生的相同误差相比，过高估计或过低估计 rs 的值都会对平均减产量造成广泛而深远的影响。当 rs/c 的值为4时，平均减产量可以忽略不计。而当 $5\% < rs \times c < 15\%$ 时，两者之间成线性关系，平均减产量也从每 $rs \times c$ 个单位上增加2%。因此，对 rs 估计的误差为5个百分点时，将会造成与预测中的平均减产量产生 $10\% \times c$ 的误差。这也就说明了通过极为认真的实行对马铃薯品种相对感染性的试验，从而将试验误差最小化是进行测量先决条件。

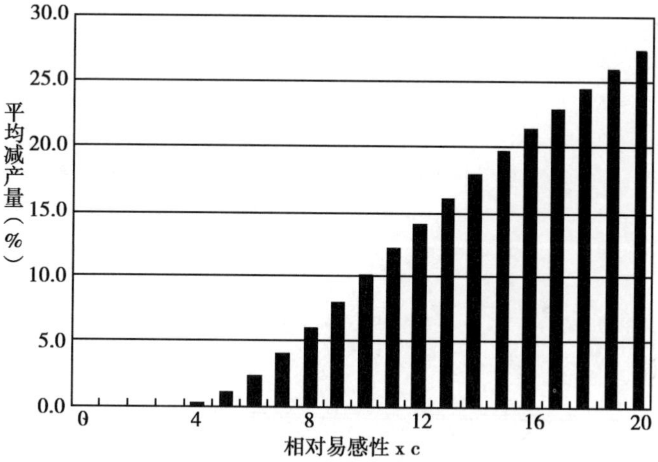

图4-6　马铃薯相对易感性与平均减产量之间的关系

注：表示当第一年未有线虫的情况下，五种拥有同样 ca_{av} 值、cM_{av} 为5个虫卵/每克土壤的作物品种而且具备部分抗性，用其预期产量的百分比对平均减产量估计 $rs \times c$ 值的变化效果所做的敏感性分析

三、相对易感马铃薯轮作计划

在运用固定轮作、胞囊线虫肆虐的种植区，且拥有相同特性的马铃薯品种的产量与其相对易感性呈负相关，但是，两者在线虫控制成本方面并无差异。因此，在其他方面都相似的品种中选择相对易感性最低的一直都是最佳选择，不是因为由线虫导致的作物损失相对更小，就是因为可以施行更短或更具收益的轮作周期。在给定的轮作条件下，单一作物的平均损

失，以及超出一定百分比损失的概率所对应的相对易感性上限仍然能处于可接受水平，这就表明测出相对易感性的上限是十分重要的。因为，一旦大范围内的线虫密度逐渐成形，在接下来的年份里，超出平均种群密度的可能性便随之增加。表 4－2 会为此提供一些基本信息。

表 4－3 呈现出取值范围在 2%～44%，且具有两种不同 Pa 3 致病型的相对易感性作物品种一览表，附带其平均产量损失与损失超过 10% 的概率。

表 4－3　不同马铃薯品种对致病型 Pa 3 两个种群的相对易感性

Pa 3 种群 1	% *rs*	平均减产量（%）								减产量 > 10% 时的概率（%）							
		a = 20								*a* = 25							
		1 : 1	1 : 2	1 : 3	1 : 4	1 : 1	1 : 2	1 : 3	1 : 4	1 : 1	1 : 2	1 : 3	1 : 4	1 : 1	1 : 2	1 : 3	1 : 4
Irene	100	58	50	41	30	58	50	42	31	100	100	100	99	100	100	100	100
Astarte	44	48	30	19	10	48	31	21	12	100	99	87	48	100	100	94	63
Producent	40	45	27	17	8	46	29	19	10	100	98	81	35	100	99	90	50
Proton	31	39	21	11	4	40	22	13	6	100	90	55	8	100	95	70	15
Karnico	30	38	20	10	4	39	21	12	5	100	88	51	6	100	94	66	11
Santé	25	33	15	7	2	34	17	9	3	100	76	27	1	100	86	40	2
Ellen	22	30	12	5	1	31	14	7	2	98	63	13	—	100	75	22	—
Atrela	20	27	10	4	—	29	12	5	1	97	51	6	—	99	65	11	—
Karna 77/281	16	21	6	2	—	23	8	3	—	92	22	1	—	96	34	1	—
Darwina	15	20	5	1	—	21	7	2	—	88	15	—	—	94	24	—	—
Seresta	2	—	—	—	—	—	—	—	—	—	—	—	—	—	—	—	—

Pa 3 种群 2	% *rs*	*a* = 20				*a* = 25				*a* = 20				*a* = 25			
		1 : 1	1 : 2	1 : 3	1 : 4	1 : 1	1 : 2	1 : 3	1 : 4	1 : 1	1 : 2	1 : 3	1 : 4	1 : 1	1 : 2	1 : 3	1 : 4
Irene	100	58	50	41	30	58	50	42	31	100	100	100	99	100	100	100	100
Astarte	40	45	27	17	8	46	29	19	10	100	97	81	35	100	99	90	50
Proton	18	25	8	3	—	26	10	4	1	96	37	2	—	98	51	4	—
Atrela	9	8	1	—	—	10	1	—	—	37	—	—	—	51	—	—	—
Santé	8	6	—	—	—	8	1	—	—	22	—	—	—	34	—	—	—
Darwina	5	1	—	—	—	2	—	—	—	—	—	—	—	—	—	—	—
Seresta	2	—	—	—	—	—	—	—	—	—	—	—	—	—	—	—	—

注：平均减产量以及在 $M_{av}/a_{av.}$ $Pi = 5$ 个虫卵/每克土壤或－≤0.5% 两个取值条件下，减产量大于 10% 时的概率。相对易感性的值是目前所有测量中的平均值，这一点与此前出版的著作有所不同

农民可以自行决定以不同的方式应用这一信息。如果想用固定轮作种

植马铃薯，且两年一次，考虑到不使用其他控制措施（如使用杀虫剂）的情况下可以节省下来的财务总数，那么就可以选择自己计划接受的产量损失风险；也可以选择能满足需求或具有相对易感性的品种，因为这类品种还有其他的培养特点，所以可以保证纯收益的最大化。另外一种可能就是，考虑选择哪一种作物轮作可以使农民所种植的最喜爱的马铃薯品种在有无附加控制措施的条件下都可以实现最大纯收益。

举一个实例：作者研究一下，在6年的时间里，是否具有18%的相对易感性品种在拥有作物轮作比例为1：2时的纯收益要比该比例为1：3时的更好。两种情况中都没有使用杀虫剂。作者认为马铃薯的纯收益为47%，而这一数字在某一特定无寄主作物中则占总收益的27%（Kwantitatieve 信息，1990—1991，IKC-agv & PAGV）。根据表4－3可知，马铃薯作物在比例为1：2的轮作中遭受的平均减产量为8.1%，而在1：3的作物轮作中却为2.5%。这两种轮作中的唯一差异就是利用轮作比例为1：2的马铃薯作物被换成额外附加的非寄主作物。因此，另外三种非寄主作物可以在计算中予以忽略。在接下来的3年里，运用比例为1：2的轮作所获得的平均纯收益为3×（47%－8.1%）/3＝38.9%；而运用的这一比例为1：3时，平均纯收益为［2×（47%－2.5%）＋27%］/3＝38.7%。因此，选用具有18%相对易感性的马铃薯品种，且两年种植一次还是享有微弱优势的。然而，如果选用具有再大一些易感性的品种种植，则作物的轮作比例就应当调整至1：3为好。

第七节　相关的结论与讨论

将公式（1）和公式（2）结合起来，并采用相应参数的频率分布，就可以计算出不同相对减产量的概率。如果对采样区内种群密度的这些相关估算与实际密度十分接近，那么就可以使用抽样结果将其代入公式（1）中。当运用适宜的轮作方式种植具有高抗性的马铃薯品种时，抽样数据就有些老旧了，因为达到最大种群密度 M 的程度时，产量增长幅度却达不到这么高，而产量的增幅是可以平衡抽样与控制成本的。已经计算出的马铃薯白色胞囊线虫 a 和 M 的频率分布，以及不存在马铃薯寄主时种群密度的下降率现已通过荷兰植物保护研究所在20个农民种植区进行的多年田间试验中得以证实。

对于大多数马铃薯品种与线虫致病型组合来说，公差极限 T 的值为2

个虫卵/每克土壤；而公式（2）中有两个十分重要的变量，一个是 P，取值为：先前马铃薯作物的 Pf 值［公式（1）中的输出值］与 c 值（作物轮作因数）的乘积；另一个变量为 m。在荷兰，当 m 的取值为 0.4 时，可以利用从经济角度上具有趣味性的种群密度对产量损失进行可靠性预测。获取相关的计算结果后，农民可以利用某一特定种植频率下的已知相对易感性，评估马铃薯品种种植的风险，从而将根据马铃薯纯收益带来的最高财务收益与自己种植区内其他可供选择的作物进行组合，实现这一概率的最大化。

敏感性分析表明，对作物品种相对易感性的准确估计要比同等情况下对 a 和 M 值［公式（1）］的精确估计重要得多。当筛选具备部分抗性的马铃薯时，应当重点放在稳定实验误差上。目前，荷兰植物保护研究所正在测试 40 多种作物品种对抗多种白色胞囊线虫致病型 Pa 3 种群的相对易感性，这些致病型种群的含毒性从中等到偏高不等。作物栽培与蔬菜大田栽培研究所（PAGV）选用了其中一些品种进行了田间试验。

如上所述，使用咨询系统前，农民需要在心里进行重新定位，因为他们仍然倾向于以获得最大化产量为目标，从而更易于选择安全性最大的品种，因为他们认为其实际产量损失不可预知。因此，使用杀虫剂处理就看似成为一个十分必要的保险手段。但是，在荷兰，根据法定措施，目前对杀虫剂的使用频率仅限于 4 年 1 次。

此外，频繁地使用杀虫剂导致植物种群改变自身来适应土壤，其结果则是加速了熏蒸剂的分解速度（Smelt 等，1989a，b）。因此，化学控制不能再继续被当作“保险”使用，来对抗在工业生产下进行马铃薯种植的产区因线虫攻击而造成的产量损失，这时的种植频率比例普遍为 1∶2。

农民应当努力优化收益而非产量，不仅要减少使用杀虫剂，也要获取更多利益。对此，咨询系统会提供必要的信息，旨在让农民运用更具收益的控制方法。但是，是否会使用这一咨询系统还取决于马铃薯种植者对建议的接受程度。使用咨询系统提供信息的主要动力仍是想要防止因禁止使用杀虫剂以后的几年，种植的马铃薯产量减少这一需求。

参考文献

Andersson S. 1987. The decline of the yellow potato cyst nematode Globodera rostochiensis under non-host crops and under resistant potato cultivars - some preliminary results. Viixtskyddsnotiser 51: 145 – 150.

Andersson S. 1989. Annual population decline of Globodera rostochiensis in the absence of host plants. Nematologica 34: 254.

Cole C S, Howard H H. 1962. The effect of growing resistant potatoes on a potato-root eelworm population – a microplot experiment. Annals of Applied Biology 50: 121 – 127.

Den Ouden H. 1960. Periodicity in spontaneous hatching of Heterodera rostochiensis in the soil. Report of the Fifth International Symposium in Plant Nematology. Nematologica, Supplement II: 101 – 105.

Den Ouden. 1963. Hatching and the rate of multiplication of Heterodera rostochiensis populations of different ages. Nematologica 9: 231 – 236.

Den Ouden H. 1970. De Afname van de Bevolkingsdichtheid van het Aardappel cysteaaltje bij Afwezigheid van Aardappelen. Jaarverslag 1969. Instituut voor Plantenziektenkundig Onderzoek, Wageningen, The Netherlands, page 133.

Den Ouden H. 1974a. The multiplication of three pathotypes of the potato root eelworm on different potato varieties. Netherlands Journal of Plant Pathology 80: 1 – 6.

Den Ouden H. 1974b. Afname van de Bevolkingsdichtheid van het Aardappel cysteaaltje bij Teelt van Resistente Aardappelen en bij Braak. Jaarverslag 1973. Instituut voor Plantenziektenkundig Onderzoek, Wageningen, The Netherlands, page 96.

Forrest J M S, Holliday J M. 1979. Screening for quantitative resistance to the white potato cyst nematode Globodera pallida. Annals of Applied Biology 91: 371 – 374.

Greco M, Di Vito A, Brandonisio I, Giordano I. 1982. The effect of Globodera pallida and G. rostochiensis on potato yield. Nematologica 28: 379 – 386.

Huijsman C A. 1961. The influence of resistant potato varieties on the soil population of Heterodera rostochiensis Woll. Nematologica 6: 177 – 180.

Jones F G W, Parrott D M, Perry J N. 1981. The gene-for-gene relationship and its significance for potato cyst nematodes and their Solanaceous hosts. Pages 23 – 36 in Zuckerman, B. M., Rohde, A. R. (Eds.) Plant Parasitic Nematodes, Vol. III. Academic Press, Inc., New York,

USA.

IKC-agv, PAGV. 1990. Kwantitatieve Informatie voor de Akkerbouw en de Groenteteelt in de Vollegrond, Bedrijfssynthese 1990 - 1991. Publicatie nr. 53.

Phillips M S. 1984. The effect of initial population density on the reproduction of Globodera pal/ida on partially resistant potato clones derived from Solamum vernei. Nematologica 30: 57 -65.

Phillips M S, Trudgill D L. 1983. Variations in the ability of Globodera pallida to produce females on potato clones bred from Solanium vernei or S. tuberosum SSP. Andigena CPC 2802. Nematologica 29: 217 -226.

Phillips M S, Trudgill D L. 1985. Pot and field assessment of partial resistance of potato clones to different populations and densities of Globodera rostochiensis. Nematologica 31: 433 -422.

Phillips M S, Trudgill D L, Evans K, et al. 1987. The assessment of partial resistance of potato clones to cyst nematodes at six different test centres. Potato Research 30: 507 -515.

Schomaker C H, Been T H. 1992. Sampling strategies for the detection of potato cyst nematodes: Developing and evaluating a model. Pages 182-194 in Gommers F J, Maas P W Th (Eds.) Nematology from Molecule to Ecosystem. Dekker and Huisman, Wildervank, The Netherlands.

Seinhorst J W. 1967. The relationships between population increase and population density in plant parasitic nematodes. II. Sedentary nematodes. Nematologica 13: 157 -171.

Seinhorst J W. 1970. Dynamics of populations of plant parasitic nematodes. Annual Review of Phytopathology 8: 131 -156.

Seinhorst J W. 1982a. The relationship in field experiments between population density of Globodera rostochiensis before planting potatoes and the yield of potato tubers. Nematologica 28: 277 -284.

Seinhorst J W. 1982b. Achtergronden van aaltjesbestrijding (3). Economische aspecten van vruchtwisseling en chemische bestrijding. Bedrijfsontwikkeling 13: 494 -500.

Seinhorst J W. 1984. Relation between population density of potato cyst nematodes and measured degrees of susceptibility (resistance) of resistant potato cultivars and between this density and cyst content in the new generation. Nematologica 30：66 –76.

Seinhorst J W. 1986a. The development of individuals and populations of cyst nematodes on plants. Pages 101 – 117 in Lamberti F, Taylor C E (Eds.) Cyst Nematodes. Plenum Press, New York, USA/London, UK.

Seinhorst J W. 1986b. Agronomic aspects of potato cyst nematode infestation. Pages 211 – 228 in Lamberti F, Taylor C E (Eds.) Cyst Nematodes. Plenum Press, New York, USA/London, UK.

Seinhorst J W. 1986c. Effect of nematode attack an the growth and yield of crop plants. Pages 191 –210 in Lamberti F, Taylor C E (Eds.) Cyst Nematodes. Plenum Press, New York, USA/London, UK.

Seinhorst J W. 1993. The regulation of numbers of cysts and eggs per cyst produced by Globodera rostochiensis and G. pallida on potato roots at different initial egg densities. Nematologica 39：104 –114.

Seinhorst J W, Oostrom A. 1984. Comparison of multiplication rates of three pathotypes of potato cyst nematodes on various susceptible and resistant cultivars. Mededelingen Faculteit voor Landbouwwetenschappen, Rijksuniversiteit Gent 49 (2b)：605 –611.

Seinhorst J W, Oostrom A, Been T H, et al. 1995. Relative susceptibilities of eleven potato cultivars and breeders' clones to Globodera pallida pathotype Pa3, with a discussion of the interpretation of data from pot experiments. European Journal of Plant Pathology (in press).

Seinhorst J W, Been T H, Schomaker C H. 1993. Partiele resistentie in de bestrijding van aardappelcysteaaltjes Globodera spp. I：Bepaling van de graad van resistentie. Gewasbescherming 24 (1)：3 –11.

Smelt J H, Teunissen W, Crum S J H, et al. 1989a. Accelerated transformation of 1, 3- dichloropropene in loamy soils. Netherlands Journal of Agricultural Sciences 37：173 –183.

Smelt J H, Crum S J H, Teunissen W. 1989b. Accelerated transformation of the fumigant Methylisothiocyanate in soil after repeated application of Metham-sodium. Journal of Environmental Sciences and Health, Part B. Pesticides, Food Contaminants and Agricultural Wastes 24 (5)：437 – 455.

第五章 马铃薯晚疫病的成因分析

由真菌致病疫霉引起的马铃薯晚疫病是马铃薯作物最严重的疾病，此病害可以完全摧毁马铃薯作物。人们可通过对作物叶子频繁地使用杀菌剂而控制疫病，但自由使用农药也正获得越来越多的关注。块茎产量的损失取决于疫病所破坏叶子量，反过来，叶子的破坏量又取决于疫病侵入的时机及病叶增加率。马铃薯疫病模型尝试利用影响疫病的诸多因素来阐明疫病发病起始和发病过程。马铃薯晚疫病有性繁殖产生卵孢子，此种繁殖方式对于该真菌的田间生存和基因型多样性的增加可能非常重要。然而，此疾病多由其无性繁殖周期导致，其无性繁殖周期是由孢子萌发、叶片感染、叶片定殖、产生孢子、孢子扩散及存活所组成。马铃薯晚疫病无性繁殖周期的不同阶段易受多种环境因素的影响，这些环境因素很大程度上取决于天气条件、基因型及病原（马铃薯晚疫病菌）和寄主的生理特征。大多疫病的影响因素间存在复杂的相互作用，影响难以量化。此外，许多影响因素不断变化并在作物水平和垂直方向的影响各不相同，所以马铃薯晚疫病现实模型的开发较为困难。

第一节 简介

晚疫病（疫病）是马铃薯作物最严重的疾病，能彻底摧毁作物。1845年6月疫病最早出现在欧洲的比利时并迅速蔓延。短短几个月内，马铃薯晚疫病摧毁了整个爱尔兰马铃薯作物（Bourke，1965），而马铃薯是大多数人的主要食物，因此数百万人或死于饥饿或移民逃荒。疫病是由病原真菌致病疫霉引起，此病原真菌可感染并杀死马铃薯植株的叶、茎及块茎。虽然马铃薯植株的基因抗病性日益重要，但自19世纪以来，杀菌剂已被用于疫病防治，今天人们也在频繁使用杀菌剂以控制疫病。农药的自由应用近年来一直受到质疑，人们担忧农药对人和环境有潜在的有害影响。因此作者希望用最少剂量的杀真菌剂而能有效防治疫病。人们通常每7～14d

对马铃薯作物茎叶喷施杀真菌剂以防治疫病，所以在马铃薯生长季节可能有多达 10 种不同防治方式（Krause 等，1975）。但是，人们早就知道疫病发病日期和发病程度尤其取决于天气。疫病与低温、潮湿的气候条件相关联，并且只有潮湿叶片才会感染病原菌（Rotem 等，1971）。依赖于天气的晚疫病的破坏本质和它的实现过程，为晚疫病预测系统的发展提供了动力，这种系统可以用来为杀虫剂的使用提供时间表。Van Everdingen（1926）根据温度、雨量、云量和露水时间制作了第一个预测疫病发病的系统。随后，又有很多人建立或者完善了新的预测系统（Harrison，1992）这也使得疫病的预测要比其他植物疾病的预测更加先进。

预报系统带动了计算机程序的开发如 BLITECAST（Krause 等，1975），这些程序的开发推进了之后描述病害发展的计算机模拟（Bruhn 和 Fry，1981）。随着系统的复杂性不断增加，一些系统现在考虑残留杀菌剂活性和马铃薯晚疫病对杀菌剂的耐受性（Levy 等，1991），其他系统根据天气预报（Raposo 等，1993）。最近，学者把描述疫病发生的计算机模型与作物生长的模型联系起来用以模拟寄主和病原的相互作用，提供病害进展与作物产量的估计（Van Oijen，1992）。

计算机模拟可用于评估改变一个或多个因素对疫病危害程度及其对马铃薯作物生长和产量的影响。虽然计算机模拟模型计算结果更加快速并且只需要相比于田间试验更少的劳动力投入，模拟的准确度依赖于对所有影响因素真实的评价及观测参数应用的完整性。由于疫病影响因素还没有被人们完全了解，计算机模型包含近似值、假设和猜测，所以其潜在的错误较多。本章以综合现有知识的视角，考虑了影响晚疫病发生的众多互作因素，发展了计算机模型。

第二节 马铃薯晚疫病菌的生命周期

有关致病疫霉生命周期及其导致相关病害周期的知识对于理解疫病发生如何受各种因素影响至关重要。致病疫霉的生命周期如图 5－1 中所示，这在很大程度上不言而喻。周期中的无性繁殖部分，即孢子囊的产生，游动孢子的游离和萌发产生菌丝，此过程在发病过程中占主导地位。有性阶段产生的可以在田间长周期存活的游动孢子需要 A1 和 A2 交配型。最近才确定了在中美洲以外它的重要性，但在欧洲没有证据表明有性繁殖会在田间频繁发生（Drenth，1994）。有性生殖过程中的遗传重组为晚疫病菌基因型多样性提供了一种机制，以便它可以更容易地适应逆境，例如，提高抗药性或克服宿主抗性。拟有性机制也可以发生基因重组（Leach 和 Rich，1969），但其在增加马铃薯晚疫病基因多样性的过程中可能仅起到次要作用。

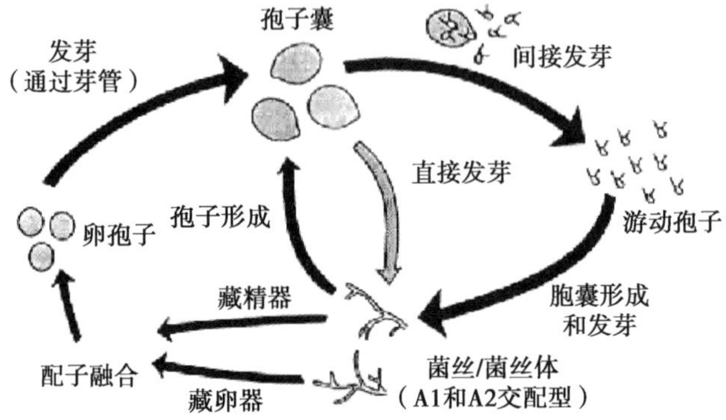

图 5－1 致病疫霉的生命周期

第三节 马铃薯晚疫病发展进展介绍

疫病周期（图 5－2）反映了病原体发生在宿主上的生命周期，叶片侵染、病斑膨胀、产生孢子、孢子扩散和存活，此过程在很大程度上决定了疾病的发展。

孢子囊源于染病块茎长出的枝条，往往聚集成堆（Van der Zaag，1956），容易被风吹到健康作物上，形成初级接种物。疫病发病通常是由一个单一的孢子囊或几个孢子囊引起叶片病变而引起。病变扩大，病原体会在侵染后几天形成孢子囊（图 5－3），并最终在单个病灶可形成约 10^6 孢子囊（Harrison 和 Lowe，1989）。孢子囊随风雨传播并会侵染更多的叶片（Fitt 和 Shaw，1989），所以无限蔓延的疫病遵循复利模式（Van der Plank，1963），感染疫病的叶片呈爆炸性增长（图 5－4）。侵染周期和病变扩展示于图 5－4，呈现出不规则的近“S”形的病变扩展曲线。作者尝试使用疫病数学模型和计算机模拟来描述该曲线，但其形式是由多种因素确定，一些因素有利于其扩展，一些则不是（图 5－5）。被冲入土壤中的孢子可以侵染块茎，使其成为传染源。

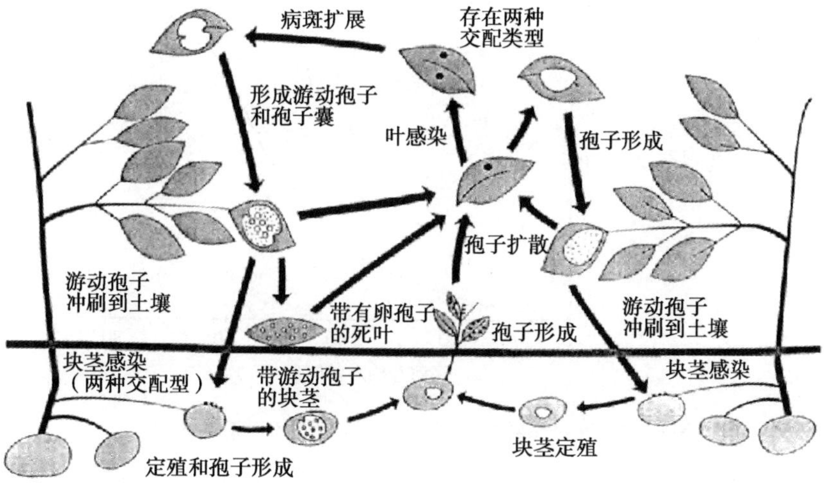

图 5－2　马铃薯晚疫病的病害周期

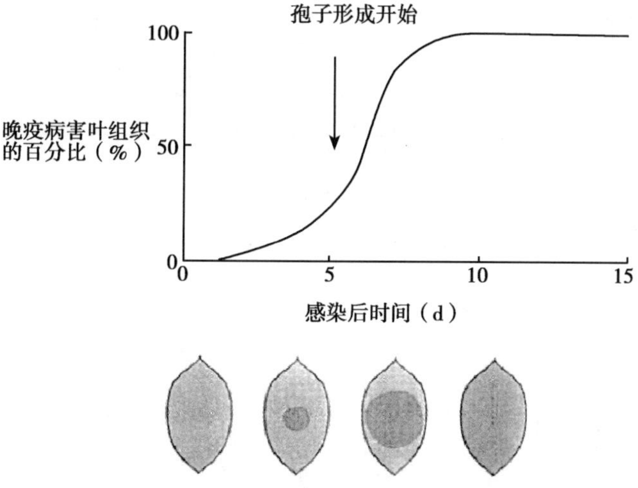

图 5－3　恒定条件下马铃薯叶上单病斑面积的增加

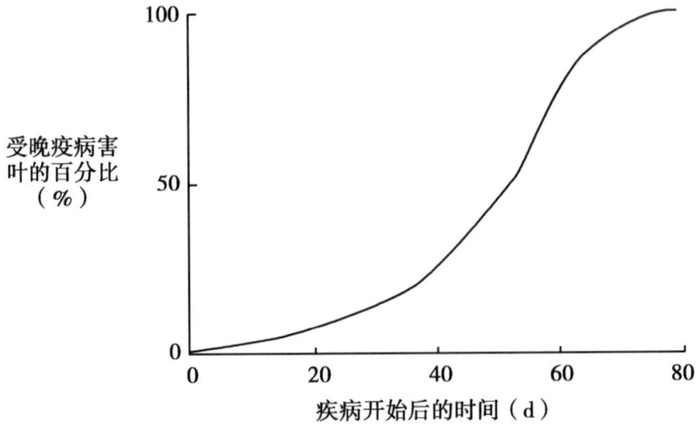

图5－4　整株马铃薯总叶片病斑面积的假设增加曲线

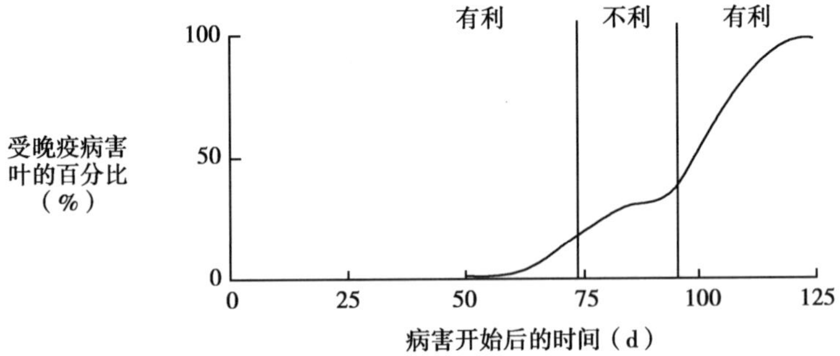

图5－5　有利和不利条件下马铃薯晚疫病的进展曲线

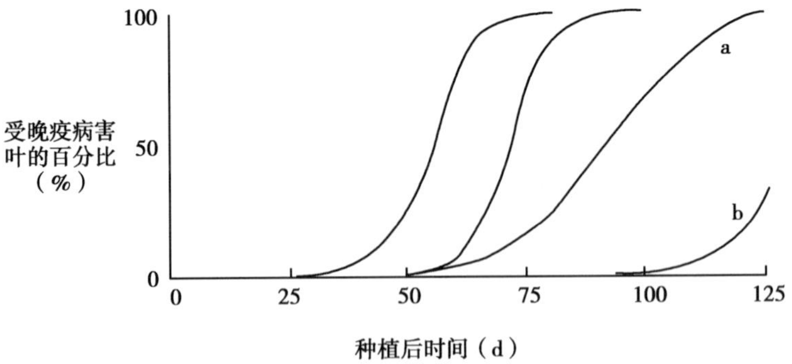

图5－6　（a）马铃薯晚疫病发病时间和（b）发病速率对病害发展影响的曲线

第四节 马铃薯晚疫病对马铃薯生长和单产的影响

疫病对马铃薯作物生长和块茎产量的影响取决于病害扩展曲线（图5-6）。在植株未衰老情况下，未感染组织的光合作用潜力不会受到其他感染病原体部位的影响，收获产量的损失几乎完全是由于病原体导致的叶面积减少（Haverkort 和 Bicamumpaka，1986；Van Oijen，1990；1991a，b）。Van Oijen（1991b）认为病害扩展曲线下面积与产量密切负相关。显然，如果疫病发展缓慢，或者大量块茎已经长成，则收获损失将相对小（图5-6）。但是，染病块茎会在存储时频繁发生腐烂（Dowson 和 Rudd Jones，1951），所以疫病也可引起收获后的损失。

第五节 影响马铃薯晚疫病菌无性繁殖周期的因素分析

影响疫病发生的时机及其扩展的因素可分为3大类：环境因素（主要是由天气条件决定）、生物因素和杀菌剂。生物因素可分为宿主，即马铃薯植物；病原体，即致病疫霉。但是，不同生物因素之间复杂相互作用对晚疫病的影响，不能被过度强调。为了解各影响因素如何影响疫病，有必要研究的它们之间的相互作用，尽管结果研究可能与特定的田间条件相关。

田间环境的动态性使得问题进一步复杂化。天气、宿主和病原体连续变化，病原体基因型也常常是连续变化的。天气的变化有连续性，寄主和病原体的生理结构以及病原体的基因型通常也是如此。可控条件下的实验结果可能难以与田间实验结果相关联。

因为无性生活周期主导了疫病发展，并且通常是疫病的唯一决定因素，所以我将讨论各种因素对致病疫霉侵染和定殖叶片、产生孢子、孢子扩散和孢子存活的影响。

一、马铃薯晚疫病自然环境因素

（1）温度。孢子囊可以通过直接形成芽管或间接释放孢子的方式发芽，但直接发芽不能造成叶感染（Schober 和 Ullrich，1985）。虽然游动孢子在温度接近0℃时也会缓慢地释放，但间接发芽的最佳温度在12～16℃之间（Crosier，1934；Yamamoto 和 Tanino，1961）。在较高温度下的孢子囊发芽比例增加，虽然间接发芽温度远高于20℃。游动孢子形成胞囊和开始发芽前的能动时间高度依赖于温度。当温度接近0℃时，它们的游动时间为24h，但24～25℃时，仅为15～20min。游动孢子对干燥非常敏感，脱离水几乎立即死亡。它们在侵染植物以前依赖植物表面的残留水分存活，所以温度和水分持续时间共同决定游动孢子是否发生侵染。Crosier（1934）和 Rotem 等（1971）观察到孢子在3～28℃之间可以萌发，21～24℃之间游动孢子萌发最快。芽管21℃伸长最快，接种2～3h 即可发生侵染。

根据菌丝生长将定殖芽管从植物组织中分离不现实，而事实上，菌丝与芽管生长需要类似的温度，2～4℃菌丝缓慢生长，28～30℃停止生长。（Zan，1962）。即使在菌丝最适生长温度下，病原体接种后至少2～3d，肉眼才能看见病灶。孢子囊的产生（芽孢形成），可认为是菌丝生长改变的过程，有着与其萌发和组织定殖类似的最低温度，但孢子形成的最高温度为25～26℃，低于其他生理过程的最高温度。经研究，适宜孢子囊形成及其最终丰富程度的温度为16～22.5℃。新鲜孢子囊可以从病灶移除旧孢子后数小时在最适条件下形成。

温度可能不直接影响孢子传播，但温度是晚疫病菌存活的主要影响因素。可以杀死病原菌的低温条件同样会彻底杀死马铃薯作物，因此在马铃薯生长季节，此低温条件与晚疫病发展无关。因此温度对晚疫病菌存活的影响主要是高温影响。孢子囊在空气中的存活取决于多种因素的相互作用，这些因素主要包括温度、湿度、辐照和孢子的生理状态，由诸如其年龄和含水量因素决定。例如，由辐射改变所引起的温度变化，影响了孢子囊相对湿度和失水速率，及其对高温潜在破坏性影响的敏感性。孢子囊在高于25℃的空气中迅速死亡（Smith，1915），并且 Wallin（1953）报道，20℃环境减弱了晚疫病菌生存能力。但是，病原菌在不同温度下的生存能力高度依赖于湿度。Rotem 和 Cohen（1974）发现，孢子囊沉积的叶表面通常由于蒸腾作用有很高的湿度，并在不利条件下有良好存活。有关温度直接影响游动孢子存活的信息很少，但植物表面的干燥，会导致游动孢子失水，此过程本身受温度影响。马铃薯晚疫病菌一旦在植物组织存活，就会对高温有较高的耐受性。病原菌可能在高于30℃时不会增长，但其可以在较高温度下存活，即使是短时间内温度高于40℃，其依然可以在天气变得更有利时重新开始增长（Rotem 和 Cohen，1974），虽然暴露于高温可能会削弱其产孢能力（Kable 和 MacKenzie，1980）。了解叶和茎的温度比环境温度高较多非常重要，特别是当叶子直接暴露于阳光下（Ansari 和 Loomis，1959）。

中等土壤温度有利于块茎感染病菌，高于18℃较为不利于其感染病菌（Sato，1979）。温度和其他因素（如植株表面水分和湿度量保持时间）之间的相互作用过程中，当其他因素为有利因素时，温度限制作用较大；当其他因素为不利因素时，温度限制作用较小（Rotem 等，1971）。温度通过影响马铃薯作物进而间接影响马铃薯晚疫病的无性繁殖（本章后面讨论），并会导致杀真菌剂活性丧失。

（2）湿度。湿度是晚疫病发展的一个主要的决定因素，但人们不可能精确控制湿度，或测量发生感染和产孢区域的湿度，因为没有可以测量邻近植物表面湿度的小型设备。蒸腾加湿叶片临近的空气，因此，虽然作物上方的空气相对干燥，但靠近叶的空气湿度接近饱和。叶表面湿度的递减等级取决于水蒸气的蒸发速率，而水蒸气的蒸发速率反过来又取决于空气的流通速度。此外，被日光加热的叶片又会加热相邻空气并降低其相对湿度（Relative Humidity，RH），结果导致对叶旁湿度的估算更加复杂。这些作用的最终结果是不知道湿度是如何影响植物表面微生物。然而，孢子囊和游动孢子仅在有液态水的情况下才会发芽，所以湿度通过影响表面水分干燥速率而间接影响晚疫病菌侵染。

一旦马铃薯晚疫病在马铃薯植株内成功侵染，80%～100%的相对湿度对晚疫病菌定殖无影响（Harrison 和 Lowe，1989）。较低的湿度可能会使得严重感染的组织较为干燥，结果限制了病原体用水并减缓其增长速度。

有大量报道指出仅在高湿度下孢子囊才会形成（如 De Weille，1963），但其对临近植物表面的具体要求条件仍未知。Harrison 和 Lowe（1989）证明了在决定单叶上形成的孢子囊数量过程中，环境湿度和气流速度存在相互作用。当环境湿度为90%～100%时，低气流速度下有大量孢子囊形成，但在85%或80%相对湿度下则不然；在100%相对湿度和较高的气流速度下也有大量孢子囊形成，但在80%～95%的相对湿度下则不然。这些结果表明，只有病原体周围湿度处于或非常接近饱和时，孢子囊才会形成。Harrison 和 Lowe 也调查了80%和100%交替相对湿度环境下的产孢量。形成的孢子囊的数量介于每个相对湿度形成的孢子数量之间，并高于90%相对湿度下所形成的孢子数。

Hirst（1958）认为，湿度变化有助于孢子囊扩散传播，虽然他观察到的相对湿度的变化很大程度上是温度变化的结果。Rotem 和 Cohen（1974）认为，低湿度会提高孢子囊扩散传播。湿度对孢子的存活有较大影响。晚疫病菌的孢子囊可在饱和空气中存活数天（Rogoshin 和 Filippov，1983），并能在叶面良好存活，但在干燥空气生中会迅速丧失生存能力（Rotem 和 Cohen，1974），高温条件较低温条件下更是如此。Warren 和 Colhoun（1975）认为，孢子囊的存活依赖于其周围的水膜，无水膜条件下，孢子囊在95%相对湿度的空气中，仅5min就会丧失生存能力。Minogue 和 Fry（1981）提出的证据表明，部分干燥的孢子囊的萌发能力取决于随后的补水率。虽然只有少数的孢子会在快速补水后发芽，但存活率较低。关于孢子囊存活与环境条件的关系，几乎可以确定取决于孢子年龄（本章后面讨论）。在没有液态水的条件下，游动孢子会迅速死亡（Lapwood，1968），并且没有不同湿度下生存率定量数据。

（3）降水和露水。大雨、小雨、雾、喷灌和结露都会导致叶片潮湿，潮湿的叶片为晚疫病菌侵染的先决条件。露水通常随着夜间气温的下降而形成，其在叶片上形成的速率与温度、环境湿度和风速有关（Collins 和 Taylor，1961），相反，叶表面水的沉积速率和蒸发也与温度（部分由照射强度决定）、环境湿度和风速有关。这些气象变量因此相互作用，以决定表面水分持续时间，以及孢子囊释放游动孢子、游动孢子萌发及感染宿主的可能性。表面水分持续时间与接种密度和温度进行相互作用，决定孢子的发芽和侵染是否发生。当达不到最佳温度时，因为温度偏离最佳，更长时间的潮湿环境是孢子发芽的必要条件；但潮湿期较短时，高孢子量抑制了其侵染能力（Rotem 等，1971）。接种后3h内，孢子发芽时遇到叶片短时间的干燥环境，会大幅度的降低病变发生数量；但在接种3h之后叶片干燥环境对病变发生影响较小，据推测是因为病菌侵染已更顺利地完成（Hartill 等，1990）。随后，表面水分可能对植物纤维定殖影响很小或没有影响。虽然孢子囊形成于降雨或露后，但此过程并不需要液态水（Harrison 和 Lowe，1989）。与叶片周围较高的湿度相关的叶表面水分无疑有助于孢子形成，而不是液态水本身。叶上大量的水确可以抑制致病疫霉孢子形成（Rotem 等，1978）。

（4）降水和风速是影响晚疫病菌孢子囊扩散的主要因素。大雨、小雨和喷灌可将叶和茎上孢子冲洗至土壤中，这些孢子在土壤中在数周内都具有感染性（Lacey，1965），并可能侵染块茎（Lapwood，1977），或飞溅到叶上以引发病变。降水强度和持续时间决定其发生程度。孢子囊很容易被纳入叶上的水中。当水滴撞击到叶或茎上的孢子囊时，便会产生含有孢子的飞溅水滴，从而完成病原体的短距离传播（MacKenzie 等，1983）。Fitt 等（1986）发现，病叶上的真菌孢子在无条件下通过水滴飞溅最远可传播 1m。孢子数量与随病源距离的增加呈指数下降趋势，并且随着高度的增加也迅速下降。只有不到 10% 孢子会飞溅到 300 多毫米的高度。相比于小雨滴，大雨滴会传播分散更多的孢子。因此，降水的持续时间和强度及水滴粒度分布对于孢子传播范围很重要。Hirst（1958）认为，大约有一半的作物叶子的枯萎病病变起源于雨水从一片叶子滴落到另一片叶子上。叶片上一小部分的含孢子水滴可以被分散至气流中。风在致病疫霉扩散传播中的作用将在下面部分讨论。雨水对于空气中孢子囊的除去非常重要（Hirst，1953）。

（5）风。就目前所知，风并不影响晚疫病菌的组织定殖，但可间接地通过逐渐蒸发的表面水分而影响马铃薯地上部分感染病菌。风通过影响叶片周围湿度进而影响孢子形成，叶片附近的湿度取决于茎叶量、单叶的几何形状和叶朝向。树冠的下部茎叶更加荫蔽，此处风速低于树冠顶部，所以通常从地面随着高度的增加，湿度呈梯度降低，这体现在树冠上部具有产生较少孢子的倾向。然而，气流随着风速的增加，可降低湿度和温度的梯度变化，从而在较高的风速时冠层内的环境变得更均匀。紧邻作物顶部的风速也随时间变化—风通常是阵风，并在空间上取决于树木和建筑物的阻力和地形。多种因素动态互动决定了作物内空气流动、表面水分的持续时间和适合孢子形成和萌发的环境条件，使得难以从气象数据量化已形成的孢子和发生的侵染。某些环境可能有利于田间某些地块的芽孢形成和侵染，而对另外一些地块则不然。风的主要作用是传播孢子囊，但几乎没有孢子仅靠风分离，即使是在高达 15 m/s 风速下情况也是如此（Stepanov，1935；Hirst，1958）。风可以吹走含孢子的被雨水飞溅的液滴，进行株内或邻地传播（Hirst 和 Stedman，1960），结果常常导致晚疫病的扇形传播。风是孢子可以长距离传播的唯一途径。一旦孢子进入作物上方的空气，包含孢子囊的液滴在强风下飞行数千米高度达 1km（Robinson，1976）。风对于晚疫病远距离传播的重要性，当然也取决于孢子的侵染能力和益于孢子侵染的凉爽、湿润、阴暗的条件。

（6）辐射。太阳辐射光谱对致病疫霉有直接和间接的影响，并且是减弱孢子囊存活的重要因素（Rotem 等，1970）。间接影响是辐射加热了植物、土壤和临近空气，降低了相对湿度并提高蒸发速率。Bashi 等（1982）认为高温提高了死亡率，充分膨大的孢子囊暴露于光下受到的伤害低于部分干燥的所受的伤害。有关太阳辐射对马铃薯晚疫病菌无性增殖直接影响的信息很少。紫外辐射在低强度时可以增加能够发芽并侵染马铃薯作物的孢子囊比重，但在高强度下可以杀死孢子（De Weille，1963；1964）。有关可见光辐射对孢子囊的产生及对定殖作用的影响的报道相互矛盾，可能仅仅是由于宿主的生理变化。

二、马铃薯晚疫病生理影响因素

（1）遗传抗性。马铃薯植株对于疫病有两种形式的遗传抗性，分别由主要和次要基因控制，虽然它们之间的区别有时并不清楚。主要和次要基因经常在马铃薯植株中一起出现。主基因抗性也被称为物种特异性抗性，通常在侵染阶段起作用，能有效地引起植株对晚疫病总的抗性。到目前为止，至少 11 种主要的抗性基因（R1 ~ R11）已被确定，每个基因都可引起植株对晚疫病的免疫反应。然而，对于每个寄主抗性基因，病菌都具有相应致病基因（致病基因或因子 1 ~ 11）可完全克服寄主抗性基因，使得病菌具有易感性。分离的具有致病基因 1（即可以克服马铃薯抗性基因 R1）的晚疫病菌株系称为种族 1 等。一株马铃薯植株可能没有任何抗性基因（RO），其易受所有种族晚疫病菌的侵染，同样其也可具有一个或多个抗性基因。同理，晚疫病菌也可能没有致病基因（种族 0，Race 0），或有一个或多个致病基因。只有晚疫病种族与马铃薯作物基因型匹配，才会发生疫病。然而，致病疫霉容易突变以克服抗性基因，这一过程很短暂（Malcolmson，1969），所以马铃薯抗性基因控制疫病的能力受到一定影响，虽然许多马铃薯品种具有多个抗性基因。有证据表明，并非所有的抗性基因基表达程度相同。特别是抗性基因 10（R10）有时不完全表达（Wastie 等，1993），使得无致病基因 10 晚疫病菌株系可侵染寄主组织，尽管是以一个被减弱的速率侵染。主基因抗性的表达部分取决于马铃薯植株生理状态，其生理状态可能取决于生长环境和植株龄期（Stewart，1990）。虽然叶和块茎的抗性都由抗性基因控制，但这两者的抗性相关性较差（Stewart 等，1992）。例如，抗性基因 R2 和 R3 可以为叶提供无致病因子 2 和无致病因子 3 致病疫霉的免疫，但其可能会在块茎上生长。相比之下，抗性基因 R1 可以为叶片和块茎都提供免疫（Roer 和 Toxopeus，1961）。这些差异可能是由于生理、生化因子的作用，例如只有块茎会产生倍半萜类植物抗毒素（Rohwer 等，1987），或是由于其他遗传因子的参与所致。

学者认为次要基因抗性是由许多基因控制，其效果可以叠加。它可以减少可侵染叶片的游动孢子的比例，但其主要作用是降低定孢子植率（Colon 等，1992），所以次要基因抗性并没有赋予植株免疫力。但不同抗性组分的相对重要性取决于品种。多基因控制下次基因抗性要比主基因抗性更持久，并具有使用最小量的杀菌剂达到长期控制疫病潜力（Shtienberg 等，1994）。但主基因抗性和次基因抗性有时难以区分（格斯和 HohI，1988）的效应，并且有证据表明两种形式的抗性之间也存在协同作用（Darsow 等，1987）。

最近几年越来越多的证据表明某些品种的次基因抗性的表达受环境的强烈影响（Simmonds 和 Wastie，1987）。这种基因型-X-环境的相互作用由 Harrison 等（1994）发现，他们确定了 3 个环境因素（光照、光照强度和温度），每个因素能有效地打开或关闭叶片的次基因抗性。根据不同品种，抗性可在一个或多个环境因素间切换，或不受任何影响。主基因抗性也可能会被类似环境影响的可能性也不能排除。次基因抗性也可以决定块茎对晚疫病的抗性（Bjor 和 Mulelid，1991）。再次，叶和块茎抗性可能不密切相关。一旦发生侵染，无论主或次基因抗性似乎都对晚疫病菌孢子有直接影响。

叶冠层结构在很大程度上由遗传因素决定，影响晚疫病的易感度，例如，通过阻碍空气循环致使叶片缓慢变干及保持树冠高湿度。

（2）生理。马铃薯植株的生理状态对晚疫病的易感性有很大影响。尽管人们对其所涉及的机制知之甚少，但某些影响侵染和定殖的因素已被确定。马铃薯植株的生理状态通常间接影响产孢，仅反映了病菌定殖程度。

由生长条件、温度和生长期的长度决定的植株生理年龄，影响叶的易感性。幼小植株的叶片是高度易受侵染的。随着叶片长大，抗侵染性增加，后又变得易于感染（Stewart，1990）。Carnegie 和 Colhoun（1982）报道大多数的晚疫病抗性品种的易感性差异巨大，这是不言而喻的。大多有关植株龄期对疫病易感性差异的研究都没有区分易感性差异对病菌侵染、定殖和产孢的影响（例如 Darsow 等，1988）。

叶位置通常易与叶龄混淆，也影响易感性。Fry 和 Apple（1986）认为，老叶是最容易受侵染的，但 Warren 等（1971）发现，成熟植物茎中部区域的叶抗性最强，对于幼小植株叶位置不影响其易感性。Populer（1978）报道，叶位置对易感性的影响取决于品种。疫病对叶组织的定殖率都类似地取决于叶片的位置和龄期。Carnegie 和 Colhoun（1982）报道，病变由植物顶部到底部的扩张速度大致呈线性增加，但这些差异不存在于幼苗植株。正如叶片，块茎的龄期会影响其对疫病的易感性（Pathak 和

Clarke，1987）。马铃薯品种成熟早晚特征是改变与叶片生长有关差异的疫病发病率的另一个影响因素（Van Oijen，1991b）。早熟品种一般都较晚熟品种更少受到疫病侵染（Kolbe，1982），可能是因为作物生长往往已经完成并在病菌未引起大范围疾病前已收获。作物抗侵染、组织定殖或产孢子的能力可能不受品种成熟早晚的影响。

寄主植物的矿质营养可能影响其对疫病的易感度，Rotem 和 Sari（1983）认为叶片受侵染、病菌定殖和病菌产孢都受到寄主营养状况的影响。生长在高氮环境的植株叶片和块茎易受到疫病侵染，并且具有基因型差异（Main 和 Gallegly，1964）的植株病变会有更高的扩增速率（Carnegie 和 Colhoun，1983；Kurzawinska，1989）。Awan 和 Struchtemeyer（1957）报道，大量施用磷或钾可减小病变的面积，并且磷对此具有更大的影响。叶片的营养水平可通过潜在形成孢子囊数目改变病斑直径。Borys（1964）发现，矿质营养特别是氯离子和硫酸根离子，强烈地影响叶片的易感度，影响大小取决于叶龄和品种。有些矿质营养，尤其是氮元素，对疫病易感度可能是间接的影响，通过促进茎叶生长或推迟叶片衰老未达到。马铃薯植株的碳水化合物水也影响植株对疫病的易感度，但其重要性尚不明确（Grainger，1979）。Carnegie 和 Colhoun（1980）报道，水分胁迫下的马铃薯植株对致病疫霉的抵抗力增加，但他们没有任何有关这种现象的解释。

（3）病毒感染。感染病毒的马铃薯植株的叶（Pietkiewicz，1974）和块茎（Fernandez de Cubillos 和 Thurston，1975）对疫病的抵抗力增加。病毒增加了马铃薯植株对致病疫霉侵染和定殖抵抗力，延缓了孢子囊形成，降低了孢子囊数量。然而，Richardson 和 Doling（1957）研究表明，感染马铃薯卷叶病毒的卷叶表面比未感染的平面叶片的湿度保持时间更长，有助于晚疫病侵染，所以感染此病毒的田块晚疫病灶更多，虽然实验室测试感染病毒的植株叶片比未感染病毒的更不容易感染晚疫病。

三、表面微生物菌群

自然存在的定殖于块茎表面的细菌最近被证明对晚疫病菌起拮抗作用（SA Clulow，个人通信），但自然发生在叶面的微生物其无论是具有促进或者抑制作用的影响效果还未知。

四、生物因素——马铃薯晚疫病基因型

致病因子在克服叶片或块茎抗性基因对晚疫病的抗性的重要性已经说明。马铃薯晚疫病菌可通过突变（Denward，1970）或体细胞变异（Caten

和 Jinks，1968）获得新的致病因子。即使没有有性生殖，一起生长的两个不同株系的菌落 Leach 和 Rich，1969；Malcolmson，1970）或通过不同的抗性基因单菌株寄主可产生新的病菌株系（Graham 等，1961）。田间环境对作物上致病疫霉株系有强烈的选择压力，所以病菌群体结构在作物生长季可能会发生变化。

马铃薯晚疫病的侵染能力是一个与致病毒力无关的特性，取决于病菌株系并决定着病菌感染率、病菌对叶和块茎的定殖能力和在叶片上的生命周期（接种和产孢的周期）（Latin 等，1981）。侵染能力如大多数的生理参数类似都难以量化，且相当不稳定，其在多个世代中可以增加或减少（Jinks 和 Grindle，1963）。Caten（1970）认为，田间环境会对病菌的快速生长和大量产孢产生定向选择，并提供了特定品种适应性的证据，支持 Latin 等（1981）的报告，次要基因抗性和侵染能力相互作用决定了晚疫病危害的严重程度。还表明可能发生晚疫病菌适应具体品种和疫病的微效基因抗性可能最终成为侵蚀。Bjor 和 Mulelid（1991）使用不带抗性基因块茎实验结果也说明了晚疫病会对特定马铃薯品种产生适应性并且次要基因对晚疫病的抗性能力最终被削弱。Spielman 等（1992）将病原体健康的各个组分与疫病发展进行关联，他们的结论是疫病发展与孢子容量关系最为密切与侵染频率关系最小。马铃薯晚疫病可以自身发生改变以适应某一区域的气候。例如埃及菌株孢子形成的最佳温度比英国的要高若干度（Shaw，1987）。这种可能是基因决定的差异，可能是某些报道晚疫病菌对温度敏感性存在明显差异的原因。

有大量关于马铃薯晚疫病菌（如 Holmes 和 Channon，1984）对杀菌剂耐受性的报道。田间马铃薯晚疫病菌种群尤其是生长季后期通常都异于他类，处于既包括杀菌剂敏感和杀真菌剂耐受类型的种群的动态平衡中，使用杀真菌剂可改变这种平衡，使得环境有利于耐杀菌剂的菌株（Levy 等，1983）。在无杀真菌剂的环境中，致病疫霉对甲霜灵的抗性菌株表现出较强的竞争力，经过 8 ~ 10 个产孢周期，其比例从 10% 增加到 100%（Kadish 和 Cohen，1988）。这种较强的竞争力可归因于病菌对叶片快速定殖，而不是侵染能力或产孢能力的差异。然而，耐甲霜灵菌株可能无法像敏感株一样在块茎存活（Kadish 和 Cohen，1992）。

虽然环境影响的事实有生理基础，但有关马铃薯晚疫病菌的生理状态对晚疫病发生可能的影响的信息较为缺乏。此外，马铃薯晚疫病的基因型对其生理状态有很大的影响，因此环境影响、遗传和生理影响都相互关联。与老化有关的生理和生化变化可能使病原体更容易受到不利条件的伤害，例如老旧孢子囊可能会比新孢子囊更迅速的失去生存能力。研究人员观察到，老旧孢子囊比新孢子囊释放游动孢子的速度慢。因此，老旧孢子囊侵染叶片需要更长时间的潮湿条件。Bashi 等（1982）报道，一天中晚些时候释放的孢子比早些时候的更具传染性。De WeiHe（1963）认为，新形成的孢子囊在发芽前需要一段成熟时间。随着成熟芽孢老化，其发芽过程可能对环境的要求变得更严格，使得游动孢子只能在很窄的温度范围内被释放。晚疫病菌老化或营养状况对其寄主定殖和产孢的影响是推测结果。

可有效侵染的孢子囊数量即接种量，显然对病害发生有重大影响。无接种孢子便无病害病发生可言。在一系列特定条件下，叶片上疫病的发生量与接种量呈正比，但也取决于可供侵染和定殖的无病叶片的量。在任何特定时间内，病菌侵染和定殖取决于疾病发生率和茎叶生长。在疫病发生初始阶段，接种物依赖于被感染茎的数量和周围情况，但有时在作物本身或在其他田里的自播植物（Van der Zaag，1956）。接种孢子也取决于适宜孢子形成和孢子扩散的天气条件和风向。随着疫病的发生，在作物上新形成的孢子囊占据主导地位，虽然外源孢子在引入新种族、菌株或致病疫霉的新交配型方面依然很重要。在空气中的孢子囊浓度呈现显著的昼夜周期性，上午晚些时候通常会达到最大值（Hirst，1953），这与适宜侵染的环境条件一起将会影响新的病灶发生量。块茎受侵染的程度明显地取决于叶上形成的孢子囊量。

五、杀菌剂

虽然本章的讨论范围仅限于疫病的“自然”生态，但几乎普遍应用的为控制疫病的杀菌剂对减缓疫病发展的效用也应简单考虑。许多因素相互作用影响杀真菌剂的效果。最重要的是杀真菌剂的毒性，其取决于晚疫病耐药性、施药频率及其在作物冠层内的分布、杀菌剂类型（如系统性或外层保护性）、杀菌剂再分配及其活性降低速率（取决于降水、温度以及杀真菌剂性质）和新叶子的生长速率。再次，其中的一些因素如杀真菌剂的分布和动态变化难以量化。

第六节　影响马铃薯晚疫病菌有性繁殖周期的因素分析

致病疫霉有性生殖在晚疫病流行病学中的潜在重要性已引起研究人员

重视，但有关影响卵孢子产生及发芽因素的信息较为有限。而当这两种交配型的菌丝成长一起卵孢子自由地产生，它们也可以在实验室中形成为任一交配型菌株的自花受精的结果（Skidmore 等，1984；Campbell 等，1985）。目前尚不清楚其田间自体受精的重要性（如果出现过）。Chang 和 Ko（1990）声称，致病疫霉分离物交配型 A1 的可能改变成型 A2 在甲霜灵的存在下生长时。由于笔者知道的交配型的改变没有其他报告，该报告可能违反常理。

Harrison（1992）和 Drenth（1994）发现，卵孢子可以在 5～25℃的温度范围内形成。卵孢子的形成数量取决于寄主品种，而易受疫病侵染或迅速腐烂叶里的孢子数量要比具有中等抗性的品种叶片数量少（Drenth，1994）。Harrison 认为卵孢子形成的速率与菌丝 1 生长速率密切相关。Smoot 等（1958）观察到卵孢子在 12～25℃可发芽。有证据表明，连续光照抑制卵孢子形成（Romero 和 Gallegly，1963），发芽对光照有要求（Shattock 等，1986）。Drenth（1994）假设雨天条件下，在或接近土壤表面释放游动孢子，这些卵孢子能够感染与土壤接触的植物部分或直接侵染树冠。卵孢子可在低至 –80℃或高达 35℃的温度下生存，但 40℃下不能生存（Drenth，1994），但有利于孢子存活率和存活时间田间环境条件仍未知。

参考文献

Ansari A Q，Loomis W E. 1959. Leaf temperatures. American Journal of Botany 46：713 –717.

Awan A B，Struchtemeyer R A. 1957. The effect offertilization on the susceptibility of potatoes to late blight. American Potato Journal 34：315 –319.

Bashi E，Ben-Joseph Y，Rotem J. 1982. Inoculum potential of Phytophthora infestans and the development of potato late blight epidemics. Phytopathology 72：1 043 –1 047.

Bjor T，Mulelid K. 1991. Differential resistance to tuber late blight in potato cultivars without Rgenes. Potato Research 34：3 –8.

Bonde R，Stevenson F J，Clark C F. 1940. Resistance of certain potato varieties and seedling progenies to late blight in the tubers. Phytopathology

30: 733 -748.

Borys M W. 1964. Influence of calcium, magnesium, chloride and sulphate nutrition on the resistance of potato leaves to Phytophthora infestans de Bary. Acta Microbiologica Polonica 13: 221 -226.

Bourke P M A. 1965. The Potato, Blight, Weather, and the Irish Famine. Ph. D. Thesis, University of Dublin, Dublin, Ireland.

Bruhn J A, Fry W E. 1981. Analysis of potato late blight epidemiology by simulation modelling. Phytopathology 71: 612 -616.

Campbell A M, Duncan J M, Malcolmson J F. 1985. Production of oospores in vitro by self fertilization in single isolates of Al mating types of Phytophthora infestans. Transactions of the British Mycological Society 84: 533 -535.

Carnegie S F, Colhoun J. 1980. Differential leaf susceptibility to Phytophthora infestans on potato plants of cv. King Edward. Phytopathologische Zeitschrift 98: 108 -117.

Carnegie S F, Colhoun J. 1982. Susceptibility of potato leaves to Phytophthora infestans in relation to plant age and leaf position. Phytopathologische Zeitschrift 104: 157 -167.

Carnegie S F, Colhoun J. 1983. Effects of plant nutrition on susceptibility of potato leaves to Phytophthora infestans. Phytopathologische Zeitschrift 108: 242 -250.

Caten C E. 1970. Spontaneous variability of single isolates of Phytophthora infestans. II. Pathogenic variation. Canadian Journal of Botany 48: 897 -905.

Caten C E, Jinks J L. 1968. Spontaneous variability of single isolates of Phytophthora infestans. I. Cultural variation. Canadian Journal of Botany 46: 329 -348.

Chang T T, Ko W H. 1990. Effect ofmetalaxyl on mating type of Phytophthora infestans and P. parasitica. Annals of the Phytopathological Society of Japan 56: 194 -198.

Collins B G, Taylor R J. 1961. Conditions governing the onset of dew on large leaves. Australian Journal of Applied Science 12: 23 -29.

Colon L T, Pieters M M J, Budding D J. 1992. Field experiments on components of resistance to Phytophthora infestans (Mont.) de Bary in wild

Solanum species. Pages 25 –29 in Rousselle Bourgeois F, Rousselle P J (Eds.) Proceedings of the Joint Conference of the EAPR Breeding and Varietal Assessment Section and the EUCARPIA Potato Section, Landerneau, France, 12 –17 Jan. 1992. INRA, Ploudaniel, France.

Crosier W. 1934. Studies in the Biology of Phytophthora infestans (Mont.) de Bary. Cornell University Agricultural Experiment Station Memoir No. 155.

Darsow U, Goebel S, Goetz E, Oertel H, Schueler K. 1987. R-Gene und relative Resistenz der Kartoffelknolle gegen Phytophthora infestans (Mont.) de Bary. Archiv fiir Ziichtungsforschung 17: 387 –397.

Darsow U, Junges W, Oertel H. 1988. Die Bedeutung der Pradisposition fur die Laborpriifung von Kartoffelblattern auf relative Resistenz gegeniiber Phytophthora infestans (Mont.) de Bary. Archiv fur Phytopathologie und Pflanzenschutz 24: 109 –119.

Denward T. 1970. Differentiation in Phytophthora infestans. II. Somatic recombination in vegetative mycelium. Hereditas 66: 35 –47.

De Weille G A. 1963. Laboratory results regarding potato blight and their significance in the epidemiology of blight. European Potato Journal 6: 121 –130.

De Weille G A. 1964. Forecasting Crop Infection by the Potato Blight Fungus. Koninklijk Nederlandsch Meteorologisch Instituut, Mededelingen en Verhandelingen No. 82.

Dowson W J, Rudd Jones D. 1951. Bacterial wet rot of potato tubers following Phytophthora infestans. Annals of Applied Biology 38: 231 –236.

Drenth A. 1994. Molecular Genetic Evidence for a New Sexually Reproducing Population of Phytophthora infestans in Europe. Doctoral Thesis, Landbouwuniversiteit Wageningen, Wageningen, The Netherlands.

Fernandez de CubilJos C, Thurston H D. 1975. The effect of viruses on infection by Phytophthora infestans (Mont.) de Bary in potatoes. American Potato Journal 52: 221 –226.

Fitt B D L, Creighton N F, Lacey M E, et al. 1986. Effects of rainfall intensity and duration on dispersal of Rhynchosporium secalis conidia from infected barley leaves. Transactions of the British Mycological Society 86: 611 –618.

Fitt B D L, Shaw M W. 1989. Transports of blight. New Scientist 123 (1677): 41 -43.

Fry W E, Apple A E. 1986. Disease management implications of age-related changes in susceptibility of potato foliage to Phytophthora infestans. American Potato Journal 63: 47 -56.

Gees R, Hohl H R. 1988. Cytological comparison of specific (R3) and general resistance to late blight in potato leaf tissue. Phytopathology 78: 350 -357.

Graham K M, Dionne L A, Hodgson W A. 1961. Mutability of Phytophthora infestans on blightresistant selections of potato and tomato. Phytopathology 51: 264 -265.

Grainger J. 1979. Scientific proportion and economic decisions for farmers. Annual Review of Phytopathology 17: 223 -252.

Harrison J G. 1992. Effects of the aerial environment on late blight of potato foliage-a review. Plant Pathology 41: 384 -416.

Harrison J G, Lowe R. 1989. Effects of humidity and windspeed on sporulation of Phytophthora infestans on potato leaves. Plant Pathology 38: 585 -591.

Harrison J G, Lowe R, Williams N A. 1994. Effects of temperature and light on non-race-specific resistance of potato leaflets to late blight. Plant Pathology 43: 733 -739.

Hartill W F T, Young K, Allan D J, Henshall W R. 1990. Effects of temperature and leaf wetness on the potato late blight. New Zealand Journal of Crop and Horticultural Science 18: 181 -184.

Haverkort A J, Bicamumpaka M. 1986. Correlation between intercepted radiation and yield of potato crops infested by Phytophthora infestans in central Africa. Netherlands Journal of Plant Pathology 92: 239 -247.

Hirst J M. 1953. Changes in atmospheric spore content: diurnal periodicity and the effects of weather. Transactions of the British Mycological Society 36: 375 -393.

Hirst J M. 1958. New methods for studying plant disease epidemics. Outlook on Agriculture 2: 16 -26.

Hirst J M, Stedman O J. 1960. The epidemiology of Phytophthora infestans. II. The source of inoculum. Annals of Applied Biology 48: 489 -517.

Holmes S J I, Channon A G. 1984. Studies on metalaxyl-resistant Phytophthora infestans in potato crops in south-west Scotland. Plant Pathology 33: 347 - 354.

Jinks J L, Grindle M. 1963. Changes induced by training in Phytophthora infestans. Heredity 18: 245 - 264.

Kable P F, MacKenzie D R. 1980. Survival of Phytophthora infestans in potato stem lesions at high temperatures and implications for disease forecasting. Plant Disease 64: 165 - 167.

Kadish D, Cohen Y. 1988. Competition between metalaxyl-sensitive and metalaxyl-resistant isolates of Phytophthora infestans in the absence of metalaxyl. Plant Pathology 37: 558 - 564.

Kadish D, Cohen Y. 1992. Overseasoning of metalaxyl-sensitive and metalaxyl-resistant isolates of Phytophthora infestans in potato tubers. Phytopathology 82: 887 - 889.

Kolbe W. 1982. Importance of potato blight control exemplified by Hofchen long-term trial (1943 - 1982), and historical development. Pflanzenschutz-Nachrichten Bayer 35: 247 - 290.

Krause R A, Massie L B, Hyre R A. 1975. Blitecast: A computerized forecast of potato late blight. Plant Disease Reporter 59: 95 - 98.

Kurzawinska H. 1989. Effect of agrotechnical factors on late blight [Phytophthora infestans (Mont.) de Bary] in selected cultivars of potato. Folia-Horticulturae 1: 63 - 73.

Lacey J. 1965. The infectivity of soils containing Phytophthora infestans. Annals of Applied Biology 56: 363 - 380.

Lapwood D H. 1968. Observations on the infection of potato leaves by Phytophthora infestans. Transactions of the British Mycological Society 51: 233 - 240.

Lapwood D H. 1977. Factors affecting the field infection of potato tubers of different cultivars by blight (Phytophthora infestans). Annals of Applied Biology 85: 23 - 42.

Latin R X, MacKenzie D R, Cole Jr H. 1981. The influence of host and pathogen genotypes on the apparent infection rates of potato late blight epidemics. Phytopathology 71: 82 - 85.

Leach S S, Rich A E. 1969. The possible role of parasexuality and cytoplasmic variation in race differentiation in Phytophthora infestans. Phytopathology 59: 1360 – 1365.

Levy Y, Cohen Y, Benderly M. 1991. Disease development and buildup of resistance to oxadixyl in potato crops inoculated with Phytophthora infestans as affected by oxadixyl and oxadixyl mixtures: experimental and simulation studies. Journal of Phytopathology 132: 219 – 229.

Levy Y, Levi R, Cohen Y. 1983. Buildup of a pathogen subpopulation resistant to a systemic fungicide under various control strategies: a flexible simulation model. Phytopathology 73: 1 475 – 1 480.

MacKenzie D R, Elliott V J, Kidney B A, et al. 1983. Application of modern approaches to the study of the epidemiology of diseases caused by Phytophthora. Pages 303 – 313 in Erwin D C, Bartnicki-Garcia S, Tsao PH (Eds.) Phytophthora. Its Biology, Taxonomy, Ecology and Pathology. The American Phytopathological Society, St Paul, Minnesota, USA.

Main C E, Gallegly M E. 1964. The disease cycle in relation to multigenic resistance of potato to late blight. American Potato Joumal 41: 387 – 400.

Malcolmson J F. 1969. Factors involved in resistance to blight [Phytophthora infestans (Mont.) de Bary] in potatoes and assessment of resistance using detached leaves. Annals of Applied Biology 64: 461 – 468.

Malcolmson J F. 1970. Vegetative hybridity in Phytophthora infestans. Nature 225: 971 – 972.

Minogue K P, Fry W E. 1981. Effect of temperature, relative humidity, and rehydration rate on germination of dried sporangia of Phytophthora infestans. Phytopathology 71: 1 181 – 1 184.

Pathak N, Clarke D D. 1987. Studies on the resistance of the outer cortical tissues of the tubers of some potato cultivars to Phytophthora infestans. Physiological and Molecular Plant Pathology 31: 123 – 132.

Pietkiewicz J. 1974. Effect of viruses on the reaction of potato to Phytophthora infestans. I. Characteristic of the reaction to Ph. infestans of plants infected with potato viruses X, Y, S, M and leafroll. Phytopathologische Zeitschrift 81: 364 – 372.

Populer C. 1978. Changes in host susceptibility with time. Pages 239 – 262 in Horsfall J G, Cowling E B (Eds.) Plant Disease. An advanced Treatise. Volume 2. Academic Press, New York, USA.

Raposo R, Wilks D S, Fry W E. 1993. Evaluation of potato late blight forecasts modified to include weather forecasts: A simulation analysis. Phytopathology 83: 103 - 108.

Richardson D E, Doling D A. 1957. Potato blight and leaf-roll virus. Nature 180: 866 - 867.

Robinson R A (1976) The components of parasite population flexibility. Pages 51 - 54 in Yaron B (Co-ordinating Ed.) Plant Pathosystems (Advanced Series in Agricultural Sciences 3). SpringerVerlag, Berlin, Germany.

Roer L, Toxopeus H J. 1961. The effect of R-genes for hypersensitivity in potato-leaves on tuber resistance to Phytophthora infestans. Euphytica 10: 35 - 42.

Rogoshin A N, Filippov A V. 1983. [Distribution and conidial viability of Phytophthora infestans (Mont.) d. By. in the air above infected potato fields]. Mikologiya i Fitopatologiya 17: 225 - 227. (Summary in: Review of Plant Pathology 63: 75.

Rohwer F, Fritzmeier K H, Scheel D, Hahlbrock K. 1987. Biochemical reactions of different tissues of potato (Solanum tuberosum) to zoospores or elicitors from Phytophthora infestans. Accumulation of sesquiterpenoid phytoalexins. Planta 170: 556 - 561.

Romero S, Gallegly M E. 1963. Oogonium germination in Phytophthora infestans. Phytopathology 53: 899 - 903.

Rotem J, Cohen Y. 1974. Epidemiological patterns of Phytophthora infestans under semi-arid conditions. Phytopathology 64: 711 - 714.

Rotem J, Cohen Y, Bashi E. 1978. Host and environmental influences on sporulation in vivo. Annual Review of Phytopathology 16: 83 - 101.

Rotem J, Cohen Y, Putter J. 1971. Relativity of limiting and optimum inoculum loads, wetting durations, and temperatures for infection by Phytophthora infestans. Phytopathology 61: 275 - 278.

Rotem J, Palti J, Lomas J. 1970. Effects of sprinkler irrigation at various times of the day on development of potato late blight. Phytopathology 60: 839 - 843.

Rotem J, Sari A. 1983. Fertilization and age-conditioned predisposition of potatoes to sporulation of and infection by Phytophthora infestans. Zeitschrift fiir Pflanzenkrankheiten und Pflanzenschutz 90: 83 – 88.

Sato N. 1979. Effect of soil temperature on the field infection of potato tubers by Phytophthora infestans. Phytopathology 69: 989 – 993.

Schober B, Ullrich J. 1985. Keimung der Sporangien von Phytophthora infestans (Mont.) de Bary auf Kartoffelblatt-und Knollengewebe. Potato Research 28: 527 – 530.

Shattock R C, Janssen B D, Whitbread R, et al. 1977. An interpretation of the frequencies of host specific phenotypes of Phytophthora infestans in North Wales. Annals of Applied Biology 86: 249 – 260.

Shattock R C, Tooley P W, Fry W E. 1986. Genetics of Phytophthora infestans: Characterization of single-oospore cultures from Al isolates induced to self by intraspecific stimulation. Phytopathology 76: 407 – 410.

Shaw D S. 1987. The breeding system of Phytophthora infestans: the role of the A2 mating type. Pages 161 – 174 in Day P R, Jellis G J (Eds.) Genetics and Plant Pathogenesis. Blackwell Scientific Publications, Oxford, UK.

Shtienberg D, Raposo R, Bergeron S N, et al. 1994. Incorporation of cultivar resistance in a reduced-sprays strategy to suppress early and late blights on potato. Plant Disease 78: 23 – 26.

Simmonds N W, Wastie R L. 1987. Assessment of horizontal resistance to late blight of potatoes. Annals of Applied Biology 111: 213 – 221.

Skidmore D I, Shattock R C, Shaw D S. 1984. Oospores in cultures of Phytophthora infestans resulting from selfing induced by the presence of P. drechsleri isolated from blighted potato foliage. Plant Pathology 33: 173 – 183.

Smith J W. 1915. The effect of the weather upon the yield of potatoes. Monthly Weather Review. United States Department of Agriculture 43: 222 – 236.

Smoot J J, Gough F J, Lamey H A, et al. 1958. Production and germination of oospores of Phytophthora infestans. Phytopathology 48: 165 – 171.

Spielman L J, McMaster B J, Fry W E. 1992. Relationships among measurements of fitness and disease severity in Phytophthora infestans. Plant Pathology 41: 317 –324.

Stepanov K M. 1935. Dissemination of infectious diseases of plants by air currents. Bulletin of Plant Protection (Leningrad), Series 2, 8: 1 –68.

Stewart H E. 1990. Effect of plant age and inoculum concentration on expression of major gene resistance to Phytophthora infestans in detached potato leaflets. Mycological Research 94: 823 –826.

Stewart H E, Wastie R L, Bradshaw J E, Brown J. 1992. Inheritance of resistance to late blight in foliage and tubers of progenies from parents differing in resistance. Potato Research 35: 313 –319.

Van der Plank J E. 1963. Plant Diseases: Epidemics and Control. Academic Press, New York, USA.

Van der Zaag D E. 1956. Overwintering en epidemiologie van Phytophthora infestans, tevens enige nieuwe bestrijdingsmogelijkheden. Tijdschrift over Plantenziekten 62: 89 –156.

Van Everdingen E. 1926. Het verband tusschen de weersgesteldheid en de aardappelziekte (Phytophthora infestans). Tijdschrift over Plantenziekten 32: 129 –140.

Van Oijen M. 1990. Photosynthesis is not impaired in healthy tissue of blighted potato plants. Netherlands Journal of Plant Pathology 96: 55 –63.

Van Oijen M. 1991a. Light use efficiencies of potato cultivars with late blight (Phytophthora infestans). Potato Research 34: 123 –132.

Van Oijen M. 1991 b. Identification of the Major Characteristics of Potato Cultivars which Affect Yield Loss Caused by Late Blight. Doctoral Thesis, Landbouwuniversiteit Wageningen, Wageningen, The Netherlands.

Van Oijen M. 1992. Evaluation of breeding strategies for resistance and tolerance to late blight in potato by means of simulation. Netherlands Journal of Plant Pathology 98: 3 –11.

Wallin J R. 1953. The production and survival of Phytophthora irifestans on tomato and potato plants. Phytopathology 43: 505 –508.

Warren R C, Colhoun J. 1975. Viability of sporangia of Phytophthora irifestans in relation to drying. Transactions of the British Mycological Society

64: 73 -78.

Warren R C, King J E, Colhoun J. 1971. Reaction of potato leaves to infection by Phytophthora infestans in relation to position on the plant. Transactions of the British Mycological Society 57: 501 -514.

Wastie R L, Bradshaw J E, Stewart H E. 1993. Assessing general combining ability for late blight resistance and tuber characteristics by means of glasshouse seedling tests. Potato Research 36: 353 -357.

Yamamoto M, Tanino J. 1961. Physiological studies on the formation and germination of sporangia of Phytophthora infestans (Mont.) De Bary. Forschungen auf dem Gebiet der Pflanzenkrankheiten (Kyoto) 7 (2): 7 -22.

Zan K. 1962. Activity of Phytophthora infestans in soil in relation to tuber infection. Transactions of the British Mycological Society 45: 205 -221.

第六章　马铃薯晚疫病模拟模型

植物病害流行病学起源于人类对传染病的研究。最古老的人类病害的数学模型起源于18世纪（Bernoulli，1760），但直至两个世纪以后人类才开始对植物病害进行数学模型分析（Van der Plank，1963）。马铃薯晚疫病在Van der Plank分析中占据突出地位，并且现有大量的关于马铃薯疫病模型的文献。20世纪80年代中期之前，马铃薯晚疫病模型主要关注病菌的生命周期，以及环境对该生命周期各个阶段的影响。近年来，人们更加关注疫病的发生在寄主生长中情况及病害造成的产量损失。现在复杂的病原体-作物组合模型已经可用。

马铃薯晚疫病模型主要服务于两个目的。首先，疫病模型可用于评估病害控制策略，特别是杀真菌剂的应用评估；其次，晚疫病模型已被用于分析品种间产量损失差异，以及各种病害发生时空的差异性。

研究人员已经开发出许多模型，功能差异很显著。模型的结构在很大程度上决定了其对参数或输入值变化的敏感性。如果研究人员更加关注模型正确的初始化和参数化操作，并对其进行综合的灵敏度分析，则模型可以得到更好的应用。

第一节　简介

由真菌致病疫霉引起的马铃薯晚疫病是马铃薯和其他茄科作物的主要传染病。疫病病斑生长和孢子形成以及孢子扩散导致新叶感染之前，感染病菌的叶子会有一段潜伏期，Harrison（1995）描述了该真菌生命周期的一些细节。马铃薯晚疫病菌的生命周期非常类似人类传染病，因此，目前的马铃薯晚疫病模型仍然与用于医学而开发的已有模型有许多相似性。本章将简要介绍植物病害流行病学模型的医学起源，再讨论新的疫病模型的发展及其前景。介绍不会严格按照时间顺序，也不会完全限于马铃薯晚疫病。自1963年以来出现的主要疫病模型按时间顺序列在表6-1中。

表 6-1 致病疫霉主要模拟模型

参考文献	关键词
Vander Plank (1963)	流行病数学分析模型
Waggoner (1968)	真菌周期气象计算机模拟模型
Sparks (1980)	茎干病变模型
Stephan 和 Gutsche (1980)	预测模型
Bruhn 和 Fry (1981)	块茎产量，杀真菌剂，品种抗性模型
Minogue 和 Fry (1983)	疫病空间分布模型
Paysour 和 Fry (1983)	地块间干扰试验模型
Fohner 等 (1984)	杀真菌剂时间安排模拟
Michaelides (1985)	天气敏感模型
Milgroom 和 Fry (1988)	真菌对甲霜灵抗性建设模型
Ferrandino (1989)	寄主生长孢子分布，疫病空间整合模型
Van Oijen (1989)	器官损伤发育抗性组成模型
Shtienberg 等 (1989)	对同步早、晚疫病杀菌剂使用安排模型
Kluge 和 Gutsche (1990)	田间有效性研究模型
Levy 等 (1991)	杀菌混合剂模型
Michaelides (1991)	植物密度高架喷灌模型
Van Oijen (1991b)	通过叶层疫病向上蔓延及抗性模型
Van Oijen (1992a)	寄主成熟类耐性组成模型
Van Oijen (1992b)	初始化、参数化、多参数变化
Raposo 等 (1993)	气候预测整合模型
Shtienberg 等 (1994)	杀菌剂安排模型

参考文献按时间顺序排列，并给出了每个马铃薯晚疫病模型的特点，或相对于旧模型的改进。此表并没有列出迄今为止的所有模型

第二节 SEIR 模型的发展历程

传染病最古老的数学模型由 Bernoulli (1760) 设计。虽然其研究病害的名字令人不悦，但他的研究目的很简单：Bernoulli 使用他的模型评估天花的疫苗接种策略。

近两个世纪后第一个模型出现了，其结构类似于传染病剂的生命周期，因此该模型可称为第一个病害模拟模型（Kermack 和 McKendrick，1927）。该病害模型由两个关联的微分方程组成，可应用于固定大小人口的传染性病害。模型区分了易感、感染和移除个体，易感个体变为感染个体的速率与易感和感染个体密度的乘积呈正比。而后，植物学文献中将模

型的比例常数称为日增殖因子（DMFR）。因此，比例常数是传染性繁殖体的增殖速率、分散性和感染性的量度。以恒定的相对速率移除传染性个体表明病害感染性、分离或死亡的原因。去除率的倒数是患病个体的平均感染周期。因此，感染周期被隐含地假定为指数分布。

Kermack 和 McKendrick 的模型被称为第一个“SIR 模型”——以传染病流行范围内三种人群的英文首字母命名，（Hethcote，1976）。而后，又分了第四类别：已经暴露于该病害但尚未感染的个体（Anderson 和 May，1982）。感染开始前的期间称为潜伏期。图 6－1 给出了 SEIR 模型结构的示意图，包括最常见的转换速率项。

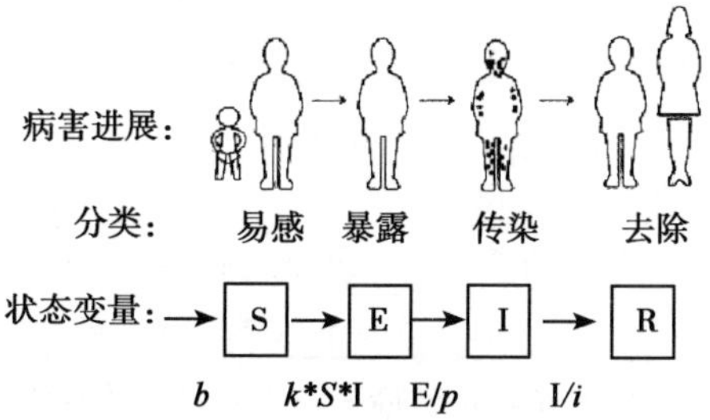

图 6－1　SEIR 模型的结构

注：病害进展通过易感个体（S）向后经过暴露（E）、感染（I）和恢复或移除（R）。该图显示了过渡率常用的术语，以及它们对增殖率（b）、接触频率（k）、潜伏期（p）和感染期（i）的依赖

第三节　SEIR 模型植物感染机理

Van der Plank（1963）对植物病害流行病学进行了数学分析。他将 SIR 和 SEIR 模型应用于植物病害，包括马铃薯晚疫病。然而，SEIR 模型中的潜伏期和感染期不是指数分布，对于所有病害单位是常量（即病变）。因此，他的模型被定义为时间延迟微差分方程，后来被称为“paralogistic”方程（Zadoks 和 Schein，1979）。

“paranogistic”方程的初步分析由 Van der Plank（1963）完成。分析遇到了阻碍，即 paralogistic 方程在分析方面比原始的人类流行病学 SEIR 模型更难处理，即使它们有相似的动态（Jeger，1986）。如果省略术语，可以推导出一些分析结果，这将使“paralogistic”变为“对指数”SEI 模型，仅可描述流行病的初始阶段。Oort（1968）通过改变潜伏和感染周期以及日增殖因子进行了初始阶段指数速率的灵敏度分析，并发现指数速率在潜伏期的变化波动最大。测试的参数和一些其他参数通常被称为“抗性组分”，它们共同决定了寄主对病原体的抗性水平（表6－2）。事实上，Oort的分析是植物病害流行病学中数学模型的第一次应用，用于测试病害发展对抗性组分水平的敏感性。遗憾的是，模型参数的初始设置并不对应流行病的真实值，但是Zadoks（1971；Rabbinge等，1989）开展了更现实和广泛的数值分析，其1971年的分析和1989年的日增殖因子分析指出了病害潜伏期的重要性。

表6－2　植物传染病抗性五个组成部分的定义

抗性组分		流行病学意义
传染率	（e:%）	孢子囊降落在植物组织上引发新的病变
潜伏期	（p：d）	感染与病变扩张和孢子形成开始之间的时间间隔
病斑生长速度	（g：m/d）	病斑径向扩张率
传染期	（i：d）	病灶部位孢子形成时间
产孢强度	（s：m^2·d）	单位病斑孢子孢子囊的产生率

注：日增殖因子（DMFR）是更高水平的参数，其代表孢子形成强度、孢子分散效率和传染性，未列出

Wagoner（1968）提出了马铃薯晚疫病的第一个计算机模拟模型，广泛模拟了环境条件对抗性组分的影响。他表示大量的晚疫病信息可以被集成到一个模拟模型中，但没有将他的模型应用于任何实际或科学领域。然而，Waggoner的复杂模型成为之后植物病害建模者的范例，作物病害模拟模型通常有众多的参数并只能进行数值分析。

第四节　潜在及感染期可替换分布

如前所述，假设潜伏期和感染期在人类病害SEIR模型中呈指数分布，而在第一植物病害SEIR模型的两阶段中病变之间应该有零方差。研究人员随后介绍了替代分布。Berger和Jones（1985）在他们的模型中的潜伏期使用了分布式延迟，但保持传染期不变。分布式延迟方法提供了一个潜伏期分布范围，从阶梯函数到近似正态分布。Berger和Jones（1985）在其模型中使用了4个延迟间隔，因此非常接近正态分布，这与真菌性叶病潜伏期的现实分布极为接近（Shaner，1980）。Knudsen等（1987）对在潜伏期和传染期的预测模型中应用了分布式延迟。

目前，研究人员仍然不清楚潜伏期和传染期的分布在流行病模型中的重要性。Van der Plank（1963）认为潜伏期和传染期的分布对计算病害发展几乎没有影响，而 Berger 和 Jones（1985）则持相反观点。

马铃薯晚疫病的茎病变代表了具有极长感染期的特殊病变，虽然通常其孢子生成强度较低。此病变在恶劣天气条件下作为特殊流行病学的生存机制被纳入了 Sparks 的预测模型（1980）。

第五节　马铃薯器官损伤生长模拟

一、病变生长建模

Zadoks（1977）提到了在对真菌生命周期各个阶段流行病学重要性分析中应该包括的 5 个组成部分，即抗性组分：感染效率、潜伏期、感染期、孢子形成强度和病变生长速率（表 6－2）。然而，迄今所讨论的模型共同特征是没有将病变生长速率作为单独的抗性组分，这可能有历史原因：在早期的人类病害模型中没有病变生长的类似物，人们假设病变从其潜伏期开始占据固定的病变区域，例如，Michaelides（1985）在其致病疫霉的模拟模型中使用0. 3cm^2 的固定病变面积。Michaelides 仔细处理了孢子囊扩散，因此，复杂化了其模型参数值的确定，同时过度简化病变生长而使其模型缺少现实应用性。

在流行病学模型中引入病变生长速率对于不确定的病害病变生长非常重要，即没有固定数值大小的最终病变，如马铃薯晚疫病。这种模型适用于马铃薯晚疫病：晚疫病病变具有恒定的径向生长速率（Gees 和 Hohl，1988；Van Oijen，1989），使得病变区域是时间的二次函数而不是指数函数。Van Oijen（1989）将可预测动态变化病变值分布的扩展亚模型引入马铃薯晚疫病模型中，圆形病变在小叶上呈泊松分布，而它们的直径由径向生长速率改变，径向生长速率与染病叶片未发病区域呈一定比例。

大多数植病病变不是随机分布，因此，模型中最好使用负二项分布代替泊松分布（Wagoner 和 Rich，1981）。Lapwood（1961）分析表明，晚疫病病变即使在马铃薯叶和小叶中也不是随机分布，在远端小叶和小叶的尖端及边缘发现病变相对较多。Wagoner 和 Rich（1981）还建议放弃病变形成率与易感和感染部位密度乘积之间的比例关系方法。这种非线性发病率在人类流行病学中研究较多（Li，1987）。

二、空间异质性

流行病学模型通常模拟“一般流行病”，即 Zadoks 和 Schein（1979）对流行病所作的定义——由均质分布的初始疾病在空间中均质发展而成。然而，现在更多的模型逐渐地纳入了流行病发展的空间因素。Minogue 和 Fry（1983）对马铃薯晚疫病的空间动力学进行了建模，发现疫病以恒定速度扩展，并取决于寄主抗性和杀真菌剂的使用情况。Paysour 和 Fry（1983）使用模型计算实验插值对马铃薯晚疫病的干扰水平。他们检查了绘图形状、大小和距离的插值影响，并表明正方形图形将插值干扰最小化。Ferrandino（1989）也提出了马铃薯晚疫病的空间模型，并用此模型检验病害的空间分布和产量损失之间的数量关系。通过模型研究，他得出高度聚集模式病害使得块茎产量损失最大的结论。

病害在作物上的垂直分布的结论在流行病学中具有一定的重要性。马铃薯晚疫病通常在下叶层开始，后逐渐蔓延到顶部冠层（Lapwood，1961；Van Oijen，1991c）。研究人员尚不清楚这是否由于较老的叶片有较高的易感度，低层更益于感染的微气候条件或低层沉积较多的孢子囊造成。Björling 和 Sellgren（1955）发现流行病发生的初期和盛期，中部和底部叶子上的孢子囊数量是顶部叶子的 2～4 倍。Van Oijen（1991b）讨论了一个疫病感染逐渐向上传播的初期模型，他表示抗性品种更多地通过降低病菌在叶层内的扩散速率以延迟病原体向上扩散。

三、寄生主体生长

直到近期，才有几个疫病模型能以真实的方式模拟寄主生长。许多模型简单地将寄主叶面积视为时间的强制函数（例如 Bruhn 和 Fry，1981；Michaelides，1985）。根据未感染叶面积的逻辑函数，叶面积向渐近线靠近的方式更加复杂。（Berger 和 Jones，1985；Van Oijen，1989）。

近年来，通过更适用的子模型，引入了寄主生长的模型，成为新模型开发的趋势。几位模型开发者开发出了寄主生长子模型，其是基于 Monteith（1977）概述的作物生长速率和绿叶面积截获的光能量之间的线性关系原理而生产。Ferrandino（1989）、Michaelides（1991）和 Van Oijen（1992a）都是根据这个原则构建了马铃薯晚疫病模型，关系的斜率即光能利用效率，不受晚疫病影响（Haverkort 和 Bicamumpaka，1986；Van Oijen，1991a）。因此，真菌病原不会损害寄主光合作用（Van Oijen，1990），马铃薯作物生长模型不需要设计用于计算病菌感染对叶光合作用参数影响的复杂子程序（Rossing 等，1992）。

马铃薯晚疫病模型的一个重要研究对象是寄主大小和病害传播之间的关系。大多数模型表明敏感叶面积减少到某一阈值以下，染病组织的面积就不再增加，一些易感组织在流行病结束时将不受感染（Kermack 和 McKendrick，1927；Vander Plank，1963）。免受病害侵袭的寄主部分尺寸取决于影响流行率的所有抗性组分的非线性函数，因此寄主帮助自身逃过病害侵袭和寄主抗性特征之间有密切联系（Van Oijen，1989）。

人类传染病的早期模型，即 Kermack 和 McKendrick（1927）的模型是确定性模型（Bailey，1975）。然而，随机性逐渐地被纳入模型，并且为现在大多数人类流行病学模型所采用（Becker，1979）。相比之下，大多数植物流行病害模型仍然是确定型（Gilligan，1985）。尽管涉及大量的病害单元，因为非线性的病态系统（Rouse，1991），这些模型的结果可能不同于随机模型。Teng 等（1977）和 Sall（1980）提出了分别用于模拟大麦叶锈病和葡萄白粉病的早期随机模型。在这些模型中，一系列的抗性组分参数值取自统一或正态概率分布。而最近开发的模型的可变模式说明孢子分散是随机处理的结果，例如 Ferrandino（1989）的晚疫病模型。随机模型可估算病害进展速率变化，如果该模型用于病害预测，这种估算非常适用，因为它们对预测的可靠性设置了边界。然而，如果该模型用于阐明其组分在系统中的作用或用于评估育种目的抗性组分的重要性，则不需要这种估算，因为它模糊化了个体组分与病害进展速率之间的关系，因此使得主要组分的识别复杂化。

第六节　马铃薯晚疫病模拟模型在杀菌剂使用中的应用

为了防止致病疫霉，定期喷洒杀真菌剂耗费大量人力、资金且会危害环境（Harrison，1995）。为使这项措施更好地执行，基于历史数据和预期天气数据，许多为预测疫病流行发展的数学模型得以建立。除少数模型外（Kluge 和 Gutsche，1990），这些模型大多数是经验性模型，直接将喷药量与天气变量联系起来，没有明确为真菌群体动态建模，更没有建立其与马铃薯的关系，因此，这些模型超出了本章所讨论的范围，针对此问题，最近有相关评论（Bajic，1988）。然而，众多模拟模型已经按照各自的应用方式应用于杀真菌剂的规划；这些模型，除了可做预报外，研究人员也可反复使用模拟模型来评估杀菌剂使用规划策略的有效性（Fohner，1984；Shtienberg，1994）。遗憾的是，这些模拟研究得出的一般结论是经常、每周或每两周不喷洒杀菌剂所节省经济收益较少。显然，晚疫病是一种暴发性病害，当天气预报仍然不可靠时，人们无法承担不进行喷雾所带来的风险。

模拟模型在指导杀真菌剂使用中的特殊作用是通过评估杀真菌剂系统的耐药性风险如甲霜灵实现（Milgroom 和 Fry，1988；Levy 等，1991）。Levy 及其同事（1991）模型研究表明，使用混合杀真菌剂可能会延迟病原群体抗性的积累。

第七节　马铃薯晚疫病模拟模型在抗性育种过程中的应用

抗性组分的直接乘积被认为是评估其对流行病发展初始指数效应的最简单模型（Zadoks，1977）。Van der Zaag（1959）发现，马铃薯栽培品种对致病疫霉局部抗性的排序与其抗性组分乘积排序相一致。然而，Van der Plank（1963）和 Oort（1968）的分析表明：这种方法认为所有组分对马铃薯总抗性具有相同的影响不合理，并且该方法并不常用。

Van Oijen（1992a，b）使用其模型确定应改善的抗性组分以培育出更多的抗性品种。模型敏感性分析显示病害发展对病变生长速率的变化最敏感，其次是感染效率和感染期，最后是潜伏期（图 6－2 中 A）。然而，此结果本身并不意味着育种者只需专注于减少病变生长速率，育种不仅取决于植物固有特性，还取决于可改造范围，即可用的遗传变异。因此研究人员进行了二次灵敏度分析，考虑了不同抗性组分的遗传变异。更现实的分析指出病变生长速率和感染效率都应成为提高马铃薯晚疫病部分抗性水平、改善育种程序的主要目标（图 6－2 中 B）。

第八节　马铃薯晚疫病模型初始化和本地化中存在的问题

当研究人员测试流行病学模型性能时，很少考虑初始接种物的状态。然而，模型初始化的方式可以多种方式影响其行为。例如，如果潜伏期不

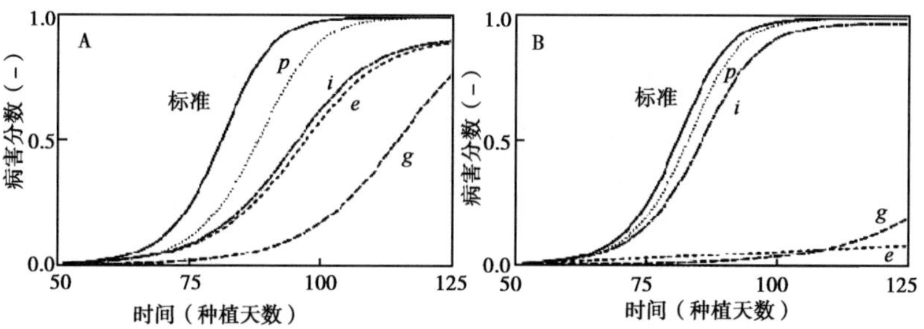

图 6-2　马铃薯晚疫病各种宿主品种的模拟发生曲线

注：（A）标准易感品种（cv. Bintje），假定具有标准值一半的部分抗性基因型感染期（*i*）、感染效率（*e*）或病变生长速率（*g*）或潜伏期（*p*）；（B）标准易感品种，其基因型中 *i*，*e*，*g* 或 *p* 的抗性值设置为最大（来源：Van Oijen，1992b）

被视为常数，那么模拟同等龄期的潜伏病变与模拟感染病变会有显著不同（Jeger，1986）。当有较多病变存在于模型初始化时，病害发展速率对感染效率变化的敏感性可能被低估（Van Oijen，1992b）。模型的时间延迟，可引起初始化的其他问题，如 Van der Plank（1963）的"paralogistic"方程式表明，病害发生水平必须在模拟周期之前定义（等于是模型中的最大延迟）。

如果模型参数与实际测量的值相差很大，则流行病学模型的参数化可能会很艰难。测量抗性组分的方法通常旨在最大限度地区分品种或处理，或易于测量。然而，为达到建模的目的，抗性组分的测量最好对应于其在病原体生命周期中的功能。因此，研究人员对测量病变生长速率相对于接种后病变测量更有实际意义。确定患病叶组织的单位面积孢子囊总产量比实际中常用的在接种后任意时间通过冲洗每个叶片或病变来定量孢子囊数目要更有效率。只有基于面积的孢子形成强度的测量，才能实现孢子形成强度与具有扩大病变真菌疾病（整合的病变生长速率）和感染期的明显分离。

实际测量和构建模型通常会将病害发生速度不同地量化，从而使模型验证复杂化。总体可见病变组织可被研究人员测量，同时潜在的、感染的和移除的组织可被模拟。模型只会使用固定大小参数模拟预测所有的病变包括隐性病变，这会导致一个特殊问题，即事实上流行病害潜在病变比可见病变占据的叶面积少得多，例如，Van der Plank（1963）"paralogistic"方程计算了总感染（潜伏＋感染＋去除）叶面积，但研究人员只能观察到感染和去除的叶面积，这是支持包括病变生长模型的进一步论证，其中潜在病变实际上可具有零或可忽略的叶面积。

模型对参数变化的敏感性通常可用作评估这些参数对疾病发生率的重要性。然而，应谨慎使用此方法，因为模型对参数变化的敏感性强烈依地赖于模型结构。因此，即使模型对参数变化的敏感性是模型的一个特征，也可能不能真实地反映出流行病的敏感性。这点可用 5 种不同的马铃薯晚疫病模型的比较进行说明（Van Oijen，1992b），在每个模型中，量化感染效率的参数都减少了 25%。当感染效率照此改变时，所有模型中的病害发生率都降低了，但降低程度大不相同（图 6 – 3）。

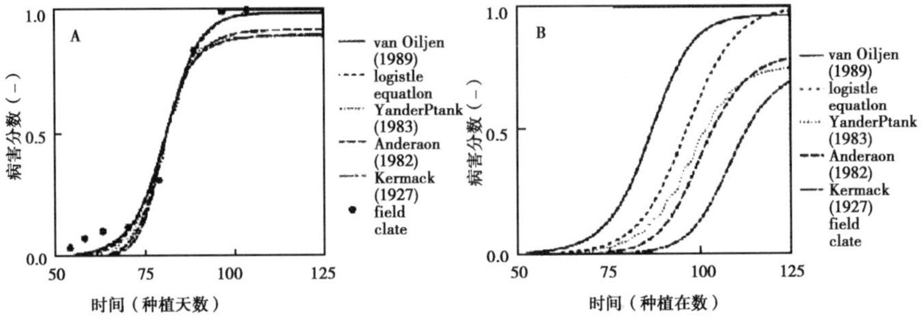

图 6 – 3　马铃薯晚疫病的 5 个不同模型的模拟发生曲线

注：（A）模型对于敏感品种 Bintje 病害发生数据的最佳拟合（田间测量 1988）；（B）使五个模型中每个模型的感染效率参数都降低 25% 的影响（来源：Van Qijen，1992b）

第九节　有关问题的结论与讨论

如前一段所述，许多建模研究都有参数化不充分和验证不足的问题。但即使已被很好验证的模型，可能仍会进一步开发更好的应用。研究人员常常缺乏对抗性成分的敏感性分析（Jeger 和 Groth，1985），尽管这样的分析已被反复强调，其对指导育种非常有益（Zadoks，1971；1977）。每当研究人员进行这种灵敏度分析时，其通常限于单参数变化，研究中也应评估同时改变多个抗性组分的结果（Van Oijen，1992b）。多参数变化分析应该考虑这样的事实：如果这些参数在遗传上或生理学上相关，在真实的系统中这些参数并不独立地变化。

病变生长率应被明确包含在可无限扩展病变的真菌性叶病模型中，其重要性已被几个含有此组分的模型说明：这些模型计算的病害发生率对病变生长速率的变化具有极大的敏感性。如果病害发生期跨越宿主生长季节，大多数真菌性叶病都是如此，特别是当评估部分抗性栽培品种时，模型会认为宿主叶面积是常量而错误地模拟疾病。对于这样的病害，如果该模型主要用于分析宿主特性及由于疾病产生产量损失的影响，其病原体对宿主生长的影响应被明确模拟。在宿主感染马铃薯晚疫病的情况下，可能需要慎重使用模型模拟块茎生长和感染状况。

尽管上面给出了警告性的说明，但显而易见，马铃薯晚疫病的模拟模型已被成功运用到许多方面。疫病模型可应用于流行病学分析、评估杀真菌剂应用策略以及鉴定改善抗性的宿主特性等诸多方面。这些领域的建模，研究人员可能旨在利用模型解释和评估疫病而不是预测，疫病模拟模型在不久的将来仍然会如此。

参考文献

Anderson R M, May R M. 1982. Directly transmitted infectious diseases: control by vaccination. Science 215: 1 053 – 1 060.

Bailey N T J. 1975. The Mathematical Theory of Infectious Diseases and its Applications. Charles Griffin & Company Ltd., London, UK.

Bajic S. 1988. A Survey on Potato Blight Phytophthora Infestans (Mont) de Bary Forecasting. The Meteorological Office, Bracknell, UK.

Becker N. 1979. The uses of epidemic models. Biometrics 35: 295 – 305.

Berger R D, Jones J W. 1985. A general model for disease progress with functions for variable latency and lesion expansion on growing host plants. Phytopathology 75: 792 – 797.

Bernoulli D. 1760. Essai d'une nouvelle analyse de la mortalite causee par la petite verole, et des avantages de l'inoculation pour la prevenir. Histoire de l' Academie Royale des Sciences (Paris) avec des Memoires de Mathematique et Physique: 1 – 45.

Bjorling K, Sellgren K A. 1955. Deposits of sporangia and incidence of infection by Phytophthora infestans on upper and lower surfaces of potato leaves. Acta Agriculturae Scandinavica V: 375 – 386.

Bruhn J A, Fry W E. 1981. Analysis of potato late blight epidemiology by simulation modelling. Phytopathology 71: 612 – 616.

Ferrandino F J. 1989. Spatial and temporal variation of a defoliating plant disease and reduction in yield. Agricultural and Forest Meteorology 47: 273 -289.

Fohner G R, Fry W E, White G B. 1984. Computer simulation raises question about timing protectant fungicide application frequency according to a potato late blight forecast. Phytopathology 74: 1 145 -1 147.

Gees R, HoW H R. 1988. Cytological comparison of specific (R3) and general resistance to late blight in potato leaf tissue. Phytopathology 78: 350 -357.

Gilligan C A. 1985. Introduction. In Gilligan C A (Ed.) Mathematical Modelling of Crop Disease. Advances in Plant Pathology 3: 1 -10.

Harrison J G. 1995. Factors involved in the development of potato late blight disease (Phytophthora infestans). Pages 215 - 236 in Haverkort A J, MacKerron D K L (Eds.) Potato Ecology and Modelling of Crops under Conditions Limiting Growth. Kluwer Academic Publishers, Dordrecht, The Netherlands.

Haverkort A J, Bicamumpaka M. 1986. Correlation between intercepted radiation and yield of potato crops infested by Phytophthora infestans in central Africa. Netherlands Journal of Plant Pathology 92: 239 -247.

Hethcote H W. 1976. Qualitative analyses of communicable disease models. Mathematical Biosciences 28: 335 -356.

Jeger M J. 1986. Asymptotic behaviour and threshold criteria in model plant disease epidemics. Plant Pathology 35: 355 -361.

Jeger M J, Groth J. 1985. Resistance and pathogenicity: Epidemiological and ecological mechanisms. Pages 310 - 372 in Fraser R S S (Ed.) Mechanisms of Resistances to Plant Diseases. Martinus Nijhoff Publishers, The Hague, The Netherlands.

Kermack W O, McKendrick A G. 1927. A contribution to the mathematical theory of epidemics. Proceedings of the Royal Society of London, Series A 115: 700 -721.

Kluge E, Gutsche V (1990) Prediction oflate blight of potato with the help of simulation modelsResults 1982 - 1988. Archiv fiir Phytopathologie und Pflanzenschutz 26: 265 -281.

Knudsen G R, 'Spurr H W, Johnson C S. 1987. A computer simulation model for Cercospora leaf spot of peanut. Phytopathology 77:

1118 -1121.

Lapwood D H. 1961. Potato haulm resistance to Phytophthora infestans. III. Lesion distribution and leaf destruction. Annals of Applied Biology 49: 704 -716.

Levy Y, Cohen Y, Benderly M. 1991. Disease development and buildup ofresistance to oxadixyl in potato crops inoculated with Phytophthora infestans as affected by oxadixyl and oxadixyl mixtures: Experimental and simulation studies. Journal of Phytopathology 132: 219 -229.

Liu W M, Hethcote H W, Levin S A. 1987. Dynamical behavior of epidemiological models with nonlinear incidence rates. Journal of Mathematical Biology 25: 359 -380.

Michaelides S C. 1985. A simulation model of the fungus Phytophthora infestans (Mont) De Bary. Ecological Modelling 28: 121 -137.

Michaelides S C. 1991. A dynamic model of the interactions between the potato crop and Phytophthora infestans. EPPO Bulletin 21: 515 -525.

Milgroom M G, Fry W E. 1988. A simulation analysis of the epidemiological principles for fungicide resistance management in pathogen populations. Phytopathology 78: 565 -570.

K P, Fry W E. 1983. Models for the spread of disease: some experimental results. Phytopathology 73: 1 173 -1 176.

Monteith J L. 1977. Climate and the efficiency of crop production in Britain. Philosophical Transactions of the Royal Society of London, Series B 281: 277 -294.

Oort A J P. 1968. A model of the early stage of epidemics. Netherlands Journal of Plant Pathology 74: 177 -180.

Paysour R E, Fry W E. 1983. Interplot interference: A model for planning field experiments with aerially disseminated pathogens. Phytopathology 73: 1 014 -1 020.

Rabbinge R, Zadoks J C, Bastiaans L. 1989. Population models. Pages 83 -97 in: Rabbinge R, Ward S A, Van Laar H H (Eds.) Simulation and Systems Management in Crop Protection. Simulation Monographs 32. Pudoc, Wageningen, The Netherlands.

Raposo R, Wilks D S, Fry W E. 1993. Evaluation of potato late blight forecasts modified to include weather forecasts: A simulation analysis.

Phytopathology 83：103 – 108.

Rossing WAH，Van Oijen M，Van der WerfW，et al. 1992. Modelling the effects of foliar pests and pathogens on light interception，photosynthesis，growth rate and yield offield crops. Pages 161 – 180 in Ayres P G (Ed.) Pests and Pathogens：Plant Responses to Foliar Attack. Bios Scientific Publishers，Oxford，UK.

Rouse D I. 1991. Stochastic modelling of plant disease epidemic processes. Pages 647-665 in Arora D K，Rai B，Mukerji K G，Knudsen G R (Eds.) Handbook of Applied Mycology. Volume 1：Soil and Plants. Marcel Dekker，New York，USA.

Sail M. 1980. Uses of stochastic simulation：grape powdery mildew example. Zeitschrift fUr Pflanzenkrankheiten und Pflanzenschutz 87：397 – 403.

Shaner G. 1980. Probits for analyzing latent period data in studies of slow rusting resistance. Phytopathology 70：1 179 – 1 182.

Shtienberg D，Doster M A，Pelletier J R，Fry W E. 1989. Use of simulation models to develop a lowrisk strategy to suppress early and late blight in potato foliage. Phytopathology 79：590 – 595.

Shtienberg D，Raposo R，Bergeron S N，Legard D E，Dyer A T，Fry W E. 1994. Incorporation of cultivar resistance in a reduced-sprays strategy to suppress early and late blights on potato. Plant Disease 78：23 – 26.

Sparks W R. 1980. Blight：A computer model relating the progress of potato blight to weather. Unpublished Agricultural Memorandum No. 899. National Meteorological Library，Bracknell，UK.

Stephan S，Gutsche V. 1980. Ein algorithmisches Modell zur Simulation der PhytophthoraEpidemie (SIMPHYT). Archiv fur Phytopathologie und Pflanzenschutz 16：183 – 191.

Teng P S，Blackie M J，Close R C. 1977. A simulation analysis of crop yield loss due to rust disease. Agricultural Systems 2：189 – 198.

Van der Plank J E. 1963. Plant Diseases：Epidemics and Control. Academic Press，New York，USA.

Van der Zaag D E. 1959. Some observations on breeding for resistance to Phytophthora infestans. European Potato Journal 2：278 – 286.

Van Oijen M. 1989. On the use of mathematical models from human epidemiology in breeding for resistance to polycyclic fungal leaf diseases of crops. Pages 26 – 37 in Louwes K M, Toussaint H A J M, Dellaert L M W (Eds.) Parental Line Breeding and Selection in Potato Breeding. Pudoc, Wageningen, The Netherlands.

Van Oijen M. 1990. Photosynthesis is not impaired in healthy tissue of blighted potato plants. Netherlands Journal of Plant Pathology 96: 55 – 63.

Van Oijen M. 1991a. Light use efficiencies of potato cultivars with late blight (Phytophthora infestans). Potato Research 34: 123 – 132.

Van Oijen M. 1991b. Identification of the Major Characteristics of Potato Cultivars which Affect Yield Loss Caused by Late Blight. Ph. D. Thesis, Wageningen Agricultural University, Wageningen. The Netherlands.

Van Oijen M. 1991c. Leaf area dynamics of potato cultivars infected to various extent by Phytophthora infestans. Netherlands Journal of Plant Pathology 97: 345 – 354.

Van Oijen M. 1992a. Evaluation of breeding strategies for resistance and tolerance to late blight in potato by means of simulation. Netherlands Journal of Plant Pathology 98: 3 – 11.

Van Oijen M. 1992b. Selection and use of a mathematical model to evaluate components of resistance to Phytophthora infestans in potato. Netherlands Journal of Plant Pathology 98: 192 – 202.

Waggoner P E. 1968. Weather and the rise and fall of fungi. Pages 45-66 in Lowry W P (Ed.) Biometeorology. Oregon State University Press, Corvallis, USA.

Waggoner P E. 1990. Defoliation, disease and growth. Pages 149-180 in Goudriaan J, Penning de Vries F W T, Rabbinge R, Van Keulen H, Van Laar H H (Eds.) Theoretical Production Ecology: Hindsights and Perspectives. Simulation Monographs. Pudoc, Wageningen, The Netherlands.

Waggoner P E, Rich S. 1981. Lesion distribution, multiple infection, and the logistic increase of plant disease. Proceedings of the National Academy of Sciences USA 78: 3292 – 3295.

Zadoks J C. 1971. Systems analysis and the dynamics of epidemics. Phytopathology 61: 600 – 610.

Zadoks J C. 1977. Simulation models of epidemics and their possible use in the study of disease resistance. Pages 109 – 118 in International Atomic Energy Agency (Ed.) Induced Mutations Against Plant Diseases: Proceedings of a Symposium. Vienna, Austria.

Zadoks J C, Schein R D. 1979. Epidemiology and plant disease management. Oxford University Press, New York, USA.

第七章　马铃薯黄萎病菌的生命周期和生态学

第一节　马铃薯黄萎病菌简介

大丽轮枝菌引起的植物病害在全球范围内广泛分布，对很多国家的经济造成重要影响（Pegg，1984）。大丽轮枝菌应该是造成马铃薯枯萎、使马铃薯早熟及造成马铃薯损伤早期死亡的主要因素。在所有双子叶植物宿主中，大丽轮枝菌主要侵染的作物是马铃薯、棉花、茄子、西红柿、薄荷和橄榄。大丽轮枝菌通过微菌核（MS）的方式生存，易存活，并且广泛侵染宿主，使其在很多农业土壤中特有（Powelson，1970）。

大丽轮枝菌是土壤入侵或根部寄生的真菌（Powelson，1970）。这些真菌均具有一个在宿主植物中寄生的阶段（图 7－1 中Ⅰ），以及宿主死亡后的腐生阶段（Powelson，1970）（图 7－1 中Ⅱ）。这两个阶段将分开讨论。

第二节　马铃薯黄萎病菌土壤种群动态研究

土壤中微菌核群落动态示意图如图 7－1 中Ⅱ所示。土壤中微菌核数量根据土壤中耕作历史构成和密度有所不同。在微菌核萌芽后，菌丝可能侵染植物根部。由于其腐生能力较弱，大多数菌丝在侵染植物根部之前死亡。下面将分别介绍微菌核的存活（图 7－1 中 a）、萌发（图 7－1 中 b）以及植物根部定殖的不同阶段。

一、马铃薯黄萎病微菌核存活研究

微菌核的存活（图 7－1 中 a）

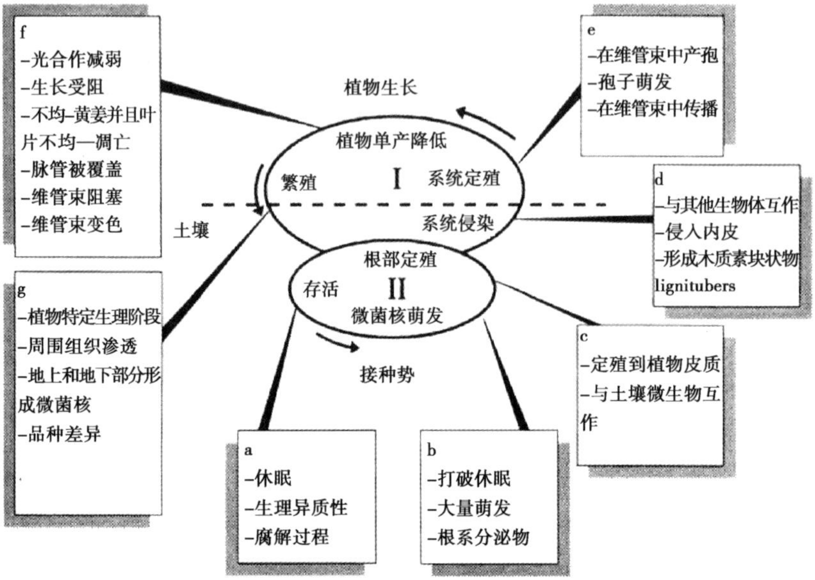

图 7－1 大丽轮枝菌生命周期示意

注：分为宿主寄生阶段 I 以及植物死亡后在土壤的腐生阶段 II，文本为具体说明。MS 表示微菌核

Ben-Yephet 和 Szmulewich（1985）的研究表明，大丽轮枝菌微菌核在田间比在实验室存活的时间长，20～25 °C，空气干燥的土壤样品中保存的微菌核在 5 年后，检测不到大丽轮枝菌。而作物轮作 7 年后 4% 的原始菌群密度依然可以存活。Wilhelm（1955）发现大丽轮枝菌可以在田间没有宿主存在的情况下存活 14 年。真菌在田间能够长期存活可能是由于其具有定殖的能力，并可以在包括单子叶植物在内的几乎所有植物根系产生新的微菌核（Martinson 和 Horner，1962）。Itoh 等人（1989）检测了大白菜种植后受感染的土壤中微菌核的密度。研究发现随着温度的升高，微菌核数量下降加快，符合一级反应函数。5～31℃ 温度范围内，半保留期微菌核数量和温度存在线性关系。

大丽轮枝菌微菌核在空气干燥的条件下生存最好（Coley-Smith 和 Cooke，1971）。虽然存活机制尚未阐明。但是机体黑色素的存在与生存有着明显的关系（Bloomfield 和 Alexander，1967）。

温度可能通过直接影响微菌核的萌发（Coley-Smith 和 Cooke，1971）以及植物材料的分解促进菌体释放，从而产生间接影响（Hancock 和 Benham，1980）。温度和分解的影响存在相关性，分解是多个过程的结果，并且在一定温度范围内，温度升高将加速大多数生物学过程。

微菌核从定殖的植物残体上释放至土壤并最终影响土壤的菌体密度需要持续至少一年的时间。1967 年，Evans 等人研究了两种栽培棉花的土壤，研究发现：在全生长季，土壤中微菌核的密度一直下降，但在收获季节，当定殖的棉花植株发生机械损伤引起了新鲜微菌核的释放，土壤中微菌核的密度会再次升高。当棉花秸秆返田后，土壤接种体密度会进一步升高。Huisman 和 Ashworth（1976）观察了 8 个具有不同种植历史的商业田块的大丽轮枝菌土壤接种密度变化情况，每 2 个月观察 1 次，共观察了 3.5 ~4 年。研究发现：持续栽培棉花的土壤，大丽轮枝菌土壤接种密度会持续升高。当棉花种植 1 年后，接种密度通常会快速升高，第 2 年不论轮作的作物是否易感病，接种密度往往会更高。微菌核从植物残体的缓慢释放可能导致了土壤中接种密度升高的延迟。侵入植物体内的微菌核与释放到土壤中微菌核相比如何存活，目前仍不清楚。据猜测，植株体内的微菌核死亡率更高，因为分解的植物残体内以及周边微生物较活跃。

二、马铃薯黄萎病微菌核萌发研究

大丽轮枝菌由于繁殖体的内生属性，发育延迟时，微菌核没有实质性休眠（Pegg，1974），但在微菌核萌发时需要从宿主或非宿主（Schreiber 和 Green，1963；Mol，1994）的根系分泌物中获取外源营养物质（Emmatty 和 Green，1969）（图 7 -1 中 b）。

外源营养物质的浓度可由土壤中施加的有机改良剂改变，有机改良剂一般影响土壤微生物活性和土壤溶液中可溶性化合物的浓度。已有研究表明，土壤微生物活性增加可以降低微菌核的萌发比例以及根际大丽轮枝菌芽管的生长（Jordan 等，1972）。在温室和田间试验中，施用切碎的大麦或燕麦秸秆在一定程度上可以降低土壤中的棉花黄萎病菌的接种密度以及马铃薯植株的发病率（Tolmsoff 和 Young，1959；Harrison，1976）。有机添加物的作用尚未明确，但有学者认为有机添加物有助于根际富集的微生物分解产生一些抗真菌物质（Curl 和 Truelove，1986）。然而，根际微生物菌群与病原体之间的相互作用可能更重要，因为土壤中产生的抗生素物质可作为有效抑制剂，在足够长时间内未降解，这一点仍然会产生疑问（Curl 和 Truelove，1986）。

微菌核可以多次萌发。1971 年，Farley 等人研究表明，每当用蔗糖溶液或水浇灌土壤，微菌核就会萌发并产孢，这种规律可持续直至第 9 次。但发芽率和生殖管的数目随发芽次数的增加而逐渐降低。田间受大丽轮枝菌感染的作物与人工接种感染的作物可能表现出不同的性状（Menzies 和 Griebel，1967）。除非该土壤有几年没有种植易感染作物，大丽轮枝菌均以微菌核部分嵌入分解的植物残体的形式侵染作物，接种可能主要是以 MS 的嵌入部分分解植物残体的形式出现。一些微菌核可能已经完成了 1 个萌芽周期而凋亡，而其他微菌核可能在耕种过程中被暴露出来，当其附近宿主幼苗根系生长时，微菌核得以大量萌发（Menzies 和 Griebel，1967）。

三、马铃薯黄萎病菌在植物根部定殖的研究

大丽轮枝菌在植物根系中的定殖（图 7－1 中 c）。微菌核萌发的菌丝可以穿透植物根的皮层，进入木质部导管，从而到达易感宿主的全身（Powelson，1970）。大丽轮枝菌微菌核主要从靠近根尖以及根毛区的部位穿透皮层进入植物体内（Fitzell 等，1980；Gerik 和 Huisman，1988）。距离根尖大于 1cm 处的定殖密度似乎是常数（Gerik 和 Huisman，1988）。大多数菌落在根表面消毒时被去除，这表明大丽轮枝菌定殖限于在根皮层表皮部位（Evans 和 Gleeson，1973）。

第三节　宿主—病原体相互作用机理

宿主—病原体互作示意图见图 7－1 中 I。当定殖到植物根部后，真菌跨过植物物理屏障进入维管束系统。当真菌侵染到植物全身后，诱发植物产生抗性反应，产量下降，最终在植物的碎片上产生微菌核。下文将集中介绍植物根部的系统性侵染（图 7－1 中 d），植物的全身定殖（图 7－1 中 e），植物响应和产量损失（图 7－1 中 t）以及大丽轮枝菌的再生（图 7－1 中 g）。

一、马铃薯根部的系统性感染研究

植物根部的系统性侵染（图 7－1 中 d）。当真菌入侵时，表皮的内切向壁会膨大（Bell，1973）。大丽轮枝菌菌丝周边的植物根部外皮层细胞偶尔会出现胶状封闭区域。这些结构被称为木质管（Griffiths，1973）。木质管是由从原生质体挤压到细胞壁和细胞膜之间的囊泡形成。当真菌开始穿透细胞壁，囊泡聚集形成木质管。随后，真菌细胞壁被木质管包围，而宿主质膜能确保完好。在大多数情况下，侵入的菌丝被降解掉，但在少数

情况下，会有部分菌丝穿透细胞壁，并且通过皮层细胞，向宿主内部渗透（Bell，1973）。

成功穿透皮层的真菌在内皮层处会遇到第二个屏障（Bell，1973）。木质管没能阻止的大多数菌丝会在内皮层被阻止；只有少数菌丝能跨过该屏障进入维管束系统，到达导管。菌丝也可以通过定殖在幼苗根尖而到达维管束系统，但也需经过内胚层。植物根部的全身性侵染决定了植物作为宿主的适宜性。例如，敏感型薄荷（薄荷属）的根具有更广泛的全身性侵染特征，而其根皮质感染的数量与抗性薄荷品种相类似（Lacy 和 Horner，1966）。

大丽轮枝菌对植物根部的侵染可能受其他生物的影响。通常情况下，马铃薯早期死亡可以由土壤微生物共同作用导致，包括：大丽轮枝菌（*V. dahliae*）、球异皮线虫属（*Globodera* spp.）、根腐线虫属（*Pratylenchus* spp.）、根结线虫属（*Meloidogyne* spp.）、炭疽病菌（*Colletotrichum* coccodes）、镰刀菌属（*Fusarium* spp.）以及立枯丝核菌属（*Rhizoctonia solani*）（Scholte，1989；Scholte 和 s'Jacob，1989）。已经与马铃薯早期死亡相关的致病因子包括胡萝卜软腐欧文氏菌（*Erwinia carotovora*）和马铃薯病毒 X（Potato Virus X）（Rouse，1985）。目前研究已表明马铃薯早期死亡是由大丽轮枝菌和内寄生线虫共同导致。Russet Burbank 和 Kotcon 等人（1985）在对马铃薯栽培品种微区试验研究中，并没有确定炭疽病菌（*C. cocodes*）和立枯丝核菌（*R. solani*）的单独作用，还是与大丽轮枝菌或根腐线虫共同作用所致。然而，对于马铃薯栽培品种 Amethyst，大丽轮枝菌和炭疽病菌共同作用导致的减产几乎是大丽轮枝菌单独作用的 2 倍（Scholte 等，1985），并且 Scholte 和 s'Jacob（1989）发现了根结线虫（*Meloidogyne* spp.）或落选短体线虫（*P. neglectus*）、立枯丝核菌（*R. solani*）以及大丽轮枝菌（*V. dahliae*）3 因子互作对几种马铃薯品种产量的影响。

目前的推测是，大丽轮枝菌和根腐线虫的互作机制导致了症状的协同表达以及马铃薯产量损失；Green（1981）认为该机制可能是线虫进食的伤口，使真菌更容易进入维管束系统，或者他本人提出的一种生理机制——植物可能由于一种易位的物质会变得更易感染真菌。Wheeler 等人（1992）比较了大丽轮枝菌和根腐线虫共同作用的马铃薯减产模型和单一因素造成的马铃薯减产模型，研究表明线虫存在时会造成更加严重的减产。然而，这一结论并没有考虑马铃薯早期死亡表达的变异性和环境条件的混杂效应。在评估他的机械模型时，Johnson（1992）描述了不包括线虫在内的多种生物胁迫因子引起的马铃薯产量损失。他的结论是，多种胁迫因素造成的产量损失小于单独作用造成的损失之和。这通过大丽轮枝菌和致病疫霉之间的竞争性脱叶得以说明。

二、马铃薯植株定殖研究

植物的全身定殖（图7－1中e），分生孢子在维管束内的生产过程仍不清晰。几个作者提出的假设是在大丽轮枝菌侵入维管束系统后，直接开始产生分生孢子（Howell，1973；Schnathorst，1981）。孢子通过简单的分生孢子梗或出芽产生（Tolmsoff，1973），并被动地通过维管束系统分布到植物全身。分生孢子可以发芽并穿透导管壁（Garber，1973；Tolmsoff，1973）。研究需要关注在维管束系统中的孢子形成和分布过程。例如，木质部中营养物浓度的影响以及上述两个过程对植物活力的影响导致了植物茎干中大丽轮枝菌微粒的起始数目，但这一点还未有实验证明。

在敏感性和耐受性较强的棉花品种中，定殖就是病原体穿过内皮层到达木质部（Garber，1973）。入侵的导管数量是测量黄萎病严重程度的一个较好的标准，因为入侵导管的数量与全身感染的数量以及从定殖点穿过皮质的菌丝数量有关。

如果马铃薯植株的叶柄为绿色，那么侵染的大丽轮枝菌的分布则局限于木质部，但在具有严重疾病症状的植物中，木髓、形成层以及皮层均被侵入（Garber，1973）。在叶子中，感染可以局限于单个羽片，或整个叶片（Garber，1973）。

在马铃薯中，大丽轮枝菌侵染的症状难以与正常衰老区分开，并且最初的病症也可能只是减缓生长（Street 和 Cooper，1984；Haverkort 等，1990）。早期叶片症状可能表现为少数植物低位叶片出现单边黄萎，随后可能出现一些整体小叶或叶子的枯萎，但是下位叶单边叶片枯死的症状更为典型（Isaac 和 Harrison，1968）。

真菌会刺激植物在维管束细胞中产生类似于软木脂的涂层、侵填体和凝胶。侵填体是堵塞在导管中的薄壁细胞（Newcombe 和 Robb，1988），凝胶是通过初生壁和胞间层组分膨胀过程中的穿孔板、端壁和纹孔膜产生（Van der Molen 等，1977）。这可能导致直接在原发感染位点上方的维管束系统中的导管闭塞（Van der Molen 等，1977；Harrison 和 Beckman，1982；Newcombe 和 Robb，1988）。制备切片样品时，通常可以看到茎基部的导管变为浅棕色（Isaac 和 Harrison，1968）。当导管中存在病原体时，会分泌特定化学品例如萜烯醛，它的累积可以降低真菌病原体的繁殖能力

(Harrison 和 Beckman，1982)。随着木质部中定殖的病原体的增多，导管可能会被菌丝体堵塞（Garber，1973)。在马铃薯植株中，从根尖到茎顶部均会被堵塞。

感染的马铃薯植株在块茎化之前并没有任何症状（Busch 和 Edgington，1967)，这表明枯萎症状只有在宿主处于发育晚期才会显著（Busch 等，1978)，通过改变光周期防止块茎化，从而防止或减轻症状出现(Busch 和 Edgington，1967)。该结果与之前栽培晚熟品种和对大丽轮枝菌的抗性关联的结果相一致（Busch 等，1978)。

Kotcon 等人（1985）将大丽轮枝菌发生率与根生长的减少、叶片重量以及块茎产量相关联。感染的植株比叶面积降低（叶面积/叶组织干重)、叶片重量比例升高（叶系统的干重/整株植株的干重）以及叶面积比升高(叶片面积/ 植株的总干重)，且在干燥条件下，相对生长速率以及叶片生长速率均会降低（干重增量/单位叶面积/每周）　(Harrison 和 Isaac，1969)。

Bowden 等人（1990）和 Haverkort 等人（1990）研究表明，初始大丽轮枝菌引起光合作用的下降是由气孔闭合引起，低气孔导度与低叶水势相关（Haverkort 等，1990)。在马铃薯中，大丽轮枝菌引起光转化效率的降低以及气孔导度更为明显的下降，导致了内/外 CO_2 比率降低，并且在相似的气孔导度下具有更高的净光合作用（Haverkort 等，1990)。净光合作用的减少似乎不是造成干物质产量减少的主要原因，其作用不超过 10%。在黄萎病导致叶片冠层早期和快速衰老的部分，光截获量的降低可能是比光合作用减少更为重要的受损原因（Haverkort 等，1990)。

虽然茎中维管束之间互相连接，但在叶柄中并没有发生。因此，叶柄束的阻塞比茎中维管束的成比例阻塞更具破坏性（Garber，1973)。大丽轮枝菌引起的马铃薯植株根长较短，降低了水分供应，并导致叶片症状的出现（Kotcon 等，1984)。据研究表明，根表面积和体积也会受到大丽轮枝菌的影响。

三、马铃薯黄萎病菌的繁殖研究

随着被侵染的植株变老，真菌会渗透到周围组织并在坏死组织中形成微菌核（Powelson，1970)。没有证据表明，除了由植物碎片产生的接种体之外还有其他接种体出现过暂时性的数量增长。被大丽轮枝菌定殖的马铃薯茎部将会充满微菌核。每厘米茎部可含有 8 000 ~ 20 000个活的微菌核，且据研究，田间每克土壤中高达 1 000个微菌核，将再次侵染马铃薯，也就是说每一棵植物的根所占据的土壤，大约有5 000万个微菌核（Menzies，1970）。不同的马铃薯栽培品种，每克茎组织中微菌核的产量从7 000～9 000个不等（Slattery，1981）。在大多数被大丽轮枝菌感染的植物当中，形成微菌核均需要水分（Powelson，1970）。然而，在温带地区，马铃薯干燥衰老期间，即使没有降雨，湿度也通常足够确保大量微菌核的形成。微菌核的形成受外部因素的影响很大。以色列内盖夫地区秋季种植的马铃薯的微菌核产量比春播马铃薯大约高100倍（Ben-Yephet和Szmulewich，1985）。秋冬季节凉爽和潮湿的条件使植物干燥缓慢，有利于微菌核的形成，或更凉爽的天气有利于真菌在植物组织中更好的存活。

Ioannou等人（1977）研究了田间不同灌溉和漫灌情况下，土壤中番茄病株残体中微菌核的形成。漫灌处理期间几乎不产生微菌核，因为土壤中O_2浓度低，而CO_2浓度高。排水后，O_2和CO_2的浓度迅速恢复到正常大气水平，微菌核开始形成。在漫灌处理后10d、20d和40d后，最终产生的MS数量分别是非灌溉处理平均值的90%、44%和46%。

新菌体的产生主要在马铃薯植株的地上部分（Ben-Yephet和Szmulewich，1985；Mol，1994）。控制大丽轮枝菌的直接方法是干扰MS的形成和传播。在实践中，主要是在病害进入土壤或在其释放接种物之前除去或破坏病株残体。但由于费用、设备缺乏以及农民不情愿破坏有机物质，这种方法没有得到广泛采用。控制病原体繁殖的另一种可行性方法是使用机械茎秆处理技术。与施用除草剂相比，马铃薯植株地上部分的再生数量在机械切碎茎秆后下降率高达80%（Mol，1994）。

已有研究表明，在没有全部感染的植物根部可产生微菌核。在小麦根部的少部分产生了大量的微菌核（Krikun和Bernier，1990）。考虑到小麦的总根长以及发现的微菌核数量（每1.3mm×0.4mm的根长中高达100个），会向土壤中释放出大量的病原菌（Krikun和Bernier，1990）。控制杂草对于限制大丽轮枝菌的繁殖非常重要（Woolliams，1966；Johnson等，1980）。大丽轮枝菌的宿主包括许多常见的杂草和天然植物。因为在作物植株中，被侵染的杂草植物的症状并不总是很明显。

第四节　马铃薯黄萎病菌宿主特异性研究

大丽轮枝菌的适应性，导致宿主特异性，对可能的抗性或耐受机制研究造成困扰。Zilberstein等人（1983）研究表明，大丽轮枝菌在琼脂上的萌芽以及微菌核对茄子、马铃薯和番茄的致病性受生长培养基和宿主来源的影响。大丽轮枝菌分离菌株的毒力取决于宿主物种、地理起源、宿主栽培种基因型的易感度以及植物器官（Zilberstein 等，1983；Michail，1989）。从易感品种中分离的大丽轮枝菌只能侵染这些品种。从抗性品种中分离的菌株可以侵染易感以及中等抗性的品种（MichaiI，1989）。

微菌核变异体的产生可能使真菌可以适应新的宿主物种品种或新的环境条件（Tolmsoff，1973）。薄荷分离株最初对番茄并没有致病性；然而，在侵染一次番茄后，分离株变得对番茄更具致病性，从而丧失了对薄荷的致病性。在一个连续播种同一作物的地区分离的致病菌株往往具有相似的特征，且都对该地区的作物具有高致病力，但通常对于其他可能感染的物种具有弱毒力（Vigoroux，1971）。因此，形成了“优先”和“随机”宿主的概念。目前，已研究得知轮枝孢属具有很大的变异性（Vigoroux，1971）。

Puhalla 和 Hummel（1983）在大丽轮枝菌中发现了 16 个不同的营养体亲和群。Joaquim 和 Rowe（1990；1991），使用另一种方法，将营养亲和群的数量减少到 4 个。来自两个不同营养亲和群的分离株对马铃薯的致病性存在显著差异（Joaquim 和 Rowe，1991），因此，在土壤中，可以建立具有不同性质个体的群体。作物的种类似乎是栽培土壤中大丽轮枝菌种群的性质和数量的一个决定性因素（Vigoroux，1971；Tjamos，1981），因此，品种选择、作物轮作以及栽培实践都不足以防止作物受到感染，但它们仍然是控制疾病严重程度的关键因素。

参考文献

Bell A A. 1973. Nature of disease resistance. Pages 47 – 62 in Verticillium Wilt of Cotton. Proceedings of a Work Conference, National Cotton Pathology Research Laboratory College Station USDA Agricultural Research Service, Texas, USA.

Ben-Yephet Y, Szmulewich Y. 1985. Inoculum levels of Verticillium dahliae in the soils of the hot semi-arid Negev region ofIsrael. Phytoparasitica 13: 193 – 200.

Bloomfield B J, Alexander M. 1967. Melanin and resistance of fungi to lysis. Journal of Bacteriology 93: 1 276 – 1 280.

Bowden R L, Rouse D I, Sharkey T D. 1990. Mechanism of photosynthesis decrease by Verticillium dahliae in potato. Plant Physiology 94: 1 048 - 1 055.

Busch L V, Edgington LV. 1967. Correlation of photoperiod with tuberization and susceptibility of potato to Verticillium albo-atrum. Canadian Journal of Botany 45: 691 -693.

Busch L V, Smith E A, Njoh-Elango F. 1978. The effect of weeds on the value of rotation as a practical control for Verticillium wilt of potato. Canandian Plant Disease Survey 58: 61 -64.

Coley-Smith J R, Cooke R C. 1971. Survival and germination offungal sclerotia. Annual Review Phytopathology 9: 65 -92.

Curl E A, Truelove B. 1986. The Rhizosphere. Advanced Series in Agricultural Sciences 15. Springer-Verlag, Berlin, Germany.

Emmatty D A, Green R J. 1969. Fungistasis and the behaviour of the micro sclerotia of Verticillium albo-atrum in soil. Phytopathology 59: 1 590 -1 595.

Evans G, Gleeson A C. 1973. Observations on the origin and nature of Verticillium dahlia colonizing plant roots. Australian Journal of Biological Sciences 26: 151 -161.

Evans G, Wilhelm S, Snyder W C. 1967. Quantitative studies by plate counts of propagules of the Verticillium wilt fungus in cotton field soils. Phytopathology 57: 1 250 -1 255.

Farley J D, Wilhelm S, Snyder W C. 1971. Repeated germination and sporulation of micro sclerotia of Verticillium albo-atrum in soil. Phytopathology 61: 260 -264.

Fitzell R, Evans G, Fahy P C. 1980. Studies on the colonization of plant roots by Verticilliumdahliae with use of immuno fluorescent staining. Australian Journal of Botany 28: 357 -368.

Garber R H. 1973. Fungus penetration and development. Pages 69 - 77 in Verticillium Wilt of Cotton. Proceedings of a Work Conference, National Cotton Pathology Research Laboratory College Station USDA Agricultural Research Service, Texas, USA.

Gerik J S, Huisman O C. 1988. Study offield-grown cotton roots infected with Verticillium dahlia using an immunoenzymatic staining technique.

Phytopathology 78: 1 174 – 1 178.

Green R J. 1981. Fungal wilt diseases of plants. Pages 1-24 in Mace E, Bell A A, Beckman C H (Eds.) Fungal Wilt Diseases of Plants. Academic Press, New York, USA.

Griffiths D A. 1973. An electron microscopic study of host reaction in roots following invasion by Verticillium dahliae. Skokubutsu Byogai Kenkyu 8: 147 – 154.

Hancock J G, Benham G S. 1980. Fungal decay of buried cotton Gossypium-hirsutum stems. Soil Biological Biochemistry 12: 35 – 42.

Harrison M D. 1976. The effect of barley straw on the survival of Verticillium albo-atrum in naturally infested field soil. American Potato Journal 53: 385 – 394.

Harrison N A, Beckman C H. 1982. Time-space relationships of colonization and host response in wilt-resistant and wilt susceptible cotton (Gossypium) cultivars inoculated with Verticillium dahliae and Fusarium oxysporum f. sp. vasinfectum. Physiological Plant Pathology 21: 193 – 207.

Harrison J A C, Isaac I. 1969. Host-parasite relations up to the time of tuber initiation in potato plants infected with Verticillium spp. Annals of Applied Biology 64: 469 – 482.

Haverkort A J, Rouse D I, Turkensteen L J. 1990. The influence of Verticillium dahliae and drought on potato crop growth. 1. Effects on gas exchange and stomatal behaviour of individual leaves and crop canopies. Netherlands Journal of Plant Pathology 96: 273 – 289.

Howell C R. 1973. Pathogenicity and host-parasite relationships. Pages 42-46 in Verticillium Wilt of Cotton. Proceedings of a Work Conference, National Cotton Pathology Research Laboratory College Station USDA Agricultural Research Service, Texas, USA.

Huisman O C, Ashworth L J. 1976. Influence of crop rotation on survival of Verticillium albo-atrum in soils. Phytopathology 66: 978 – 981.

Ioannou N, Schneider R W, Grogan R G. 1977. Effect of flooding on the soil gas composition and the production of micro sclerotia by Verticillium dahliae in the field. Phytopathology 67: 651 – 656.

Isaac I, Harrison J A C. 1968. The symptoms and causal agents of early-dying disease (verticillium wilt) of potatoes. Annals of Applied Biology

61: 231 -244.

Hoh S, Komoda H, Monma T, Amano T. 1989. Development offield diagnosis system (FDS) for preventing continuous cropping injury of crop. 12. Study of factors related to the development of a prediction model of Verticillium yellows in Chinese cabbage. Bulletin of the National Agriculture Research Center, Japan 16: 33 -53.

Joaquim T R, Rowe R C. 1990. Reassessment of vegetative compatibility relationships among strains of Verticillium dahliae using nitrate-nonutilizing mutants. Phytopathology 80: 1 160 -1 166.

Joaquim T R, Rowe R C. 1991. Vegetative compatibility and virulence of strains of Verticillium dahliae from soil and potato plants. Phytopathology 81: 552 -558.

Johnson K B. 1992. Evaluation of a mechanistic model that describes potato crop losses caused by multiple pests. Phytopathology 82: 363 -369.

Johnson W M, Johnson E K, Brinkerhoff LA. 1980. Symptomatology and formation of micro sclerotia in weeds inoculated with Verticillium-dahliae from cotton Gossypium-hirsutum. Phytopathology 70: 31 -35.

Jordan V W L, Sneh B, Eddy B P. 1972. Influence of organic soil amendments on Verticillium dahliae and on the microbial composition of the strawberry rhizosphere. Annals of AppliedBiology 70: 139 -148.

Kotcon J B, Rouse D I, Mitchell J E. 1984. Dynamics of root growth in potato fields affected by the early dying syndrome. Phytopathology 74: 462 -467.

Kotcon J B, Rouse D I, Mitchell J E. 1985. Interactions of Verticillium dahliae, Colletotrichum coccodes, Rhizoctonia solani, and Pratylenchus penetrans in the early dying syndrome of Russet Burbank potatoes. Phytopathology 75: 68 -74.

Krikun J, Bernier C C. 1990. Morphology of micro sclerotia of Verticillium dahliae in roots of gramineous plants. Canadian Journal of Plant Pathology 12: 439 -441.

Lacy M L, Horner C E. 1966. Behaviour of Verticillium dahliae in the rhizosphere and on roots of plants susceptible, resistant, and immune to wilt. Phytopathology 56: 427 -430.

Martinson C A, Horner C E. 1962. Importance of non-hosts in maintaining the inoculum potential of Verticillium. Phytopathology 52：742.

Menzies J D. 1970. Factors affecting plant pathogen population in soil. Pages 16 –21 in Toussoun.

T A，Bega R V，Nelson P E（Eds.）Root Diseases and Soil-Borne Pathogens. University of California Press，Berkeley，USA.

Menzies J D，Griebel G E. 1967. Survival and saprophytic growth of Verticillium dahliae in uncropped soil. Phytopathology 57：703 –709.

Michail S H. 1989. Fusarium and Verticillium wilts of cotton. Pages 199 – 217 in Agrihotri V P.

Singh N，Chaube H S，Singh U S，Dwivedi T S（Eds.）Perspectives in Phytopathology. Today and Tomorrow's Printers & Publishers，New Delhi，India.

Mol L. 1994. Control of Verticillium dahliae by catch-crops and haulm killing techniques. Pages 223 –228 in Struik P C，Vredenberg W J，Renkema J A，Parlevliet J E（Eds.）Plant Production on the Threshold of a New Century. Kluwer Academic Publishers，Dordrecht，The Netherlands.

Newcombe G，Robb J. 1988. The function and relative importance of the vascular coating response in highly resistant，moderately resistant and susceptible alfalfa infected by Verticillium alboatrum. Physiological and Molecular Plant Pathology 33：47 –58.

Pegg G F. 1974. Verticillium diseases. Review of Plant Pathology 53：156 –182.

Pegg G F. 1984. The impact of Verticillium diseases in agriculture. Phytopathologia Mediterranea 23：176 –192.

Powelson R L. 1970. Significance of population level of Verticillium in soil. Pages 31-33 in Toussoun T A，Bega R V，Nelson P E（Eds.）Root Diseases and Soil-Borne Pathogens. University of California Press，Berkeley，USA.

Puhalla J E，Hummel M. 1983. Vegetative compatibility groups within Verticillium dohliae. Phytopathology 73：1 305 –1 308.

Rouse D I. 1985. Some approaches to prediction of potato early dying disease severity. American Potato Journal 62：187 –193.

Schnathorst W C. 1981. Life cycle and epidemiology of Verticillium. Pages

81-111 in Mace M E.
Bell A A, Beckman C H (Eds.) Fungal Wilt Diseases of Plants. Academic Press, New York, USA.

Scholte K. 1989. Effects of crop rotation and granular nematicides on the incidence of Vertlcillium dohliae Kleb. and Colletotrichum coccodes (Wallr.) Hughes, in potato. Potato Research 32: 377 -385.

Scholte K, s'Jacob J J. 1989. Synergistic interactions between Rhizoctonia solani Kuhn, Verticillium dohliae Kleb., Meloidogyne spp. and Pratylenchus neglectus (Rensch) Chitwood & Oteifa, in potato. Potato Research 32: 387 -395.

Scholte K, Veenbaas-Rijks J W, Labruyere R E. 1985. Potato growing in short rotations and the effect of Streptomyces spp., Colletotrichum coccodes, Fusarium tabacinum and Verticilium dahlia on plant growth and tuber yield. Potato Research 28: 331 -348.

Schreiber L R, Green R J. 1963. Effect of root exudates on germination of conidia and microsclerotia of Verticillium albo-atrum inhibited by soil fungistatic principle. Phytopathology 53: 260 -264.

Slattery R J. 1981. Inoculum potential of Verticillium - infested potato cultivars. American Potato Journal 58: 135 -142.

Street P F S, Cooper R M. 1984. Quantitative measurement of vascular flow in petioles of healthy and Verticil/ium-infected tomato. Plant Pathology 33: 483 -492.

Tjamos E C. 1981. Virulence of Verticillium dahliae and Verticillium alboatrum isolates in tomato seedlings in relation to their host origin and the applied cropping system. Phytopathology 71: 98 -100.

Tolmsoff J. 1973. Life cycles of Verticillium species. Pages 20-38 in Verticillium Wilt of Cotton. Proceedings of a Work Conference, National Cotton Pathology Research Laboratory College Station USDA Agricultural Research Service, Texas, USA.

TolmsoffW J, Young R A. 1959. The influence of crop residues and fertilizer on the development and severity ofVerticillium wilt of potatoes. Phytopathology 49: 114.

Van der Molen G E, Beckman C H, Rodehorst E. 1977. Vascular gelation: a general response phenomenon following infection. Physiological

Plant Pathology 11：95 –100.

Vigoroux A. 1971. Hypothesis to explain the anomalous pathological behaviour of some Verticillium isolates. Page 31 in Proceedings of the International Verticillium Symposium. Wye College，London，UK.

Wheeler T A，Madden LV，Rowe R C，Riedel R M. 1992. Modelling of yield loss in potato early dying caused by Pratylenchus penetrans and Verticilliurn dahliae. Journal of Nematology 24：99 –102.

Wilhelm S. 1955. Longevity of the Verticillium wilt fungus in the laboratory and field. Phytopathology 45：180 –181.

Woolliams G E. 1966. Host range and symptomatology of Verticillium dahliae in economic，weed，and native plants in interior British Columbia. Canadian Journal of Plant Science 46：661 –669.

Zilberstein Y，Chet I，Henis Y. 1983. Influence of microscl erotia source of Verticilliurn dahliae on inoculum quality. Transactions of the British Mycological Society 81：613 –617.

第八章　综合作物毛长模型与流行病模型预测种薯单产和病毒感染

植物病毒传播媒介的病理系统包括元病毒、传病媒介与植物，受人类及自然环境的影响。大部分病毒流行病模型强调传病媒介与病毒的重要性，但却忽略了植物的作用。

EPOVIR 模型是首个与作物生长子模型共同作用的病毒流行病模型，后者常与土壤水平衡子模型组合应用。这两种子模型经常用在计算薯块产量及大小、薯叶的生理龄期三方面，而干旱胁迫压力经常应用在估量病毒易感度（抗老化性）方面，最终，需计算出被冠层覆盖的土壤部分在种薯产地所占的传病媒介比。由于病毒传播的速率是植物生理学与物候学模拟的作物子模型所呈现出的函数，流行病就应当在人或环境引起的植物生长变动中出现相应的反应。正是依据这一实例，作者将展示种植密度如何对病毒感染造成影响。

EPOVIR 模型已经被纳入到“TuberPro”决策支持系统中，这一系统旨在分析按大小预测块茎产量以及块茎对马铃薯 Y 病毒与卷叶病的感染情况。该模型支持种薯生产过程中实现碎茎日期的最优化。而这一系统会估算出预期的种产量以及病毒感染处于种子鉴定允许界限内的可能性。两种要素的组合会为产区提供种子平均产量的预期认证，实现后续优化目标。

第一节　简介

植物病毒传播媒介的病理系统包括元病毒、传病媒介与植物，且受人类及自然环境的影响（Robinson，1976）。大部分预测系统与病毒流行病模型主要关注的焦点是传病媒介、病毒以及两者之间的关系（如 Marcus 和 Raccah，1986；Kisimoto 和 Yamada，1986；Miyai 等，1986；Madden 等，1990；Kendall 等，1992），但却忽略了植物在病毒感染、染色体易位及其对传病媒介群的影响与作用，最多仅用部分常用参数和函数进行描述（如 Sigvald，1986；Ruesink 和 Irwin，1986；Van der Werf 等，1989）。然而，准确地描述植物生理学与物候学在流行病进程中发挥的作用以及外部环境对这一系列关系的影响，作物生长子模型的应用就必不可少。作物模型已经同真菌流行病模型一起应用于植物病理学中（Rouse，1988），但就作者所知，植物模型从未与病毒流行病模型共同应用过。

马铃薯作物子模型与病毒流行病模型结合，从而预测种薯的产量及病毒感染情况。作物子模型本应与土壤水平衡子模型组合，说明次优情况下的供水状况。这两个子模型虽然影响流行病模型，但是仍为独立存在的模型。作者并未研究因病毒造成的减产原因，因为作者主要感兴趣的是种薯生产，因病毒感染造成的产量减少因素所占比例较少。但是，经济因素对病毒感染造成的影响甚深，因为一旦种批被驳回，将会切断农民所有收入。作者已经研发出种薯决策支持系统，可以预测出块茎的病毒感染情况，特别需要指出的是，这一系统还可以预测出病毒感染超出一定质量等级允许界限的风险，作者将这一系统称之为“TuberPro”系统（Solanum tuberosum prognosis，即马铃薯栽培预断），帮助实现种薯生产过程中碎茎日期的最优化（Nemecek 等，1994）。

本章的目标是展现将作物子模型应用于病毒感染模型以及“Tuber-Pro”系统的原因及方法。

第二节　马铃薯流行病毒模型“EPOVIR”概述

“EPOVIR”模型（epidemiology of potato viruses 即马铃薯病毒流行病学）是“TuberPro”系统的核心，这一模型旨在从一个产区，模拟作物从出苗到碎茎过程中作物的生长情况以及病毒感染状况。第一版详细描述出自 Nemecek（1993），主要由 4 种子模型组成（图 8－1）。

（1）接种子模型（继 1986 年 Irwin 和 Ruesink 的研究后），主要根据产区的传病媒介丰度计算传病媒介强度、传病媒介倾向以及传病媒介行为；吸虫塔的捕获数可以估计传病媒介丰度。无翅蚜虫传病媒介只传播卷叶病毒。这一子模型表示传病媒介的作用以及病毒和作物之间的关系。

（2）感染子模型决定了马铃薯 Y 病毒与卷叶病毒分别对植物与块茎的感染状况。此外，还可以通过病毒潜伏期以及植物对病毒感染的抗老化性估计传染源所占的比例。这一子模型主要表现病毒的角色及其与作物的

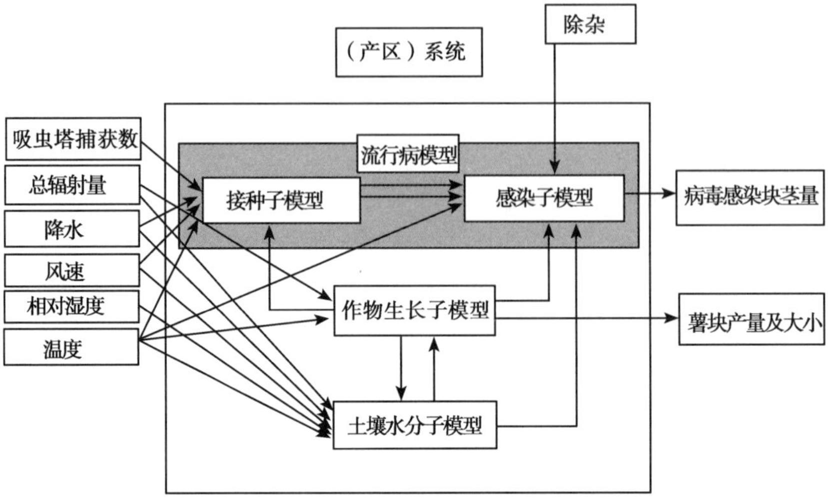

图 8－1 EPOVIR 模型结构（马铃薯病毒流行病学）

关系。

（3）作物生长子模型用于计算马铃薯的叶、茎、根和块茎的干质量，保持与冠层的生理状态相一致，并实现块茎产量的分级。

（4）土壤水分平衡子模型用于通过实际与潜在蒸发率计算出土壤含水量以及马铃薯植株的水分胁迫。

Johnson（1986；1987）等人继承并发展了作物模型。曾用于测量土壤湿度的原始模型在作者的试验中并未使用，取而代之的是土壤水分平衡子模型，这一模型是基于 Van Keulen 和 Wolf（1986）的研究成果。马铃薯作物的生长与土壤水分平衡子模型结合应用以及验证结果的完整描述出自 Roth（1995）等人之手。作物子模型适用于在瑞士栽培的各个品种，并通过块茎大小子模型实现深化，由 Nemecek 和 Derron（1994）给出了相应的验证结果。这一作物模型能够极具可靠性地重述干湿条件下收集到的实验数据。

通过使用 Model Works（Fischlin 等，1994）模拟环境，“EPOVIR”模型可以应用在 Apple 和 Macintosh™等机型的 Modula-2 编程语言中（Wirth，1985）。

第三节　马铃薯作物生长模型的应用

一、马铃薯病毒感染对生长状况的影响

除了作者对植物-传病媒介-病毒的病理系统所下的定义外，作物生长以及土壤水分平衡模型在“EPOVIR”模型中的应用主要体现在 3 个方面：计算产量、描述植物生理学对病毒增殖与易位的影响以及描述植物气候学对传病媒介行为的影响。

预测块茎产量及块茎大小当然是决策支持系统中的重要因素。作物子模型可计算出块茎总产量和任意块茎大小极值间的比值（详见图 8 –2）。

马铃薯处在生长期时，块茎不易被病毒感染（Beemster，1987）。这种现象被称之为“抗老化性”或“植株抗性”。其成因尚未查明，但可能与叶片的生理活性有关（Venekamp 等，1980）。作物子模型一直跟踪记录每日新生的叶片质量及其生理龄期（Johnson 等，1987；Roth 等，1995）。作者用“*ly*”代表低于生理龄期阈值的幼叶叶片质量（$0 \leqslant ly \leqslant 1$），从而估算出植物对病毒的易感性与龄期的相关性方程式 AS，干旱也影响这一易感性。

Wislocka（1982）对此作出了证明：干旱（次最优情况下的供水）显然会破坏部分植物的抗老化性，增加块茎的感染。作者用“*wStress*”代表水分胁迫（$0 \leqslant wStress \leqslant 1$，1 = 最理想的生长条件，0 = 不生长，详见 Roth 等，1995）从而证明水分胁迫的影响在 *AS* 方程式中也可以体现出来。

$$AS = \begin{cases} 1-(1-l_y)[1-0.5(1-wStress)] & \text{如遇干旱} \\ l_y & \text{除干旱外} \end{cases}$$

常数 0.5 是权重因数，用于表示干旱因素（土壤含水量低于临界土壤湿度，Roth 等，1995）对易感性的影响。图 8 –2 表示 *AS* 方程式特有的演化。在植物幼期，*AS* 数值为 1，不考虑干旱原因。随着植物不断成长，*ly* 逐渐减小至 0，从而导致 *AS* 值也随之递减。也就是说，与供水条件理想化的情况下相比，干旱会增加植物在这一阶段的病毒易感性。倘若没有干旱，*AS* 的值应当与 *ly* 相等。

二、马铃薯载体相互作用机制

植物不仅影响病毒感染、增殖及易位，还会改变蚜虫群落的发展及其

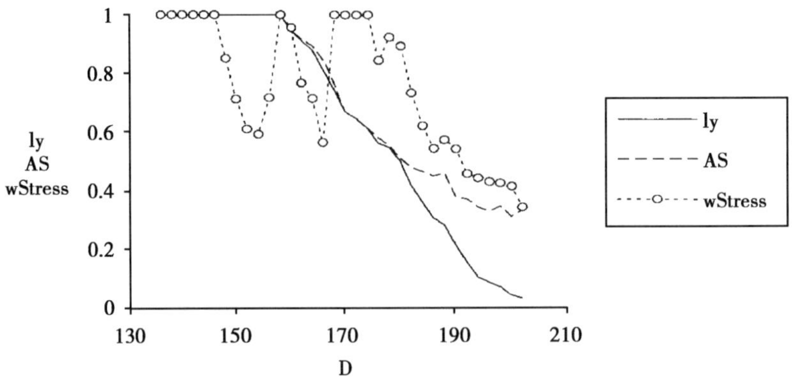

图 8－2　植物对病毒的易感性与龄期的相关性方程式

注：*AS* 是一个关于低于生理龄期阈值的幼叶叶片质量 *ly* 与水分胁迫 *wStress* 的函数，*D* 代表一年中的天数

行为，因此，间接地影响着病毒传播（Irwin 和 Kampmeier，1989；Nemecek，1993）。宿主的适合性强烈地影响着传病媒介行为与病毒传播。Nemecek（1993）已经证明载体种类并未拓殖至马铃薯植株上，而是通过行为序列增加了传病媒介的效率。仿真研究揭示出，与拓殖到植株上的传病媒介种类相比，这一特性将会导致非持续的病毒传播效率增加 2 倍。

生物气候学影响传病媒介群的行为。作者用 *R* 表示栖息率（同样时间间隔下，吸虫塔捕捉到的每平方米栖息的蚜虫数量）（Taylor 和 Palmer，1972）。随着冠层覆盖量的增加，栖息率 *R* 会逐渐减小（图 8－3）。蚜虫栖息率是根据渔网格覆盖（Derron 和 Goy，1993）的捕获量及其与在马铃薯植株上栖息之间数量的关系间接计算得来（Derron 等，1989）。冠层覆盖由作物子模型进行模拟。*R* 的平均值与冠层覆盖之间的回归分析适用于第 3、第 4、第 5、第 10 组，所有的数据点大致相近。表示最佳线性关系的回归分析中的最高值 r_2 也应用在这一模型中（图 8－3）。由于在模型计算时，冠层覆盖量是一个变量，所以这一关系很容易囊括在其中。随着冠层的变化发展，栖息率及其作用下的病毒传播也在不断变化。随后作者会对两组变量种类进行更深一步的辨别：第一组中有一个相对栖息率，平均高出第二组 8.6 倍。

作者用两个例子说明模型应用（更多实例详见 Nemecek 等，1994）。“TuberPro”系统的主要目标就是通过预测下列变量，实现判定决策：①病毒病与卷叶病导致的预期块茎感染及其变量；②一定规模层级的预期块茎产量及其变量。

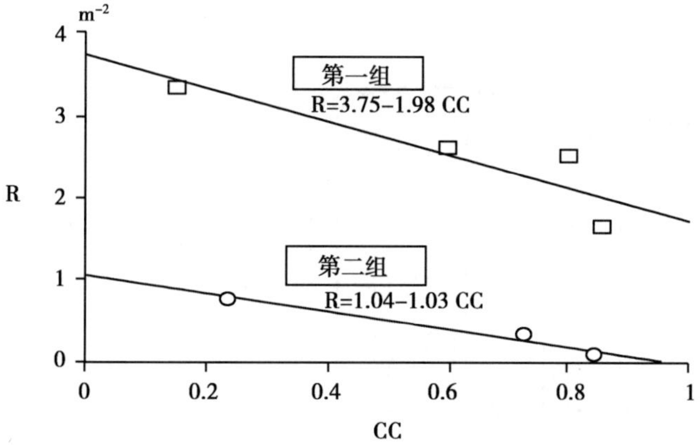

图 8－3　蚜虫相对栖息率

注：R＝马铃薯产区的蚜虫栖息率［# m^{-2} d^{-1}］与吸虫塔捕获量［#/d］作为冠层覆盖量 CC（冠层覆盖的土壤比例）函数得出的比率。包含在第一组中的传病媒介种类：蚜虫、桃蚜（Sulz）、长管蚜（Thomas）；第二组：豌豆蚜（Harris）、短尾蚜（Kalt）、指头蚜（Schrk）、禾谷缢蚜

针对病毒感染的逻辑回归转换预测通常呈正态分布。平均数和方差用来描述预测的概率分布，主要包括投入时间序列的未知发展前景所导致的不确定性（气候变量以及蚜虫的存在比）以及对某些特定参数的估值存在误差。对未知未来的投入值已经被近年来收集的数据所替代，而模拟的范围仍局限于单一领域。但是，对代表区域或全国生产的平均情况进行定义仍具可行性。类似的实例详见图 8－4。在瑞士，被马铃薯 *Y* 病毒与卷叶病感染达到 10% 的种批被认证为头等种子。因此，种批获得认证的概率 *P* 即为块茎被病毒感染≤10% 的概率，这也正是预期种批被认证的比例。种子级别的预期最高产量 *S* 于 7 月 4 日实现；*CS* 表示两个变量的乘积（$CS = S \times P$）。*CS* 的最大化早在 6 月 24 日就已达到。因此，6 月 24 日是已认证种子产量最大化的时间，也正是最佳的碎茎日期。

另一大战略旨在实现农民收入的最大化。因为块茎可以用于人类消费，所以农民也对获取更大的块茎兴趣浓厚。通常情况下，被认证的种子实现预期最大化后，预期收入的最大化也就随之实现。使用“TuberPro”系统的用户一定要自定义这一战略，从而实现自己的目标。

这一模型也可用于评估栽培技术。模拟分析研究表明，通过增加种植密度可以大幅减弱病毒感染（图 8－5）。主要原因还是冠层覆盖量的增加（图 8－3）以及土壤的稀释作用（Power，1990；1992）。这两大因素导致

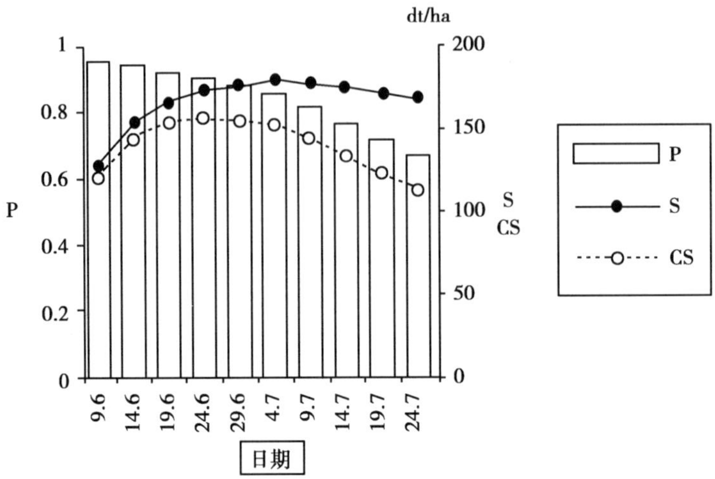

图 8－4　概率与生长日期的关系

注：被认证的种批概率为 *P*，种子级别的预期最高产量为 *S* [dt/ha]，*CS* [dt/ha] 表示种子级别预期认证的平均产量。结果来自仿真模拟运行 30 次后的平均值：三种条件（有利、正常与不利）结合过去 10 年间 6 月 14 日以后的输入数据（到目前为止，已经使用本年度数据）

单株植物的传病媒介数较低，从而会降低被感染植株比例。与拥有较低种植密度的产区相比，种植密度高可以减少单株传病媒介平均值的 58%，减少 62% 病毒感染可能性，也就是说，病毒感染减少的量甚至多于传病媒介密度。这种差异主要是由植物生长初期传病媒介丰度降低引起的，这一时期病毒传播是危害最大的（Nemecek，1993）。

第四节　有关问题的总结与讨论

作物会对自然环境以及人类带来的影响做出反应。由于马铃薯病毒的增殖完全取决于寄生植物的新陈代谢，所以任何其内在的改变都会影响病毒流行进程。例如，如果叶片生长持续的时间更长，就会延迟植物的抗老化性，从而更利于病毒传播。传病媒介与作物的关系也十分密切。像这种为病毒流行进程带来的间接影响，最好通过病毒感染模型与作物生长模型的结合共同描述。

由于经过有效验证的模型目前已可应用于多种作物，作者认为作物模型应当在模拟病毒感染方面更加广泛地使用。作物子模型的应用使整个模型更加灵活，同样也适用于其他情况，如应用于其他国家不同的生产方案

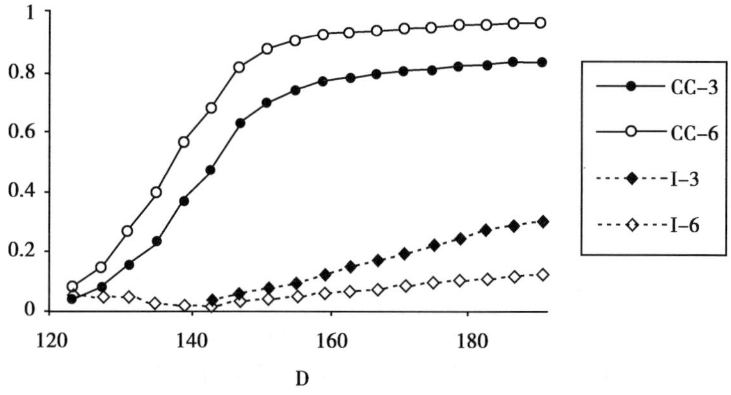

图 8 -5　种植密度对病毒传播的影响

注：*I* 表示受马铃薯 *Y* 病毒感染的块茎比例，*CC* 表示冠层覆盖的土壤比例，*D* 代表天数，数字 3 和 6 分别表示种植密度为每平方米 3 个或 6 个种子

中。将植物-病毒与植物-传病媒介的关系囊括在内，模型也变得更加翔实清楚，为验证假设、评估栽培措施提供更好的可能（Nemecek，1993）。Ruesink 和 Irwin（1986）对这一模型研究的尚未出版的版本也考虑到了冠层覆盖对蚜虫栖息率的影响。然而，笔者使用测量的时间序列作为输入值，因此，每当生长参数发生改变时，就需要进行新的实验，这在作者的研究方法中是没必要的。作者支持使用作物子模型的最后一个理由，也是最显而易见的理由是：作物子模型可以计算出块茎产量，因而可以根据不同标准优化种薯生产（图 8 -4）。同时，这一模型还可以对病毒控制措施对产量产生的影响进行评估。

将作物子模型与病毒流行模型结合也会产生一些弊端。模型整体的复杂度会大大提高，所需的各项参数不断增加，而这些参数并不适用于所有情况，这时，就需要使用系统默认值或平均值。如此一来，作物子模型就丢失了一些应有的优势。最终，模型的发展及其之间的结合就会耗时费力，尤其是还未出现可以直接使用的合适作物模型。所以，是否使用作物子模型将取决于研究目标、适合模型的可用性以及可利用的资源。

“TuberPro”系统是种薯生产强大的系统管理工具。这一系统保证了生产根据决策者选择的不同策略实现优化。农民会尽可能地保证自己一定的收入；一个种子种植机构会努力增加自身的利润；而国家主要专注于保持作物的高质量。“TuberPro”系统可以计算出不同水平最优解所必需的变量，尤其是根据特定目标确定的碎茎日期。此外，这一系统已经被证明有助于为专家、农民提供指导；有助于评估栽培措施以及病毒控制措施

（Nemecek 等，1994）。目前在瑞士，“TuberPro”系统正在经受实践的检验。由于该系统的灵活性，在合理的情况下，经过人们的努力，“TuberPro”系统应当可以应用于其他条件下。

参考文献

Beemster A B R. 1987. Virus translocation and mature-plant resistance in potato plants. Pages 116 – 125 in De Bokx J A, Van der Want J P H (Eds.) Viruses of Potatoes and Seed-Potato Production, 2nd Ed. Pudoc, Wageningen, The Netherlands.

Derron J O, Goy G. 1993. Description et mode d'emploi d'un piegepour l'etude du vol des puceronsvecteurs de virus. Revue Suisse d'Agriculture 25: 135 – 137.

Derron J O, Goy G, Genthon M. 1989. Le piegeage des pucerons ailes: potentialites et limites de differents types de pieges. Pages 71-82 in Cavalloro R (Ed.) Euraphid Network: Trapping and Aphid Prognosis. Proceedings of a Meeting of the EC-Experts' Group, Catania, Italy, 7 – 9.
November 1988. ECSC-EEC-EAEC, Brussels, Belgium.

Fischlin A, Roth O, Gyalistras D, Ulrich M, Thony J, Nemecek T, Bugmann H, Thommen F. 1994. ModelWorks 2.2: An Interactive Simulation Environment for Personal Computers and Workstations. Internal Report No. 14. Systems Ecology. ETH, Zurich, Switzerland.

Irwin M E, Kampmeier G E. 1989. Vector behaviour, environmental stimuli and the dynamics of plant virus epidemics. Pages 14 – 39 in Jeger M J (Eds.) Spatial Components of Plant Disease Epidemics. Prentice Hall, Englewood Cliffs, NJ, USA.

Irwin M E, Ruesink W G. 1986. Vector intensity: a product of propensity and activity. Pages 13 – 33 in McLean G D, Garett R G, Ruesink W G (Eds.) Plant Virus Epidemics (Monitoring, Modelling, and Predicting Outbreaks). Academic Press, Sydney, Australia.

Johnson K B, Johnson S B, Teng P S. 1986. Development ofa simple potato growth model for use in crop-pest management. Agricultural Systems 19: 189 – 209.

Johnson K B, Teng P S, Radcliffe E B. 1987. Coupling feeding effects of potato leafhopper, Empoasca fabae (Homoptera: Cicadellidae), nymphs to a model of potato growth. Environmental Entomology 16: 250 - 258.

Kendall D A, Brain P, Chinn N E. 1992. A simulation model of the epidemiology of barley yellowdwarf virus in winter sown cereals and its application to forecasting. Journal of Applied Ecology 29: 414 - 426.

Kisimoto R, Yamada Y. 1986. A planthopper-rice virus epidemiology model: rice stripe and small brown planthopper Laodelphax striatellus Fallen. Pages 327 - 344 in McLean G D, Garett R G, Ruesink W G (Eds.) Plant Virus Epidemics (Monitoring, Modelling, and Predicting Outbreaks). Academic Press, Sydney, Australia.

Madden L V, Raccah B, Pirone T P. 1990. Modelling plant disease increase as a function of vector numbers: nonpersistent viruses. Research in Population Ecology 32: 47 - 65.

Marcus R, Raccah B. 1986. Model for the spread of non-persistent virus diseases. Journal of Applied Statistics 13: 167 - 175.

Miyai S, Kiritani K, Nakasuji F. 1986. Models of epidemics of rice dwarf. Pages 459 - 480 in McLean G D, Garett R G, Ruesink W G (Eds.) Plant Virus Epidemics (Monitoring, Modelling, and Predicting Outbreaks). Academic Press, Sydney, Australia.

Nemecek T. 1993. The Role of Aphid Behaviour in the Epidemiology of Potato Virus Y: A Simulation Study. Ph. D. Thesis. No. 10086. ETH, Zurich, Switzerland.

Nemecek T, Derron Jo. 1994. Validation et application d'un modele de croissance de la pomme de terre. Revue Suisse d'Agriculture 26: 311 - 315.

Nemecek T, Derron J O, Schwarzel R, Fischlin A, Roth O. 1994. Un modele de simulation au service des producteurs de plants de pommes de terre. Revue Suisse d'Agriculture 26: 17 - 20.

Power A G. 1990. Cropping systems, insect movement and the spread of insect transmitted diseases in crops. Pages 47 - 69 in Gliessman S R (Ed.) Agroecology (Ecological Studies 78). Springer, New York, USA.

Power A G. 1992. Host plant dispersion, leafhopper movement and disease transmission. Ecological Entomology 17: 63 – 68.

Robinson R A. 1976. Plant Pathosystems. Springer, Berlin, Germany.

Roth O, Derron J, Fischlin A, Nemecek T, Ulrich M. 1995. Implementation and parameter adaptation of a potato crop simulation model combined with a soil water subsystem. In: Kabat.

P, Van den Broek B J, Marshall B, Vos J, Van Keulen H (Eds.) Modelling and Parametrization of the Soil-Plant-Atmosphere System - A Comparison of Potato Growth Models. Wageningen Pers, Wageningen, The Netherlands (in press).

Rouse D I. 1988. Use of crop growth-models to predict the effects of disease. Annual Review of Phytopathology 26: 183 – 201.

Ruesink W G, Irwin M E. 1986. Soybean mosaic virus epidemiology: a model and some implications. Pages 295 – 313 in McLean G D, Garett R G, Ruesink W G (Eds.) Plant Virus Epidemics (Monitoring, Modelling, and Predicting Outbreaks). Academic Press, Sydney, Australia.

Sigvald R. 1986. Forecasting the incidence of potato virus Y^o. Pages 419 – 441 in McLean G D, Garett R G, Ruesink W G (Eds.) Plant Virus Epidemics (Monitoring, Modelling, and Predicting Outbreaks). Academic Press, Sydney, Australia.

Taylor L R, Palmer J M P. 1972. Aerial sampling. Pages 189 – 234 in Van Emden H F (Eds.) Aphid Technology. Academic Press, London, UK.

Van der Werf W, Rossing WAH, Rabbinge R, De Jong M D, Mols P J M. 1989. Approaches to modelling the spatial dynamics of pests and diseases. Pages 89-119 in Cavalloro R, Delucchi V (Eds.) Parasitis 88. Proceedings of a Scientific Congress, 25-28 October 1988, Barcelona, Spain.

Van Keulen H, Wolf J (Eds.). 1986. Modelling of Agricultural Production: Weather, Soils and Crops. Pudoc, Wageningen, The Netherlands.

Venekamp J H, Schepers A, Bus C B. 1980. Mature plant resistance of potato against virus diseases. III: Mature plant resistance against potato virus Y^N, indicated by decrease in ribosome-content in ageing potato plants under field conditions. Netherlands Journal of Plant Pathology 86: 301 – 309.

Wirth N. 1985. Programming in Modula-2, 3rd Ed. Springer-Verlag, Berlin a. o. , Germany.

Wislocka M. 1982. Einfluss der Trockenheit vor und zu verschiedenen Zeitpunkten nach Inokulation auf den Knollenbefall der Kartoffelsorte 'Uran' mit Kartoffelvirus y. Potato Research 25: 293 – 298.

第九章 预定模型和虫害管理软件在马铃薯种植中的应用

与马铃薯农作体系集约管理所关注的内容相同，社会上对使用杀虫剂和化肥的担忧大多和农业投入品的选择、使用时间和用量有关。威斯康星州的马铃薯产业过去曾实施过一项“病虫害综合防治”计划，因为①当地种植户向来思想活跃，对开展病虫害综合防治兴趣浓厚；②作物种植所需的杀虫剂、化肥和灌溉水等多种投入品采用集约化管理；③马铃薯是一种高价值、高风险的作物；④作物种植区域通常对环境较为敏感；⑤科研基础扎实，可支持病虫害综合防治计划的实施。威斯康星州“病虫害综合防治”计划的一个重要内容是开发专用的计算机软件用以协助种植户做管理决策。计算机软件是一种有效的田间工具，可用于分析复杂的环境和作物信息，并提出具体的管理建议。威斯康星大学集结多个学科的力量，共同开发了“马铃薯作物管理”软件。软件包含多个模块，可预测作物出芽、调度灌溉、防治病虫害（病害、虫害和杂草）以及评估仓库通风需求等。自从 1989 年向种植户和病虫害综合防治顾问发布该软件以来，软件使用量就在不断增长。目前，大批种植户和顾问都在使用这一软件，影响了大约 28 300hm^2 的马铃薯种植土地。使用此软件可减少杀虫剂和灌溉水的用量，比使用软件前年均节省至少 589 万美元。然而，软件目前还只能解决种植户所面临的几个问题。研发人员正在加强软件设计，关注马铃薯林冠发育、作物营养、种块腐烂和农场信息记录等方面的问题。“马铃薯作物管理”软件还从过去的微软磁盘操作系统（MS-DOS®）转变为微软 Windows 操作系统（Microsoft Windows™）。新版软件将被命名为“WISDOM”，导入和交换数据将更为便捷，新添加数据的地理表征功能，还能针对不同地块和年份进行环境、作物或虫害数据的对比。“WISDOM”将为农业种植管理带来更多可能性，包括用于管理通常与马铃薯进行轮作的作物。目前正在开发新的模块，用于管理四季豆（灌溉调度、虫害防治、白霉病风险评估）和甜玉米（灌溉调度和虫害防治）。随着田间研究和实验室研究不断推陈出新，程序的模块化可促进未来纳入其他马铃薯和轮作作物的模块。

第一节　简介

用于加工、生鲜市场销售和种薯生产的马铃薯种植业是威斯康星经济的一项重要资产，其种植面积约 27 500 hm^2，总价值 1.2 亿美元，总生产量全美排名第四（Pratt，1993），每年为威斯康星的就业和加工业等相关农业产业所贡献的经济价值接近 3.5 亿美元。采用灌溉的马铃薯生产成本与全美平均水平相当，目前晚熟 russet 马铃薯的平均成本为 4 199 美元/hm^2，总回报为 4 900 美元/hm^2 以上。

美国中西部地区的气候条件有利于马铃薯等耐寒作物的生产，但同时也带来各式各样严重的虫害问题。在威斯康星州和美国中北部大部分地区，种植马铃薯一般需要大量使用杀虫剂来管理病虫害和除杂草；化肥用量也相对较高，特别是需要进行灌溉的沙质土壤；而且需要进行大量灌溉以获得最佳产量。15 年前，种植户要在种植于壤砂土地上的 Russet Burbank 这一当地主要作物的种块上施用杀真菌剂，叶片喷施杀真菌剂多达 12 次以防治早疫病和晚疫病，播种阶段系统性施用杀虫剂，叶面喷施杀虫剂多达 4 次，喷施除草剂 1～3 次以除草和阔叶杂草，施用茎叶干燥剂 1～2 次，每周灌溉 3 次，周总灌溉量 5cm，所含化肥用量，氮：280～392kg/hm^2，P_2O_5：134～179kg/hm^2，K_2O：336～448kg/hm^2。

20 世纪 80 年代初期，投入品的大量使用杀虫剂开始引发市民和种植户的担忧，特别是担心环境风险、种植户和消费者安全以及常用杀虫剂出现抗药性的可能性。对于大量使用杀虫剂等投入品的依赖性和与人们对马铃薯种植有关的环境关联推动了威斯康星作物管理和病虫害综合防治计划的出台。种植户、加工业等伙伴产业也加入了计划的设计，通过优化马铃薯种植过程中投入品的使用、减少或杜绝使用会给环境带来负面影响的投入品，来保持马铃薯产业的竞争力。

作物管理和病虫害综合防治计划的建立过程代表了多个领域的科研历史包括病害防治（正确地使用杀真菌剂的同时进行病害预测预报）、虫害防治（针对虫害的脆弱阶段并关注经济阈值）、杂草治理（预测作物出芽期、植株树冠发育和除草剂施用时机）、培肥措施（所需养分的用量、位置、来源和时机）和灌溉管理（将灌溉与作物需求对接）。在上述及其他关键领域的科研成果将产生庞大的数据，为解决具体生产问题提供有效建议。20世纪80年代中期，马铃薯产业联盟和威斯康星大学研究人员发现，若使研究成果发挥最大效用，必须将所得信息以更加全面和简明的形式呈现出来；当时，将计算机应用于田间试验仍处于早期探索阶段，但它是整合信息和改善信息获取性的便捷载体。

第二节　计算机应用的初步开发

早期开发的计算机软件是一个由“马铃薯病害防治”和“威斯康星灌溉调度”两个程序组成的独立软件。“马铃薯病害防治”程序利用改良后的BLITECAST方法（Krause等，1975；Stevenson，1983）解决了早疫病的预测和防治（在威斯康星每年都会出现）（Pscheidt和Stevenson，1986）和晚疫病的预测和防治（受接种体和天气影响偶有发生）。20世纪80年代初期，因早疫病和晚疫病而蒙受损失的种植户们欣然接受了这种使用计算机处理信息的模式，由此他们可以预测病害的发生，从而通过计算机优化的杀真菌剂施用调度和施用量选择来改善病害防治。“马铃薯病害防治”是进行病虫害防治的指定软件，种植户可通过它使用在自家地块收集的天气和作物信息。“马铃薯病害防治”软件十分成功，它也因此在之后开发综合性更强的计算机软件的过程中发挥了重要的基础作用。“马铃薯病害防治”程序得到应用之后，随着计算机技术不断进步，“病虫害综合防治”计划的计算机程序员能够将“马铃薯病害防治”和“威斯康星灌溉调度”两个程序合二为一，由此开发出一个综合性更强的软件——“马铃薯作物管理”。这款软件融入了多个新模块，可用于病害预测和防治、灌溉调度（Curwen和Massie，1984）、作物出芽预测、通过杀虫剂的靶向用药来进行虫害发展预测和防治以及使用经济阈值（Walgenbach和Wyman，1984a，b）、计算马铃薯贮存的通风需求量等。上述模块反映了种植户和相应领域研究人员的要求。“马铃薯作物管理”软件实现了不同模块的整合，种植户可使用自家地块中收集到的信息并用于辅助管理农田的各项决策（图9－1）。

第三节　软件使用带来的价值

根据种植户和“病虫害综合防治”项目顾问的报告，使用了“马铃

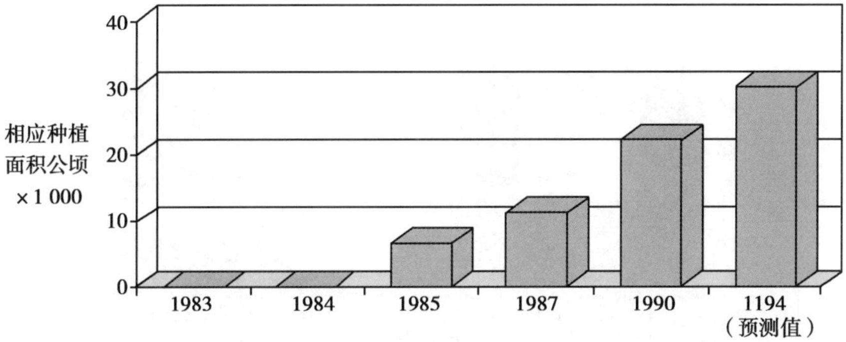

图9-1　1983年以来马铃薯产业中“马铃薯病害防治/马铃薯作物管理”软件使用情况

薯作物管理”软件的马铃薯用地面积约为16 200hm²，而威斯康星州马铃薯种植总共占地27 500hm²。邻近的几个州也在使用这个软件，覆盖面积达到12 100hm²（图9-1）。通过将种植初期阶段（发病度值累积到18之前（Krause等，1975）以及达到300个生理天数之前（Sands等，1979；Pscheidt和Stevenson，1986）的杀真菌剂施喷次数减少2~3次，该软件预计为种植户节省49美元/hm²（Connell等，1991）。使用软件后，早疫病和晚疫病防控措施的采用时间得以后延，预计为中西部地区的马铃薯种植户节省超过140万美元/年（图9-3）。如果种植户根据环境条件和累计的生理天数调整杀真菌剂用量，还有可能在种植季将用量减少15%~20%。使用“马铃薯作物管理”软件和相关技术还有其他好处：①可减少高达50%的除草剂，延迟施用时间；②通过使用经济阈值和关注最容易受到虫害而造成经济损失的生长阶段，可减少40%~50%的杀虫剂用量；③根据作物需求实施灌溉，可减少10%~15%的灌溉投入品用量；④减少20%的氮元素投入品用量（表9-1和图9-4）。使用“马铃薯作物管理”软件和相关技术可节省约208美元/hm²的费用。长期使用则效益更高，但需注意的是，由于使用软件的同时需要对土壤和植株进行组织分析，雇佣观测服务和对环境进行监测，会带来额外的成本而抵消掉部分节省的费用。在威斯康星，这笔额外的成本费用为37~49美元/hm²，在美国其他地区可能还会更高。威斯康星目前正在对种植户和使用软件的行业进行一项调查，探究全面使用病虫害和作物管理技术将产生的成本和创造的价值。调研结果还将帮助找出在采用其他管理技术的过程中可能遇到的障碍。

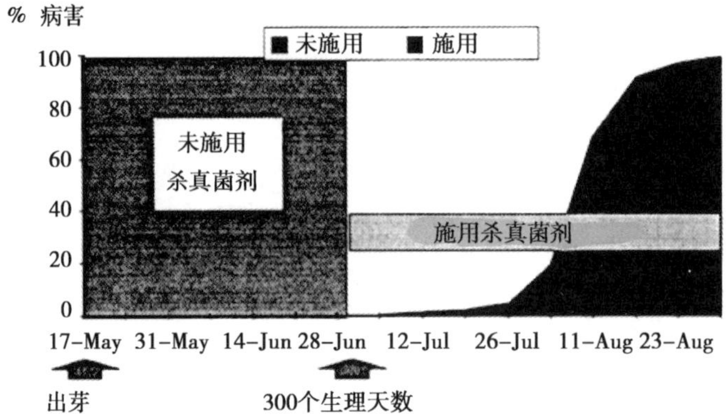

图 9-2 早疫病防控采用 300 个生理天数的施用阈值可延迟杀真菌剂的施用时间

注：300 个生理天数之后，在整个种植阶段根据环境条件每隔 5 ~ 10d 施用一次杀真菌剂。施用杀真菌剂能减缓病害发展速度

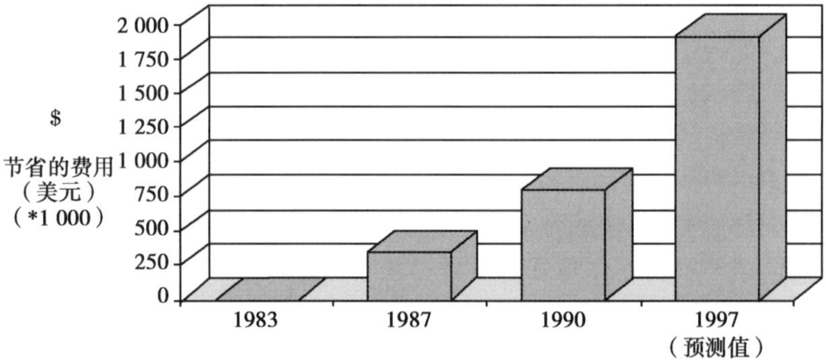

图 9-3 因应用"马铃薯病害防治/马铃薯作物管理"软件减少杀真菌剂用量而减少的费用

表 9-1 使用"马铃薯作物管理"软件和相关技术生产晚熟 Russet Burbank 马铃薯可减少的投入品分类

减少的投入品	投入品减少量/hm^2	投入品价值/hm^2
减少两次含 EBDC 的施喷	3.4kg	49.40 美元
每次含 1.7kg/hm^2 杀真菌剂，加上施喷成本（12.35 美元/hm^2）		
在保留的施喷次数中减少杀真菌剂用量		
例，EBDC 传统用量为 13.4kg/hm^2，可减少 18%	2.52kg	18.53 美元

（续表）

减少的投入品	投入品减少量/hm²	投入品价值/hm²
减少50%的除草剂	异丙甲草胺：1.7kg	56.81美元
减少一次芽后除草剂施喷		
减少芽前除草剂用量	嗪草酮：0.56kg	
减少40%的杀虫剂	2.8kg	49.40美元
取消种植阶段的系统施喷，减少1~2次叶部施喷		
减少灌溉和能源成本		
灌溉水的使用配合作物需求	13.2cm	11.12美元
减少使用20%的氮肥	56kg	22.72美元
投入品总减少量	10.9kg杀虫剂 56kg氮元素 13.2cm灌溉水	207.97美元

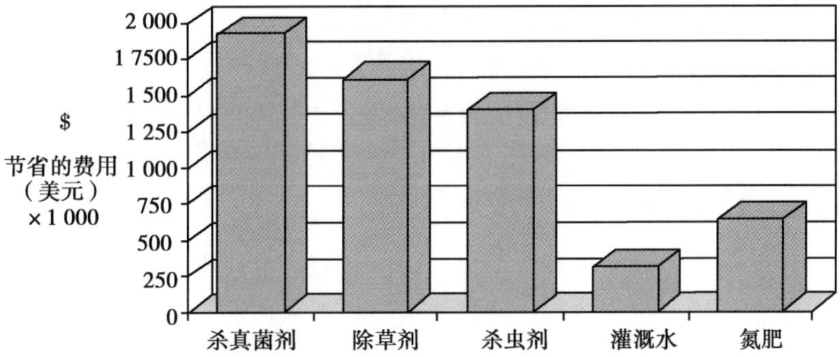

图9-4　1994年使用“马铃薯作物管理”软件和相关技术在28 300hm² 土地上减少投入品而节省的费用预估

第四节　规范虫害管理软件介绍

一、疾病管理

早疫病和晚疫病防治。叶部施喷是为了防治早疫病（茄链格孢）和晚疫病（致病疫霉菌）。在美国多个州出现了致病疫霉菌的第二种交配型和多个致病疫霉菌甲霜灵抗性菌株，更凸显了在气候条件适宜病害发展时多次施用保护性杀真菌剂的重要意义。目前在美国的威斯康星州、加拿大其他马铃薯种植区所种植的马铃薯品种都很容易患早疫病和晚疫病。针对多个马铃薯品种和新育品系的一项田间评估表明，有可能可以减少某些品种和新育品系防治早疫病所需杀真菌剂的用量，但当前可用品种的晚疫病抗病力尚未达到有意义的水平。威斯康星的田间试验已显示出品种具备晚疫病抗病力的好处，在已施喷和未施喷的地块上病害的发展都得到了抑制（图9-5）。将病害预报技术和品种叶片病害易感性信息相结合，可帮助种植户就施喷杀真菌剂的时间和用量做出知情决定。

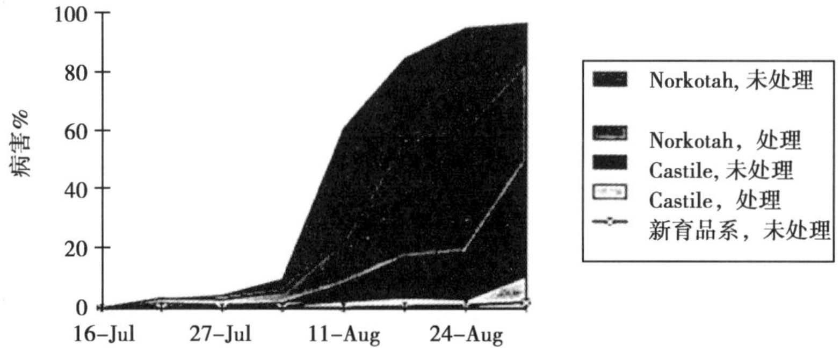

图9-5 增强抗病力和施用杀真菌剂改变了早疫病的发展情况

二、种薯块处理

种块腐烂、停工损失和植株生长强度减弱给威斯康星种植户带来重大生产风险。马铃薯生产过程中对种块进行处理和种植阶段是最关键的几个环节之一。若未能发现种块腐烂有关的隐患并及时补救，会有严重的后果，可能需要重新栽种，或导致生长迟缓，从而最终影响收获、产出、市场销售和利润率。为降低种块腐烂的风险，种植户的传统做法是先将种块切小，涂上杀真菌剂粉末，然后再进行种植。若每公顷土地种植2t种薯，且每100kg种薯的粉末涂抹量为1kg，则每1hm^{2}土地上就会施用20kg的杀真菌剂粉末。威斯康星大学一项研究表明，威斯康星州出现的种块腐烂现象一般是由马铃薯黑胫病菌等细菌引起的，且施用杀真菌剂粉末对防止种块腐烂几乎起不到任何作用。研究还显示，与种块腐烂直接相关的是种植前和种植期间种块的处理方式，以及种植期间和种植后一周内种块所处环境的温度和湿度条件。因此，种块腐烂不是偶然的；相反，种植户可以对其加以控制。田间和实验室数据为开发一个专业的系统奠定了基础，该系统可用于分析种块处理和气候条件信息，为使用者量身打造建议方案，从而以最小剂量的杀真菌剂防止种块发生腐烂。在“马铃薯作物管理”软件中新增这一专业系统将为今后采用“病虫害综合防治”技术提供新路径，还有助于进一步减少使用杀真菌剂。

三、昆虫管理

“马铃薯作物管理”软件新应用的软件提高了交互式工具的作业能力，也为种植户提供了更多机会，他们可以使用实时观测数据来改善虫害管理的相关决策，并参考借鉴其他田间和历史数据。种植户可通过调节当前的施用量阈值来解决不同品种对关键害虫的抗性不同的问题，作物在生长周期中易感性不同的问题也可用类似方法加以解决。通过调节施用量阈值来反映已知品种差异，还可减少抗性品种的杀虫剂施用量。杀虫剂用量越小，越有可能在虫害管理中引入生物防治。田间观测可将实时信息以图像形式呈现出来，包括某地区害虫和益虫的数量等，可帮助种植户对其进行生物调控，显著增大成功的几率。害虫抗药性仍然对防控马铃薯甲虫和桃蚜等害虫构成重大威胁。种植户在互联网上获得的害虫历史数据可帮助他们更有效地推进每一种害虫的管理策略，并确保对其他害虫的治理不会妨碍这些策略。

四、杂草管理

多年来，杂草管理由施喷芽前除草剂、犁地和施喷芽后除草剂（偶尔）组成。马铃薯种植中的杂草管理依赖于化学品的使用，这也催生了大量相关研究，希望找到相应办法减小化学品对环境产生的影响。最近一段时间的研究关注的是建模模拟马铃薯生长和评价遮阳对杂草生长的影响（Connell 和 Binning，1991；Raby 和 Binning，1986）。Russet Burbank 等马铃薯品种的林冠生长茂盛，出芽 9 周后的土壤表面遮阳率还能达到 95%。了解到这一点，作者就可以调整除草剂的施用时间和用量，即：仅针对出芽后的 7 ~9 周施用一次除草剂。其他实验结果表明，减少除草剂用量后，除草剂通过土壤进入地下水的可能性被大幅降低或是消除了。为了在更广的范围和更多品种（竞争力各不相同）中利用这一信息，作者着力建模模拟不同遮阳条件下杂草的生长情况。根据杂草地理分布历史数据和不同品系的遮阳率，作者对不同光照条件下不同种杂草的生长情况进行评估，结果将可用于指导在杂草出现前进行预测。杂草管理计划可利用这一信息帮助种植户做出环境友好且有经济价值的决策。

五、灌溉管理

改进灌溉管理可降低某些植株病害的影响、提高块茎质量、减小杀虫剂和氮肥沥滤的可能性。出芽到结薯期间灌溉过度似乎会加重根系感染黄萎病菌，而块茎膨大期避免水分亏缺又有可能降低发病度（Powelson 等，1993）。常见的黑星病菌（疥链霉菌）一般在湿润的土壤里比较活跃，块茎形成期和膨大期保持 80% ~90% 的土壤湿度可有效抑制发病度（Powelson 等，1993）。植株和块茎生长阶段的土壤湿度若不控制在最佳范围内，可能会出现空心、糖点和块茎畸形等问题。通过谨慎调节作物生长各个重要阶段的用水量来为灌溉计划进行微调，可帮助减少灌溉水用量，更重要的是，提高块茎收获时的质量，使其更加畅销。

六、施氮管理

氮肥的传统施用量为每公顷 280 ~ 392kg。在大多数情况下，224kg/hm^{2}氮的施用量下的产出和质量最佳（Fixen 和 Kelling，1981）。最近的一项研究进行了叶柄硝酸盐校准测试，得出的结论是，可减小早期氮肥施用量，而晚期只需根据需要施用氮肥。采用此法可减少最终沥滤到地下水中的过剩氮元素。已经有研究对多个马铃薯品种采用校准法进行分析（Kelling 和 Wolkowski，1992）。他们的后续研究显示，当马铃薯与豆科作物轮作时，这些测试还能帮助种植户精确地找出田间氮元素来源（Kelling 等，1993）。这些信息正被纳入氮肥管理模块，帮助种植户为当前种植的品种选择最佳氮肥施用时间和用量，以及在给定地块的管理选项间作出选择。该模块还将帮助种植户为所选品种做应季叶柄测试，恰当时还会提供氮肥施用量的建议。

七、昆虫和作物相互作用机制

作物选取、虫害发展和农作投入品之间的关系错综复杂，通常需要一个综合全面的方案。例如，缺氮的植株比营养充沛的植株会更容易患早疫病。缺氮和叶片过早脱落的植株的早疫病病情发展十分快速，即便多次施喷杀真菌剂也于事无补。灌溉水过量会促使氮元素从马铃薯根系中沥滤而出，导致植株缺氮。频繁和过量灌溉还会使叶子长时间处于潮湿状态从而变成疾病蔓延的温床。缺氮的植株还会过早枯萎，叶片更易受到早疫病感染，林冠密度和遮阳度都会下降。透光度增大会刺激杂草生长，导致杂草和植株争抢光照、水和营养元素，最终可能导致减产。病弱的马铃薯植株还会引来昆虫，进一步加重叶片脱落。使用指定的软件可协助种植户改善这些复杂问题的管理，获得及时有效的补救方法。

第五节　计算机模型的应用前景

一、增强计算机应用

威斯康星马铃薯产业提出新增土壤肥力和农田记录等模块的要求，这在当前微软磁盘操作系统下是不可能实现的，需要开发加强版的软件。目前，“马铃薯作物管理”软件正在向 Microsoft Windows 系统转型，新的平台叫做 WISDOM。今后，软件编程将有更高的灵活性，还为用户提供了许多新功能。

（1）改进电子表格格式，数据输入和文档间交换数据更简单。环境和作物数据可通过手动输入，呈现在屏幕显示的电子表格中。触摸左或右箭头可在不同生长日期、周、月份之间进行切换。一些种植户报告，他们会使用从同一个监测站搜集到的数据，在有不同品种、不同种植时间和不同生长速度的邻近田块的生产中指导决策。使用新软件，可在不同文件之间轻松地复制/粘贴数据，用户还可将新输入的数据设置为自动复制到指定文件中，从而节省数据输入时间。

（2）数据小结以图表形式呈现。以线形图展示种植户需要的信息，包含多个指标（例如生理天数、发病度值、日度、降水量、灌溉量、出芽后天数）随时间的变化情况。当示数超过某一特定阈值时，会对用户发出警告，从而及时采取管理措施。

（3）不同年份间数据的图像比较。某一具体地块或区域的历史数据可单独存储，并用于与当前种植季的各项生产指标进行比较。通过逐个对比，用户可对作物和病虫害管理活动做进一步微调。

（4）作物和病虫害情况记录。可记录和存储杀虫剂和肥料施用量以及病虫害观测信息，以便日后参考。记录的内容是和其他多个模块相互关联，对于反复出现的某种病虫害和实施某项具体的管理决定，软件可利用这些信息为用户量身打造应对建议，所以这些记录对未来具有重要意义。杀虫剂施用量的记录还对管理或防止病虫害抗药性的策略实施有重要作用。

（5）病虫害等作物问题的图片数据库。昆虫、病害、杂草、营养缺乏、中毒和生理问题的图片都会被扫描存储于磁盘或光盘中。用户之后还可选择存储的照片（还附有描述信息和防治建议）进行浏览和查看。这一功能将有助于对问题进行诊断，以及实施正确的文化、生物或化学补救办法。

二、计算机模型的农业应用领域

“马铃薯作物管理”软件是全面的，但其设计初衷仅针对农业生产中的一个领域——马铃薯。当前，还需要将常与马铃薯进行轮作的其他作物的“病虫害综合防治”信息纳入软件，才能在整个农作体系中为种植户提供指导。威斯康星马铃薯种植通常要进行2～3年的轮作，涉及有市场销售潜力的大宗农产品（四季豆、甜玉米、饲料玉米、豌豆）和不进行收获的覆盖作物（苜蓿、红三叶草、苏丹草、黑麦）。两种类型的轮作作物都对农业生产有好处。覆盖作物可保持土壤和养分，作为诱虫作物消灭害虫，或是为害虫的天敌提供生长空间，以及对杂草和病虫害产生异株克生效应。许多覆盖作物不会在短期内给种植户带来经济回报，但长期来看却会给农业生产的整体带来好处，因此也许具备同等甚至更高的价值。相比之下，有市场的轮作作物会在短期带来收入，也有可能带来长期效益。然而，有一些轮作作物会延续甚至加重病虫害问题，所以可能会给马铃薯种植环节带来风险。

表9－2　以马铃薯为主要作物的2～3年轮作的六种轮作安排

轮作号码	轮作的年份					
	1	2	3	4	5	6
1	马铃薯	四季豆	马铃薯	四季豆	马铃薯	四季豆
2	马铃薯	苏丹草	马铃薯	苏丹草	马铃薯	苏丹草
3	马铃薯	甜玉米	马铃薯	甜玉米	马铃薯	甜玉米
4	马铃薯	四季豆	苏丹草	马铃薯	四季豆	苏丹草
5	马铃薯	四季豆	甜玉米	马铃薯	四季豆	甜玉米
6	马铃薯	四季豆	红三叶草	马铃薯	四季豆	红三叶草

1991年，一个综合研究项目启动，探究以马铃薯为主要作物的6种具体轮作顺序（包含可市场销售作物和非市场销售作物）的短期和长期效益以及风险。该项目做了一个历时6年的长期实验，研究包含马铃薯的两年和三年轮作的6种顺序。研究找出了影响或干扰马铃薯作物预防性病虫害管理的若干因素。研究给出的信息将被整合到“马铃薯作物管理”软件当前使用的预防性和治疗性策略中。目前，扩大软件内容以包含四季豆和甜玉米模块的工作正在进行中。四季豆管理模块包含灌溉调度和日度差计算

器，可用于预测虫害发展、作物成熟度、评估白霉病发展态势，马铃薯和四季豆都会患白霉病。甜玉米模块包括灌溉水调度和虫害管理。所有这些模块都将被纳入致力于改善马铃薯和轮作作物管理的整体计划中。

第六节 有关管理软件的总结

种植户和相关产业快速采取了“病虫害综合管理”新措施，这些措施能够提高生产力、减少作物投入品、保持同其他主产区的竞争力。计算机技术为“病虫害综合管理”相关技术在田间实施提供了便捷的载体。种植户反馈显示，进一步采用相关技术的需要和机会还在涌现。这些反馈为学术界带来了灵感，很多局部研究和综合研究都正在进行中，将为当前的问题提出解决方案。“马铃薯作物管理”软件自 1989 年发布以来，内容已被大大丰富和拓展。从过去只关注单一作物，到今天覆盖整个农业生产范围，作者更加明确这项事业存在的风险和提供的机会。

参考文献

Connell T R，Binning L K. 1991. Canopy development model for weed management in potatoes. American Potato Journal 68：602（Abstract）.

Connell T R，Koenig J P，Stevenson W R，Kelling K A，Curwen D，Wyman J A，Binning L K. 1991. An integrated systems approach to potato crop management. Journal of Production Agriculture 4：453 –460.

Curwen D，Massie L R. 1984. Potato irrigation scheduling in Wisconsin. American Potato Journal 61：235 –241.

Fixen P E，Kelling K A. 1981. Potato Fertility Requirements and Recommendations：Nitrogen.

Wisconsin Potato Production. 81-DF-B，University of Wisconsin Extension Publication，Wisconsin，USA.

Kelling K A，Wolkowski R P. 1992. Interaction of potato variety and N rate on yield，quality and petiole nitrate-nitrogen levels. Proceedings of Wisconsin's Annual Potato Meetings 5：23 –29.

Kelling K A，Hero D，Grau C R，Rouse D I，MacGuidwin A. 1993. Potato responses to nitrogen following various legumes. Proceedings of Wisconsin's Annual Potato Meetings 6: 93 – 104.

Krause R A, Massie L B, Hyre R A. 1975. BLITECAST: A computerized forecast of potato late blight. Plant Disease Reporter 59: 95 – 98.

Powelson M L, Johnson K B, Rowe R C. 1993. Management of diseases caused by soilborne pathogens. Pages 149-158 in Rowe R C (Ed.) Potato Health Management. APS Press, St. Paul, Minnesota, USA.

Pratt L H. 1993. Wisconsin Agricultural Statistics. Wisconsin Department of Agriculture, Trade and Consumer Protection, Madison, WI, USA.

Pscheidt J W, Stevenson W R. 1986. Comparison offorecasting methods for control ofpotato early blight in Wisconsin. Plant Disease 70: 915 – 920.

Raby B J, Binning L K. 1986. Weed competition study in 'Russet Burbank' and 'Superior' potato (*Solanum tuberosum*) varieties with different management practices. Proceedings of the North Central Weed Control Conference 41: 28 – 29.

Sands P J, Hackett C, Nix H A. 1979. A model of the development and bulking of potatoes (*Solanum tuberosum* L.). I. Derivation from well-managed field crops. Field Crops Research 2: 309 – 331.

Stevenson W R. 1983. An integrated program for managing potato late blight. Plant Disease 67: 1 047 – 1 048.

Walgenbach J F, Wyman J A. 1984a. Colorado potato beetle (Coleoptera: Chrysomelidae) development in relation to temperature in Wisconsin. Annals of the Entomological Society of America 77: 604 – 609.

Walgenbach J F, Wyman J A. 1984b. Dynamic action threshold levels for the potato leafhopper (Homoptera: Cicadellidae) on potatoes in Wisconsin. Journal of Economic Entomology 77: 1 335 – 1 340.

附录　部分原文摘录

Crop physiological responses to infection by potato cyst nematode (*Globodera* spp.)

A.J. HAVERKORT[1] and D.L. TRUDGILL[2]

[1] *Research Institute for Agrobiology and Soil Fertility (AB-DLO), P.O. Box 14 6700 AA Wageningen, the Netherlands*

[2] *Scottish Crop Research Institute (SCRI), Invergowrie, Dundee, DD2 5DA, U.K.*

Abstract. Fresh potato tuber yields are determined by the total amount of photosynthetically active radiation intercepted by the green foliage of the crop, its conversion efficiency into dry matter, the proportion of dry matter allocated to the tubers and the tuber dry matter concentration. Reduced light interception at the beginning and at the end of the growing season, best explain yield losses following infection by potato cyst nematode (*Globodera* spp.). Retarded canopy closure during canopy establishment is due to a reduced photosynthetic rate (caused by reduced nutrient uptake rates, hormonal signalling and disrupted plant water relations), increased dry matter allocation to the roots, the formation of fewer stems and the leaves being smaller and thicker. In the second half of the growing season crops infected by potato cyst nematode senesce earlier because the reduced assimilation rate leads to reduced formation of new leaves and the reduced water and nutrient uptake is aggravated because rooting depth is less whereas water and nutrients are depleted in the rooted zone leading to increased leaf shedding. Moreover, tuber fresh yields are reduced because infection by potato cyst nematode reduces the number of tubers and increases the tuber dry matter concentration.

The short-term and long-term effects of potato cyst nematode infection on crop functioning are illustrated and their implications for plant breeding for tolerance , cultural practices and modelling are discussed.

Introduction

Crop performance may be expressed in terms of factors that set the upper limits to yield (e.g. temperature and solar radiation), that restrict yield (e.g. lack of water and nutrients) and reduce yield (e.g. pests and diseases). Those stress factors restricting and reducing yield may be abiotic or biotic in origin and several such factors may operate in the course of the growing season. For instance, emergence of potato sprouts may be hampered by unfavourable soil conditions (cold or wet) or soil-borne pathogens (e.g. *Rhizoctonia solani* and nematodes) or seed-borne (viruses) ones. Between emergence and canopy development the same stress factors affect crop growth but more hazards (night frosts or drought) are added. The canopy is subjected gradually to air borne diseases such as late blight caused by *Phytophthora infestans* and aphid-transmitted viruses. Towards the end of the growth cycle depletion of water and minerals as well as certain diseases such as potato wilting caused by *Verticillium*

A.J. Haverkort and D.K.L. MacKerron (eds.), Potato Ecology and Modelling of Crops under Conditions Limiting Growth, 167–184.
© 1995 *Kluwer Academic Publishers.*

dahliae may become increasingly important. The objective of crop management and disease control is to reduce the chance of growth being restricted by unfavourable factors.

Crop reactions to nematode attack may be partitioned into those that are short- and long-term. Short-term reactions include effects on plant and cell water relations, stomatal conductance, photosynthesis, respiration, nutrient uptake and chemical composition. To measure these effects techniques such as the pressure chamber, porometry, infrared gas analysis and isotope labeling have been applied. Longer-term effects are often expressed as growth retardation or earlier senescence and by wilting, yellowing and leaf shedding. They may be determined by crop growth analysis, measurement of intercepted radiation and radiation-use efficiency, dry matter partioning and specific leaf area. Other longer-term effects may be detected through special measurements such as crop reflectance, transpiration efficiency, stable isotope fractionation and crop mineral content.

The objectives of this chapter are to review the short-term and long-term effects of potato cyst nematodes on crop growth, development and productivity. We discuss the relevance of the knowledge of the mechanism of damage by potato cyst nematodes to modelling crop development and growth, to breeding strategies and to cultural practices. Attention is paid to the interaction of the effects of nematode infection with abiotic soil factors, especially the availability of water, nutrients and the soil pH.

Short-term effects

From measurements (Fatemy et al. 1985; Haverkort et al. 1991a) it is clear that potato cyst nematodes affect water relations in potato plants. The moisture condition of a plant cell is described by its water potential (Ψ) comprising an osmotic component (Ψo) and a pressure component (Ψpi): Ψ=Ψo+Ψpi, expressed in MPa. When plant or cell moisture content decreases because transpiration exceeds water transport through the roots (due to a high evaporative demand, a dry soil or root damage) Ψ, and Ψpi decrease as well as the relative water content of the cell or the tissue (Turner 1988). Leaf water potential may be observed with a pressure chamber or psychrometrically (Brown and Van Haveren 1972) and the osmotic potential with the aid of an osmometer. Vos and Oyarzún (1987) observing potatoes growing in moist soil found typical values of Ψ and Ψo of −0.5 and −0.8 Mpa respectively. Drought coupled with infection of roots by potato cyst nematodes can decrease leaf water potentials from −0.6 MPa to about −1.1 MPa (Haverkort et al. 1991a). Fatemy et al. (1985) found that nematode infection increased stomatal resistance, reduced photosynthetic rates and reduced water use of the plants. Exogenously applied abscissic acid temporarily increased these effects (for 2 days with cv. Pentland Dell and for 6 days with the tolerant cv. Cara); apparently the plants grow accustomed to increased abscissic acid levels and the

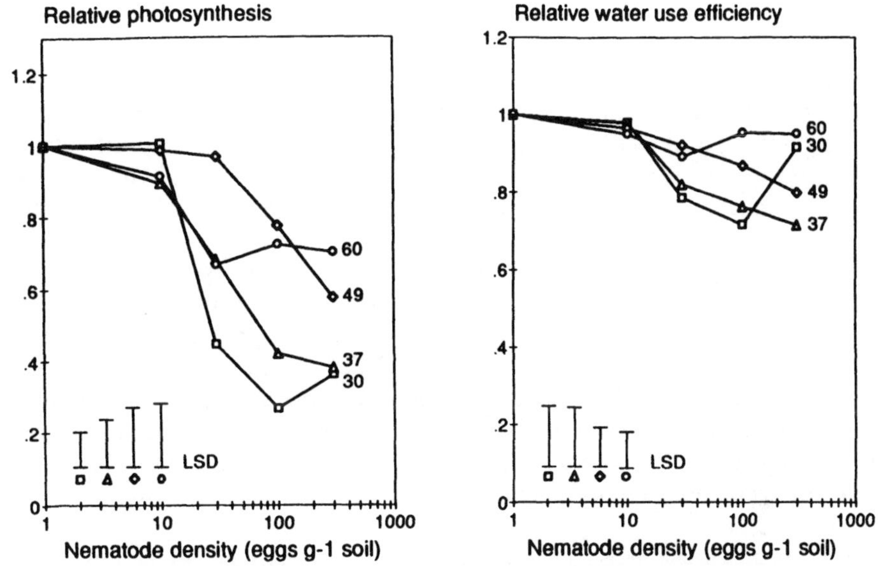

Figure 1. Photosynthetic rate and water use efficiency at 30 (□), 37 (△), 49 (◇) and 60 (○) days after planting at various nematode densities (from Schans and Arntzen 1991).

effect disappears. Water stress also increased abscisic acid levels so the initial effect of drought on stomatal regulation may be through hormonal effects.

Schans and Arntzen (1991) showed that infection by potato cyst nematodes decreased rates of photosynthesis of potato leaves. This effect was progressively more severe at higher nematode densities but at each nematode density, the effects appeared to be lesser with increasing age of the plant (Figure 1). The authors suggested that the gradual relative increase of photosynthesis is due to two factors. With time the attack decreases due to depletion of the inoculum, leading to less hormonal signalling as less infected root tips are present, and, with time the leaf area increases, leading to a possible dilution of the signal in the leaves, resulting in a reduced stomatal response. The authors did not record levels of ABA or other hormones. A simultaneous effect of potato cyst nematode attack on photosynthesis and on water-use efficiency (Figure 1) shows that not are only stomatal processes involved (as then the water-use efficiency increases) but that photochemical or biochemical processes are also implicated. Reduced uptake of nutrients is a major component as will be discussed later. Schans and Arntzen (1991) conclude that their observation of a decreasing effect of potato cyst nematodes on the water-use efficiency of individual leaves agrees with Evans' (1982) finding that whole plant water-use efficiency was decreased up to 32 days after planting but increased from 32 days onward. The reported relationships between gas-exchange and water-use efficiency from literature, however, have not yet proven unambiguously that the two mechanisms (signalling and root damage leading to disrupted water and nutrient uptake) operate. The methodology used and the results obtained are conflicting still and the processes involved need further elucidation.

Nematode attack (Haverkort et al. 1991a; Schans 1993), like water – and heat – stress can reduce the photosynthetic rate indirectly by closure of the stomata (prompt effect) or directly by a reduction of the photosynthetic capacity of the leaves (after a few days). As a side-effect of stomatal closure the assimilation/transpiration ratio increases as does the proportion of stable isotope 13C within the plant (Farquhar et al. 1982). No consensus exists on the primary site of the nematode effect on photosynthesis and whether photoreactions in the thylakoid membranes or biochemical reactions of the Calvin cycle are the most affected. In vivo fluorescence signals give information on the light use efficiency and on the rate of electron flow in the thylakoids. Combined with gas exchange measurements, analysis of fluorescence signals provides information on rate limitation of processes related to resistance to flow in the gas phase and to internal photosynthetic processes, but these approaches have not been applied to effects of potato cyst nematodes on potato.

Table 1. Effects of potato cyst nematode infection (15 living eggs per g soil) and drought (no water applied from emergence) on plant water relations (container experiment, 42 days after planting, Haverkort et al. 1991a) Data, within a row, with different letters differ at the 1% confidence level

Observation	Treatment		
	Control	Infected	Droughted
Leaf water potential (10^4 Pa)	6.9a	10.8b	11.8b
Diffusion resistance (s cm^{-1})	1.36a	1.51a	5.11b
Transpiration (ug H_2O cm^{-2} s^{-1})	11.0a	10.21a	4.09b
Water use efficiency (g kg^{-1})	7.34a	6.43b	9.19c
Leaf dry matter concentration (%)	8.5a	10.5b	12.3c
Stem dry matter concentration (%)	6.5a	8.0b	8.0b
Tuber dry matter concentration (%)	15.0a	15.2a	21.7b

The effects of infection by potato cyst nematode (*G. pallida*) and drought on the water relations of the potato plant are shown in Table 1. Nematodes and drought both reduced leaf water potentials as measured with a pressure chamber and increased dry matter concentrations of all plant parts. The effects of nematodes on stomatal diffusion resistance and transpiration were negligible whereas drought led to a strong increase in stomatal resistance and consequently to a strongly reduced transpiration rate (Haverkort et al. 1991a). Drought increased the water-use efficiency but, as was seen in Figure 1, potato cyst nematodes decreased the water use efficiency showing that, in contrast to drought, nematodes reduced photosynthesis more than transpiration.

During the later stages of growth, the water use efficiency of infected plants increased strongly, even above that of uninfected plants. Similar results were obtained by Evans et al. (1975): later in the season, the efficiency with which water is taken is taken up is impaired, intolerant cultivars suffering most leading to water stress and premature senesence. Haverkort et al. (1991a) concluded that at least two yield reducing mechanisms exist of which the relative importance varies with time. Firstly, reduced apparent assimilation rates which are not wholly related to a change in the water balance caused by the initial attack by the cyst nematodes. Secondly a reduced dry matter accumulation resulting from a decrease of water and/or nutrient uptake. Effects of drought and cyst nematodes (as observed in field experiments which will be discussed later) were often additive. Where they were not additive this was because infected plants used less water than uninfected plants leading to less water stress.

Long-term effects

Effects on morphology

The most convenient means of observing differences in crop development caused by environmental factors is to make comparative assesments of the proportion of the ground covered by the canopy as this is strongly correlated with the percentage of light interception. It is also highly highly correlated with the leaf area index (below 3), and with the proportion of the incoming infra-red radiation reflected by the canopy i.e. leaves are more reflective than the soil (Haverkort et al. 1991b). This is illustrated in Figure 2 where the proportion of infra-red reflectance of potato crops affected by drought (unirrigated), by potato cyst nematodes (unfumigated) and by night frost (7 weeks after planting) are shown. Ground cover and hence infra-red reflectance were reduced more strongly by nematodes than by drought. Night frost did not decrease the proportion of the ground covered by green leaves (data not shown) but it did reduce the canopy reflectance.

The effect of nematode attack and an early drought period on some morphological characteristics of potato plants grown in containers under a rain shelter is shown in Table 2. At the end of the dry period the leaf area of the infected plants was less than half that of the control 3181 cm^2 per plant instead of 8253 cm^2 per plant, explained because thicker (SLA decreased from 301 to 263 $cm^2\ g^{-1}$) and smaller 72 instead of 150 cm^2 per leaf) leaves were formed. The effect of drought on these morphological characteristics were similar. The plants subjected to either or both stress factors remained shorter, had lower tuber yields and lower shoot/root ratios than the control plants. Four weeks after rewatering the droughted plants, droughted and undroughted plants had similar leaf areas and weights although the plants that had been droughted before still had thicker leaves and lower leaf area ratios. The unstressed plants were senescing at 70 days after planting whereas the plants which were subjected

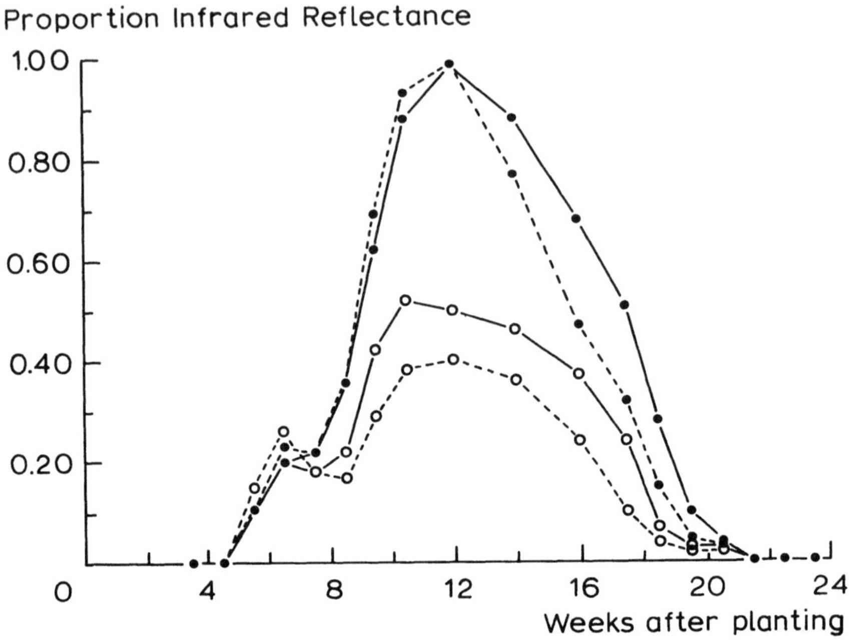

Figure 2. Proportion of infrared reflectance of potato crops subjected to drought and potato cyst nematodes. — irrigated, --- unirrigated, ● nematodes controlled, o nematodes not controlled (after Haverkort and Schapendonk 1994).

Table 2. The main effects of potato cyst nematode and drought on plant characteristics of potato cv. Mentor, 43 days after planting. Crop characteristics differed significantly $P < 0.05$ between the control and nematode or drought treatments (Haverkort et al. 1991a)

Plant characteristic	Potato cyst nematode (eggs per g soil)			
	0 moist	18.5 moist	0 dry	18.5 dry
Leaf, dry (g/plant)	27.4	12.1	20.2	10.4
Leaf area (cm^2/plant)	8253	3181	4661	2258
Leaf size (cm^2/leaf)	150	72	108	58
SLA ($cm^2\ g^{-1}$ leaf dry weight)	301	263	228	218
LAR ($cm^2\ g^{-1}$ plant dryweight)	93.1	89.0	74.2	73.1
Number of stems	5.4	4.5	5.2	4.5
Stem length (cm)	52	30	34	24
Spec. stem weight (mg cm^{-1})	62	46	62	52
Tuber dry weight (g/plant)	42	15	27	13
Number of tubers per plant	30	16	28	16
Shoot/root ratio	17.3	8.6	15.1	7.9

to an early drought stress continued to grow longer, probably because they had not yet depleted soil nutrients.

In Table 1 it was shown that similarities exist in the water relations of droughted crops and those infected by potato cyst nematodes. The same holds for the influence of the two stress factors on plant morphology. The number of tubers, however, is more strongly reduced by potato cyst nematodes (from 30 and 28 to 16 tubers per plant) than by drought The difference is only partly explained by a reduction in numbers of stems due to cyst nematode infection (from 5.4 per plant to 4.5 per plant). The shoot/root ratio in this experiment was only marginally affected by drought but strongly decreased in the presence of potato cyst nematodes, indicating that a major shift in relative dry matter allocation occurred in favour of the roots.

The recently constructed Wageningen Rhizolab (Smit et al. 1994) allows a frequent and non-destructive observation of root growth of plants subjected to various soil stress factors including nematodes and drought. Recent studies using that facility (Haverkort et al., 1994), to compare uninfected potato plants with plants infected with *Globodera pallida*, (40 eggs per g soil) showed increased root growth in the top 30 cm of soil but reduced root growth in the subsoil (Figure 3). Potato cyst nematodes initially caused the formation of less roots in the topsoil but root formation (branching) continued until 120 days after planting, whereas in the control, root formation in the topsoil stopped after 50 days after planting to allow exploitation of the subsoil.

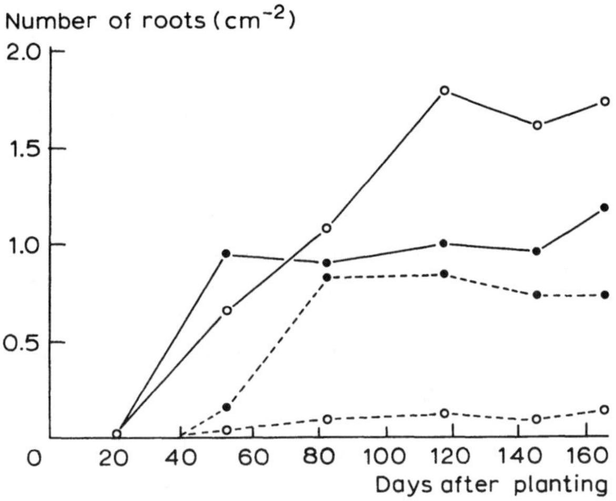

Figure 3. Mean number of roots observed per cm^2 surface of minirhizotrons with time at 0 (●) and 40 (o) living juveniles of *Globodera pallida* per g soil at 0–30 (—) and 30–100 cm (---) soil depth (Haverkort et al., 1994).

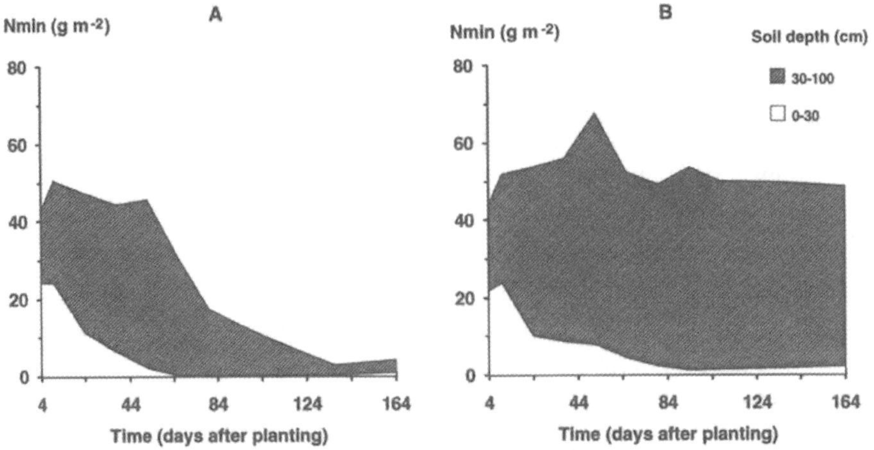

Figure 4. Total soil mineral nitrogen in the 0–30 cm and the 30–100 cm soil layers in the experiment in the wageningen Rhizotron. (A) no potato cyst nematodes present, (B) initial population of 40 living juveniles per g soil.

Effects on nutrient uptake

The total amount of soluble nitrogen (nitrate and ammonia) was determined every two weeks in the topsoil from 0–30 cm and in the subsoil from 30–100 cm soil depth (same experiment as shown in Figure 3). Without cyst nematodes, the soil profile towards crop senescence was depleted of mineral nitrogen to a depth of at least 1 m whereas, following high initial population densities there was no change in nitrogen concentration in the subsoil between 30 and 100 cm (Figure 4). The spatial heterogenity of roots (Figure 3) and nitrogen (Figure 4) in the soil (as there were hardly any roots in the subsoil so the plants could not exploit the nitrogen in that layer), caused by potato cyst nematodes apparently is an important mechanism of damage, especially in the second part of the growing season and may contribute to the earlier senescence which is observed in infected fields.

The effects of potato cyst nematode on the growth of potato were analysed in a long-term trial involving irrigation, fumigation and resistant cultivars (Trudgill et al. 1975). Based on their observations the authors hypothesized that reduced rates of top growth due to potato cyst nematodes were a consequence of root damage which decreased rates of nutrient uptake and led to chronic deficiency of the least available nutrient. The findings shown in Figures 3 and 4 are in line with this hypothesis.

This hypothesis was also tested in soils heavily infected with potato cyst nematode at two field sites and in a pot experiment by observing the interaction between rates of fertilizer and a non-fumigant nematicide. Percentage ground cover and yield of potato plants infected with potato cyst nematodes were affected much more by changes in fertilizer inputs than by nematicide (Trudgill

Table 3. Effects of nematicide and additional fertilizer on cv. Pentland Crown (pot trial 14 weeks after planting, weights in g per plant, nutrient concentrations in haulm dry matter, additional fertilizer was 2 × standard dose of N, or 2 × K or 3 × P). Weights in g per plant

Observation	With nematicide fertilizer increased				Without nematicide fertilizer increased			
	Standard	N	P	K	Standard	N	P	K
Haulm (g)	298	448*	306	251	98	265*	208*	160*
Root (g)	60	84*	77	34	11	64*	56*	22
Tuber (g)	363	372	363	357	100	158	168	142
% N	2.5	3.4*	2.5	2.4	3.3	3.5	3.3	3.3
% P	0.16	0.22*	0.26*	0.15	0.18	0.20	0.27*	0.20
% K	3.5	3.3	3.4	5.2*	2.6	3.1	2.9	4.5*

* statistically significant different ($P < 0.01$) within a nematode treatment.

1987). In one field trial the relatively tolerant cv. Cara was much less responsive to rates of either nematicide or fertilizer than the intolerant cv. Pentland Dell. Increasing fertilizer rates from 0.5 to 1.5 t per ha increased tuber yields of the unprotected cv. Pentland Dell from 0.13 to 0.82 kg per plant, compared with 1.30 to 1.60 kg per plant for those protected with a nematicide. In a pot experiment to vary P, K and N independently, Trudgill (1980) showed that nematode damage interacted strongly with the availability of N and of P. Doubling the amount of N and P more than doubled top weight in the absence of nematicide. Amounts of K had a smaller but still significant effect (Table 3). Greater effects were observed in the pot experiment than in one of the field trials possibly because of adsorption of the P in the heavy clay soil at that site.

Yield formation

As indicated before, the yield of a crop (Y) is determined by the amount of photosynthetically active solar radiation it intercepts (R), the conversion efficiency of radiation to dry matter (E), the proportion of the total dry matter in the harvested parts (H) and its dry matter concentration (D). In a formula: $Y = R\ E\ H/D$.

The value of each parameter in this simple model of crop growth is determined by a number of underlying processes. R depends on the (duration of) leaf expansion rate and light extinction by the canopy, E depends on photosynthesis and respiration rates and H depends on dry matter distribution processes related to crop development. In a series of field trials in the north east of the Netherlands in which four cultivars were either irrigated or not irrigated and either fumigated or not fumigated (leading to averages of 5 or 44 eggs per g soil) periodic harvests were made to determine the effect of drought and cyst nematodes on components of yield. The effects on infrared reflection by cv. Mentor were shown in Figure 2. Both intercepted radiation and radiation use efficiency were negatively affected by increased levels of potato cyst nematodes and drought.

Table 4. Relative values of yield components in unirrigated and unfumigated plots (mean data 1989 and 1990, values of irrigated and fumigated plots = 100) (Haverkort et al. 1992)

Year	Cultivar	Irrigated	Fumigated	Yield =	R ×	E ×	H /	D
88–90	Darwina	+	–	48	61	90	94	105
	Désirée	+	–	52	71	86	93	102
	Elles	+	–	73	86	95	102	107
	Mentor	+	–	49	57	92	100	101
89–90	Darwina	–	+	55	62	99	94	105
	Désirée	–	+	77	88	99	94	105
	Elles	–	+	80	93	90	95	101
	Mentor	–	+	73	87	97	97	111

Table 4 shows the values of intercepted radiation, radiation-use efficiency, harvest index and dry matter concentration of the unirrigated and unfumigated plots relative to those of the irrigated and fumigated controls of the four cultivars (including cv. Mentor) tested.

Apparently cyst nematodes and drought act similarly on the components of yield: yields are mainly reduced because crops intercept less solar radiation. Other factors reducing fresh tuber yields are lower efficiencies whereas reduced harvest indices and increased dry matter concentration also play a role. The cultivar least tolerant of nematode damage (Darwina) was, also least tolerant of drought and cv. Elles was most tolerant of both stress factors. Trudgill et al. (1990) found a value of 1.05 g MJ^{-1} for tolerant Cara and a value of 1.25 g MJ^{-1} for intolerant Pentland Dell which maintained ground cover for a longer period and hence intercepted more solar radiation. This phenomenon of a lower radiation use efficiency associated with a higher value of intercepted radiation is also presented in Table 4 where the unfumigated drought tolerant cv. Elles had a 5% lower E-value and a 14% lower R-value than the fumigated treatment whereas the cv. Mentor (less tolerant of potato cyst nematodes) had a 43% lower R-value when not fumigated. With drought this phenomenon is even more apparent: cv. Darwina lost only 1% of its efficiency but 38% of its intercepted radiation when not irrigated. Cultivars that are able to maintain a well developed canopy apparently are not able to keep it functioning at maximum rates as, during stress, photosynthetic rates decline whereas respiration does not decline. Beside long-term visible effects of stress factors on crop morphology and measurable effects on intercepted radiation and radiation use efficiency, stresses affect the transpiration efficiency, 13C-discrimination and mineral uptake. Haverkort et al. (1991a; Table 1) reported transpiration efficiencies of 7.34 g kg^{-1} in plants that were not subjected to stress factors but a value of 6.43 g kg^{-1} in potato plants subjected to cyst nematodes and 9.19 g kg^{-1} in plants subjected to drought.

Different cultivars react differently to potato cyst nematode infection in the development of ground cover as was shown by Mulder (1994). Comparison of

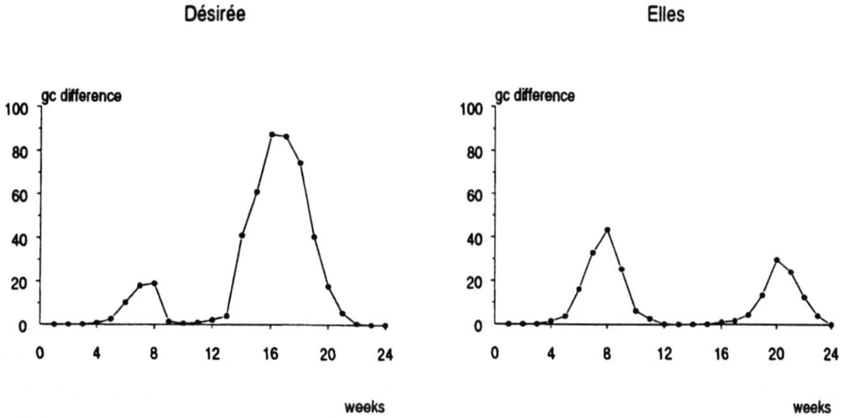

Figure 5. Difference in percentage ground cover (gc difference) between lightly and heavily infected crops of Désirée and Elles (Mulder 1994).

ground cover values of healthy crops and infected crops shows (Figure 5) that ground cover development of cv. Désirée was much less affected early in the growing season than in the later stages of crop growth, whereas with cv. Elles (tolerant cultivar), the reverse was observed.

Potato cyst nematodes and drought both decrease 13C-discrimination indicating that nematodes disrupt plant water relations (Haverkort and Valkenburg 1992). Drought and potato cyst nematodes, however, are not identical in their effects as are shown by the reduced water use efficiency following nematode infection and increased water-use efficiency following drought. Apparently, the negative effect of potato cyst nematode infection on plant processes such as photosynthesis and dry matter allocation to the roots is greater than the increase in water use efficiency. Another dissimilarity between the effects of drought and potato cyst nematodes is that drought decreases Ca-uptake whereas potato cyst nematode infection increases it (Fatemy and Evans 1986). Calcium is taken up by the root tips and is transported within the plant with the transpiration stream; a process that is enhanced where synchytia break the endodermis allowing apoplastic movement. As is shown in Figure 3, the ramification of the roots induced by potato cyst nematodes leads to many more root tips in the topsoil and hence contributes to the increased calcium uptake.

Tolerance of potato cyst nematodes

Resistance to potato cyst nematodes is expressed as the ratio of final to initial nematode population density. Resistance is of importance in reducing the nematode population density but does not necessarily reflect the degree of damage which is to be expected. Tolerance of potato cyst nematodes in one cultivar, is defined as a smaller reduction in yield at high nematode densities

Table 5. Relations between soil pH, tuber yields, dry matter and initial density of *Globodera pallida* (Haverkort et al. 1993)

pH_{KCl}	Tuber dry yield (kg m^{-2})	Initial density (eggs g^{-1} soil)
4.5	1.09	17.5
5.0	1.04	13.8
5.5	1.03	14.4
6.0	0.92	11.3
6.5	0.75	8.8
$LSD_{0.05}$	0.11	8.3

than in another cultivar. It is a relative measure. Evans and Haydock (1990) reviewed some of the (curvi-)linear relations between the initial nematode density and associated tuber yield losses conceived by authors such as Brown, Oostenbrink and Seinhorst. Although such relations may be valid when expected damage has to be forecast, they convey no information as to the mechanism of damage occurred. Evans and Haydock (1990) in their review, attributed differences in tolerance between cultivars to differences in nematode hatching, invasion and colonisation of roots, root vigour, form of the root system, local reaction of roots to nematode attack, physiological responses of plants to nematode attack, and to interactions with other organisms. They found no evidence that tolerance is consistently greater to one species of potato cyst nematode (*Globodera rostochiensis*) than the other (*Globodera pallida*). Damage due to potato cyst nematodes has been reported to be increased by biotic soil-borne factors such as *Rhizoctonia solani*, *Verticillium dahliae* and *Verticillium albo-atrum*. Abiotic soil-related factors, however, may be of equal or greater importance than biotic factors. Mulder (1994) compared relative damage due to potato cyst nematodes with that caused by soil type, soil fertility, soil water availability and soil pH. Higher soil pH-values appeared to be especially important and Haverkort et al. (1993) found a strong interaction between yield losses and pH in a light, sandy soil in the north-east of the Netherlands. Through annually liming, plots with pH levels of 4.5 to 6.5 were obtained in a reclaimed peat soil with about 22% organic matter. Potato yields (Table 5) in these plots decreased from 45 t per ha at pH 4.5 to 33 t per ha at pH 6.5, whereas the nematode population densities decreased from about 18 to 9 juveniles per g soil. In a container experiment with the same soil a strong interaction was observed between soil pH and the presence of potato cyst nematodes. Tuber yields were about 11% lower at pH 6.5 than at 4.5 in the absence of nematodes, but 44% lower when an initial population of 27 juveniles per g soil was present. Plants grew more vigourously at pH 4.5 even though initial nematode densities were higher than at pH 6.5; the more vigourous (root) growth at the lower pH values probably contributed to higher multiplication rates that were observed.

When modelling the effect of potato cyst nematodes on crop growth, development and tuber yield, such interactions should be taken into account in

Table 6. Effect of inoculation with 10 000 eggs per small pot on root length of cultivars differing in tolerance (Trudgill and Cotes 1983)

Cultivar	Nematodes	Root length (cm/pot)		
		Main	Lateral	Total
Maris Piper	not inoculated	138	384	522
(tolerant)	inoculated	275	490	766*
Epicura	not inoculated	118	298	416
(intermediate)	inoculated	87	307	397
Maris Anchor	not inoculated	179	973	1153
(intolerant)	inoculated	219	260*	480

* significantly different ($P < 0.05$) per cultivar and root class.

order to accurately estimate model parameter values of uninfected and infected crops. Breeding new cultivars with resistance to potato cyst nematodes and capable of high yields in infested soils needs screening techniques for tolerance. Evans and Haydock (1990) reviewed several direct and indirect methods of screening for tolerance in the field and in pots. Trudgill and Cotes (1983) showed that an infected tolerant cultivar had a greater total root length in pots than an uninfected intolerant cultivar (Table 6).

Arntzen (1993) found no relation between tolerance and maturity class of potato as indicated in the Netherlands Recommended list of Varieties. However, Haverkort et al. (1992) and Trudgill et al. (1990) concluded that culivars which develop abundant foliage and intercept more light (generally late-maturing cultivars) usually suffer less damage from potato cyst nematode infection than cultivars which develop less foliage and hence keep the ground covered for a shorter period (early maturing cultivars). Arntzen also found that, in the greenhouse, the reduction of the leaf area due to potato cyst nematode infection was mainly because the plants formed smaller leaves rather than because they formed fewer leaves. He reported a good correlation between reduction in total plant weight in pots in the greenhouse and yield losses in the field. The depression of growth in pots was mainly associated with the rate of hatching of the nematodes and the differences in growth of the roots, which became apparent within a few days in vitro.

Screening techniques in pots may be useful to determine the degree of tolerance of potato cyst nematodes in potato genotypes. They are not useful for crop ecological purposes with a view to mechnistic modelling of the development and growth of field grown crops. In pots the inoculum and nutrients may be depleted sooner than in the field, and fluctuations in water availability are likely to occur within a shorter time span. In pots it is often observed (Fasan and Haverkort 1991) that infected plants, after an initial depression of growth continue to grow for a longer period than uninfected plants once the inoculum is depleted as they continue to benefit from nutrients present within the soil while all the nutrients have already been taken up by the healthy plants.

Discussion

Three major processes determine crop growth and their contribution to reduced crop growth following potato cyst nematode infection is of greatest importance during two distinct phases in the growing season.

In the presence of potato cyst nematodes, the three following effects are found:

1. reduced assimilation rates,
2. dry matter allocation unfavourable to the leaves, and
3. morphological changes unfavourable to leaf expansion.

Considering the interception of light and its use, the phases of crop growth that are most crucial to yield formation are:

1. canopy establishment from emergence to canopy closure, and
2. crop senescence at the end of the season when light interception is reduced from 100% to 0%.

The three major effects on crop physiological processes operate in both those phases and the mechanism of damage to the crop is explained as follows. For maximum yields, the time between planting and emergence should be as short as possible. The time between emergence and 100% light interception should also be as short as possible then the period of 100% light interception should be as long as possible. The time between planting and emergence is not affected by potato cyst nematode (Haverkort et al. 1992) but the initial stage of crop growth between emergence and 100% light interception (canopy establishment) and the final stage of crop growth when ground cover declines from 100 to 0% (canopy senescence), are both affected.

As was shown, potato cyst nematodes lead to a direct reduction of photosynthesis. This reduction may be mediated initially by a hormonal signal such as abscissic acid. Water uptake, by infected crops is also reduced leading to symptoms of drought in the plant (stomatal closure, lowered leaf water potential and reduced transpiration). However, it is evident that rates of photosynthesis are affected more than those of transpiration. Reduced mineral uptake (nitrogen and phosphorus) may also lead to a decrease in the efficiency of the photosynthetic apparatus (rubisco) thus contributing to reduced radiation use efficiency. Later canopy closure following potato cyst nematode infection may be due mainly to disrupted water relations, whereas in the second part of the growing season assimilation is reduced because roots of infected crops penetrate less deeply into the soil and in the second half of the growing season the topsoil is becoming depleted of water and nutrients. The resulting reduction in assimilation rates leads to a reduction in the rate of appearance of new leaves (beside leaf shedding contributing to reduced radiation interception) and to reduced rates of tuber bulking due to the combined effect of reduced radiation interception and lower radiation-use efficiency. A reduction in the number of tubers formed following infection with potato cyst nematode may also contribute to the reduction of assimilation rates as fewer sinks are formed which, through feed-back mechanisms, render the source less effective.

Potato cyst nematode damage is associated with a relative increase in dry matter partitioning to the roots, as is shown by decreased shoot/root ratio and increased rooting in the Wageningen Rhizolab. Dry matter allocation favouring the roots decreases the amount available to be invested in the leaves thereby reducing the amounts of intercepted radiation and decreasing crop growth rates. The timing of tuber initiation is hardly altered by potato cyst nematodes which means that tuber initiation in infested crops occurs with smaller plants than in healthy crops, again leading to a pattern of dry matter allocation unfavourable to the leaves contributing to the later closure of the canopy and to an earlier onset of senescence.

To have a maximum effect on ground cover and radiation interception, the amount of dry matter which is allocated to the leaves should be spread over the soil surface as widely as possible. This means that a larger specific leaf area (the formation of thinner leaves) a higher leaf water content (leading to larger leaves) and more and longer stems (allowing a better spread of the leaves over the soil surface with less overlap) are the major factors contributing to light interception with a given amount of dry matter partitioned to the plant tops. Potato cyst nematodes, however, unfavourably influence these four factors. As was shown in the previous paragraphs potato cyst nematodes damaged plants have smaller specific leaf areas, lower leaf water contents, less and shorter stems resulting from disrupted water relations and decreased uptake rates of N, P an K by the damaged roots.

Finally, marketable yields and crop profitability, do not only depend on the amount of dry matter partitioned to the tubers but also to tuber size distribution. Infected crops yield smaller tubers because of reduced crop growth rates, and, because tuber water content is lower.

The relative importance of three processes in their influence on tuber yield varies. Which underlying process influences assimilation (signalling, lack of water or nutrients), partitioning (favouring the roots or the reproductive organs) and morphology (leaf thickness and stem length), depends on many factors. Between emergence and canopy senescence, lack of nitrogen is likely to become increasingly limiting for optimal photosynthesis and may gradually replace the immediate signalling effect which is likely to play a major role in reducing photosynthesis during the early stages of crop growth. Dry matter partitioning favouring the roots initially withdraws assimilates which in uninfested conditions would contribute to leaf growth. In the second half of the growing season dry matter allocation to the tubers probably plays a more significant role: tuber initiation in smaller plants leads to an earlier allocation of virtually all assimilates that are produced to the tubers and consequently leads to earlier senescence as no assimilates are left for the formation of new leaves and/or maintenance of the old ones. Thicker leaves initially lead to later closure of the canopy but the earlier reduction in light interception following potato cyst nematode infection may rather be due to the reduced number of leaves that are formed.

The relative importance of the three major processes to yield reduction following potato cyst nematode infection, vary with cultivar (degree of tolerance and physiological processes on which this tolerance is based), cultural practices (soil type, pH, soil moisture and nutrient availability throughout the season) and environmental factors that the grower cannot alter, such as temperature and radiation. The many interactions between plant growth processes and environmental and genetic factors cannot possibly be quantified within a single comprehensive simulation model of crop growth and development. Modelling, therefore will focus on part of the interactions such as, for instance, root growth and nitrate uptake as influenced by cultivar, soil type, and moisture availability. At a very high level of integration such as shown in the fresh yield formula based on radiation interception, radiation use efficiency and harvest index, models may be used in simple crop guidance systems.

Acknowledgements

D.L. Trudgill acknowledges the financial support of the Scottish Office of the Agriculture and Fisheries Department

Modelling the interaction between potato crops and cyst nematodes

M. VAN OIJEN[1]*, F.J. DE RUIJTER[1] and R.J.F. VAN HAREN[2]
[1] *Research Institute for Agrobiology and Soil Fertility (AB-DLO), P O Box 14, 6700 AA Wageningen, The Netherlands*
[2] *Research Institute for Plant Protection (IPO-DLO), P O Box 9060, 6700 GW Wageningen, The Netherlands*
* *Present address Wageningen Agricultural University, Dept of Theoretical Production Ecology, P O Box 430, 6700 AK Wageningen, The Netherlands*

Abstract. Simulation modelling of the effects of cyst nematodes on potato crop growth is still in its infancy The first attempts to simulate nematode population dynamics together with host growth were done in the 1970s For many years the models focussed on the rate of growth of the nematode population, while using oversimplified, descriptive functions for host growth and its response to nematodes Further development was hampered by lack of information on the physiological damage mechanisms involved

However, recent years have witnessed the development of sophisticated combination models that treat both pest and host in a more realistic manner It is becoming increasingly clear that cyst nematodes affect both the carbon and nutrient metabolism of their host Possible ways to incorporate the various damage mechanisms into simulation models are discussed

Introduction

The use of statistical regression models, linking crop yield to pre-planting densities of nematodes, has been reviewed elsewhere (Elston et al. 1991). In this chapter we will restrict ourselves to simulation models. In such models the physiological processes involved in the interaction of crops with nematode populations in soil and roots are simulated dynamically. We will focus mainly on the role of the models in accounting for nematode effects on host growth.

The earliest simulation models were developed in the 1970s. In a model by Ferris (1976; 1978) of *Meloidogyne arenaria* in grapevine (*Vitis vinifera*), host growth rate is temperature-driven but decreases in inverse proportion to the fraction of the root system infected by the nematodes. In contrast to this simple approach to host growth, the model included much detail of the nematode life cycle, and its dependence on various environmental conditions. This approach to plant-nematode simulation modelling, involving much complexity for the nematode but only very simple descriptive functions for host growth, remained the common approach during the following years (McSorley et al. 1982). Jones (Jones and Perry 1978; Jones et al. 1978; Jones and Kempton 1980) produced such a model for cyst nematodes, including *Globodera rostochiensis*, the golden potato cyst nematode. One of the new features of Jones' model study was the incorporation of nematode genetics, which allowed application of the model to

A J Haverkort and D K L MacKerron (eds), Potato Ecology and Modelling of Crops under Conditions Limiting Growth, 185–196
© 1995 *Kluwer Academic Publishers*

the problem of selection for virulence. The study also showed how simulation models could be used to study the effect of nematicides, resistant cultivars in various rotation schemes, and competition between nematode species on nematode populations.

For some years there remained a gap between modelers of nematode population dynamics and crop growth modelers. At a conference in 1982, nematologists expressed the wish to bridge the gap: 'if we wish to build valid plant models, we had better bring in some agronomists and work closely with them' ...'they had models of nutrition, photosynthesis and growth and carbon allocation and no one uses the models' (Freckman 1982, pp. 200–201). The need to couple pests to crop growth simulations was reaffirmed by Boote et al. (1983), who indicated how a soybean crop growth model might easily be adapted for that purpose.

However, simulation models in which a very detailed nematode submodel was coupled to an oversimplified crop model remained the trend (e.g. Bird et al. 1985; Caswell et al. 1986; Schmidt 1992; Van Haren et al. 1994). An exception was the model of Ward et al. (1985), who incorporated the dynamics of potato cyst nematodes into the sophisticated crop growth model SUCROS (Van Keulen et al. 1982). Unfortunately, they did not take full advantage of their crop growth simulation in that they restricted the effects of the nematodes to altered assimilate partitioning, while neglecting effects on photosynthesis and leaf senescence. SUCROS was later used by Den Toom (1990) to examine hypotheses about relations between populations of the ectoparasitic nematode

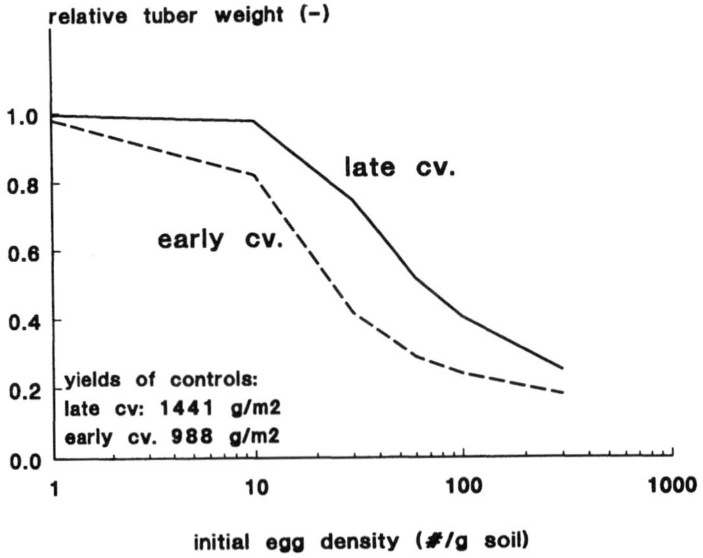

Figure 1. Simulated final tuber yields at different levels of soil nematode density, for an early and a late maturing cultivar. (Source: Schans, 1993).

Tylenchorynchus dubius and damage to *Lolium perenne*. Den Toom's model study showed that *T. dubius* probably impaired the growth of its host by obstructing water uptake.

Wallace (1987) reviewed the literature on the effects of nematodes on photosynthesis and strongly advocated incorporating the effects into existing crop growth models. This was finally done by Schans (1993) who simulated the population dynamics of potato cyst nematodes in great detail in his model. Again, SUCROS was used to simulate host growth. Schans (1991) had earlier concluded from experiments that invasion of roots by second stage juvenile potato cyst nematodes triggers an hormonal signal from root to shoot, causing stomatal closure and impairing photosynthesis. He incorporated this damage mechanism in his model by decreasing photosynthetic rate proportionately to the ratio of root penetration rate to leaf area index. By using this ratio he accounted both for stimulation of hormone production in penetrated roots and for dilution of the hormonal signal in the leaf tissue. Schans used his model to explain several observations on nematode-related damage in potato crops. His model accounted very well for the relatively high level of tolerance of late maturing cultivars (Figure 1). The physiological explanation for the tolerance by late cultivars, brought out nicely by the model, was that late cultivars produce a greater leaf area that dilutes the hormone more than in early cultivars, and that late cultivars only start tuber bulking when most of the root invasion by nematodes has ended.

Schans' model is very complex but still lacks some of the damage mechanisms identified by Haverkort and Trudgill (1995). In the remainder of this chapter we will show how a simpler model can be developed, which can give a more comprehensive treatment of damage, despite its simplicity.

Physiological effects to be incorporated in a model for potato cyst nematodes

Crop growth rate can be analysed as the product of light interception by the crop and the efficiency with which intercepted light is used to produce biomass (Monteith 1977). This type of analysis has also proved useful to study the effects of potato cyst nematodes on potato crop growth (Trudgill et al. 1990; Haverkort and Trudgill 1995). Nematodes primarily affect leaf area dynamics, i.e. light interception, but light-use efficiency is reduced as well. Haverkort and Trudgill (1995), reviewing the literature, identified four mechanisms through which cyst nematodes affect leaf area dynamics and LUE of potato crops:

1. Nematode infection reduces the specific leaf area (SLA);
2. Allocation of assimilates to growth of leaves and stems is decreased in favour of allocation to the root system;
3. Leaf senescence is accelerated;
4. Photosynthesis is affected resulting in a decreased LUE.

The extent to which these four mechanisms contribute to loss of yield will differ, so a simulation model need not necessarily contain all four. In fact, we will show

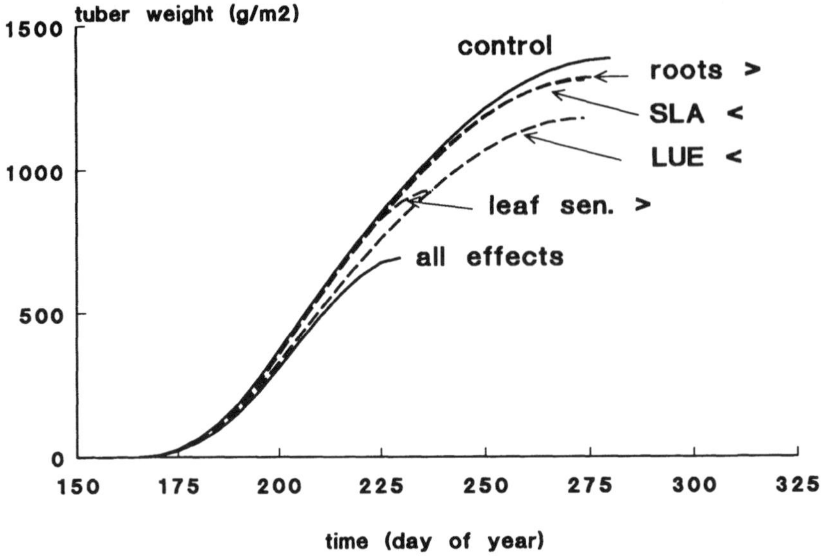

Figure 2. Simulated time course of tuber dry matter growth in a potato crop affected by various numbers of damage mechanisms: zero ("control"), one (increased allocation to roots, decreased SLA or LUE or accelerated leaf senescence: "roots>", "SLA<", "LUE<" and "leaf sen.>", respectively) or all four damage mechanisms simultaneously ("all effects").

how to use a simple simulation model to quantify the importance of the four mechanisms. For that purpose we applied the potato crop growth simulation model LINTUL (Spitters and Schapendonk 1990, where a program listing can be found). In LINTUL, crop growth rate is calculated as the product of light interception and LUE, both assimilate partitioning and leaf senescence depend on thermal time, and leaf area is calculated from leaf weight using a constant SLA. For the magnitude of the four damage effects we took a 10% reduction of SLA and LUE, and a doubling of leaf senescence rate and allocation to roots. These changes are typical for potato crops growing on heavily infested soils (Haverkort and Trudgill 1995; Van Oijen et al. 1995a). The model was not extended with a nematode submodel or with calculation of root length density, since the effects of nematodes were treated as forcing functions. A similar approach, also using standard crop growth models to identify the major damage mechanisms in a plant-nematode system, was used by Boote et al. (1993) for soybean root-knot nematodes.

The simulation results are shown in Figure 2. Obviously, acceleration of leaf senescence and, to lesser extent, reduction of LUE, account for most of the final yield loss. We conclude that the effect of nematode infection on SLA need not be incorporated in a potato cyst nematode model. Stimulated allocation of assimilates to roots also has little *direct* effect on crop growth potential (Figure 2). However, if we plan to couple the crop growth model to a simulation model of nematode population dynamics, root growth should be accurately simulated so the effect of nematodes on allocation of assimilates to roots must still be incorporated. For the same reason the effect of nematodes on the acceleration of root senescence (not discussed by Haverkort and Trudgill 1995) should also be included in the model.

A simple simulation model for potato cyst nematodes

We will now show how the damage mechanisms, listed in the previous paragraph, can be incorporated in a simple crop growth model. A preliminary version of such a model has been presented before (Van Oijen et al. 1993). For the crop growth model we choose LINTUL, already referred to above. A complete listing of this model is given by Spitters and Schapendonk (1990), but for convenience the six main equations of LINTUL are repeated here (see Table 1: Eqs. 1–6).

To calculate the rate of penetration of roots by second-stage juveniles we added three equations to the model (Table 1: Eqs. 7–9), which in fact constitute a simplified representation of the model of population dynamics given by Schans (1993). These equations are based on the following three assumptions: (1) the specific root length is constant; (2) root length is randomly distributed in the soil, and (3) rate of penetration of the root system by second-stage juveniles is proportional to both the nematode density in the soil and the rate of increase of soil volume colonized by roots.

Finally, we need four more equations to quantify the damage mechanisms selected in the previous paragraph (Table 1: Eqs. 10–13). We will consider these in more detail. Firstly, the acceleration of leaf senescence is assumed to be proportional to the ratio of crop growth rate and root length density (Eq. 10). This reflects the fact that nematode-induced reduction of root length may lead to an insufficient supply of nutrients to sustain growth and maintain leaf area (Haverkort and Trudgill 1995).

Secondly, we assume that root death is proportional to the density of recent penetrations by second-stage juveniles in the root system (Eq. 11). It is possible that later stages of the nematodes present in the roots may also contribute significantly to root damage. We did not quantify this possibility in the model since it would have involved increasing model complexity without major effects on model behaviour.

The third and fourth damage mechanisms incorporated in the model are increased allocation of assimilates to the roots (Eq. 12) and decreased LUE (Eq. 13). We assume that both mechanisms depend on the ratio of penetration of roots to the leaf area index. This represents, as in the model of Schans (1993), the hypothesis that both effects are caused by an hormonal messenger from roots to the shoot, the production of which, probably, is proportional to root penetration rate.

Obviously, the damage mechanisms could be modeled in a more comprehensive manner, by adding state variables for plant nutrient and

Table 1. Basic equations for a simple simulation model of growth of potato crops infested by cyst nematodes

(1)	LAI	$= W_l \times SLA$
(2)	$I_{intercepted}$	$= I_0 \times (1 - e^{-k \times LAI})$
(3)	CGR	$= I_{intercepted} \times LUE$
(4)	$d\ W_l/dt$	$= f_l \times CGR - Senescence_l$
(5)	$d\ W_r/dt$	$= f_r \times CGR - Senescence_r$
(6)	$d\ W_t/dt$	$= f_t \times CGR$
(7)	L_r	$= W_r \times SRL$
(8)	$V_{rhizosphere}$	$= V_{total} \times (1 - e^{-Lr \times c/Vtotal})$
(9)	Penetration	$= k_1 \times P_i \times d\ V_{rhizosphere}/dt$
(10)	$Senescence_l$	$= k_2 \times CGR/L_r$
(11)	$Senescence_r$	$= k_3 \times Penetration/W_r$
(12)	f_r	$= k_4 \times Penetration/LAI$
(13)	LUE	$= LUE_{max} - k_5 \times Penetration/LAI$
Where:	LAI	= Leaf Area Index ($m^2\ m^{-2}$)
	W_l, W_r, W_t	= Weight of leaves, roots, and tubers ($g\ m^{-2}$)
	SLA	= Specific Leaf Area ($m^2\ g^{-1}$)
	I_0, $I_{intercepted}$	= Incident and intercepted light ($MJ\ m^{-2}\ d^{-1}$)
	k	= Light extinction coefficient (–)
	CGR	= Crop Growth Rate ($g\ m^{-2}\ d^{-1}$)
	LUE, LUE_{max}	= Light-Use Efficiency under actual and optimum conditions ($g\ MJ^{-1}$)
	f_l, f_r, f_t	= Partitioning coefficients, dependent on thermal time (for f_r see also Eq. 12)
	L_r	= Root length ($m\ m^{-2}$)
	SRL	= Specific Root Length ($m\ g^{-1}$)
	$V_{rhizosphere}$, V_{total}	= Volume of rhizosphere and total rooting zone ($m^3\ m^{-2}$)
	c	= Cross-sectional area of the rhizosphere surrounding the roots (m^2)
	$Senescence_l$, $Senescence_r$	= Senescence rate of leaves and roots ($g\ m^{-2}\ d^{-1}$)
	Penetration	= Rate of penetration of roots by second-stage juveniles ($m^{-2}\ d^{-1}$)
	P_i	= Soil nematode population density (m^{-3})
	k_1, k_2, k_3, k_4, k_5	= proportionality constants (–, $m\ m^{-2}$, $g^2\ m^{-2}$, $m^2\ d$, $g\ MJ^{-1}\ m^2\ d$)

hormonal status and for nematodes in another than the second stage. However, we refrained from doing so because of the scarcity of data on these processes, the associated problems with model parameterization, and the experience that added complexity need not necessarily improve model performance.

Preliminary tests of the model

The first test of the model would be to check its ability to simulate plant-nematode relations observed in the field. For this purpose we collected

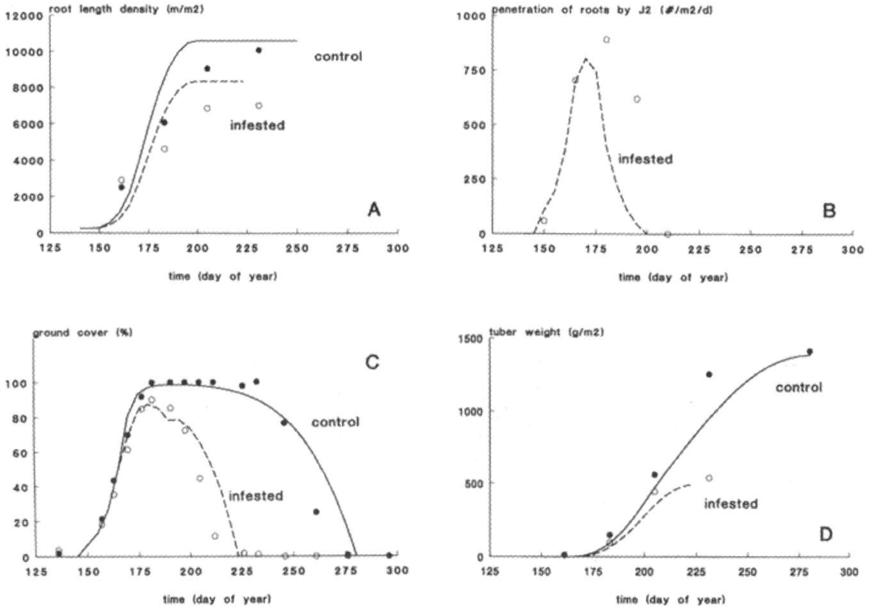

Figure 3. Simulated time courses for cultivar Darwina on infested soil and in a control treatment (points: experimental observations, lines: simulations).
A: root length; B: root penetration by second-stage juveniles; C: ground cover by green foliage; D: tuber dry matter growth.

information from a field experiment conducted in 1991 on a sandy soil infested with *G. pallida*, in the North of the Netherlands (Van Oijen et al. 1995a,b). In that experiment cultivar Darwina was subjected to the prevailing high nematode density (about 15 living juveniles cm^{-3}), and to a control treatment where most nematodes had been killed by nematicide. The model used local weather as input, and the crop growth parameters were maintained at standard values for Dutch growing conditions. The parameters that quantify the plant-nematode relationships (k_1-k_5 in Eqs. 9–13 of Table 1) were fitted to the measurements. Therefore, in this simulation only the control treatment represents an independent check of the model, or rather of its crop growth part, while the simulation of the interaction with the nematodes can only show the ability of the model to capture the complex response of the crop to the nematode in a few simple equations. The results of the simulations are given in Figure 3. We consider that the model simulates the time courses of root length, penetration of roots, ground cover and tuber growth quite well, for both the control treatment and the treatment with nematode infestation. These results do not suggest the need for a more complex model.

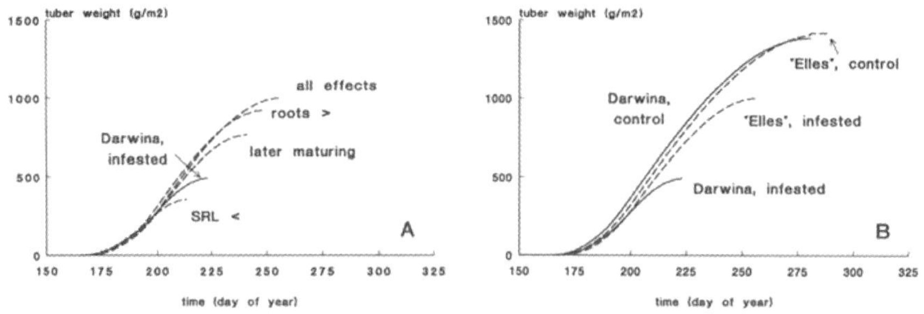

Figure 4. Simulated time courses of tuber growth for the nematode-sensitive cultivar Darwina, with or without hypothetical changes in plant characteristics. Each change corresponds to an observed characteristic of the nematode-tolerant cv. Elles.
A: the effect on growth in infested soil of changing one plant characteristic (increased allocation to roots, delayed maturation, decreased specific root length (SRL)), or all three combined; B: the effect of changing all three characteristics, simultaneously, on growth in infested soil and in a control treatment.

Application of the model to the evaluation of tolerance characteristics

As a further test of the model we analysed its ability to explain the high level of nematode-tolerance of cultivar Elles, from knowledge of plant characteristics. Elles differs from cultivar Darwina by virtue of its later maturity, its greater root-shoot ratio and its lower specific root length. In contrast to cultivar Darwina, Elles hardly suffered any loss of yield under the experimental conditions described earlier for cultivar Darwina (Van Oijen et al. 1995a). We used the simulation model to see whether the typical characteristics of cultivar Elles fully accounted for its maintenance of yield. The results of the simulation are shown in Figure 4. It seems that allocating a greater proportion of biomass to roots is indeed advantageous, as is the late maturation. Regarding the effect of late maturation, our simple model thus confirms the results of the complex model of Schans (Figure 1). The low specific root length of cv. Elles, on the other hand, only increases loss of yield. When all three characteristics are combined in one model run, we see that loss of yield is much reduced compared to cultivar Darwina. However, the simulated tolerance of nematodes is still only partial, whereas the measurements for cv. Elles showed complete tolerance. The difference between measurements and simulations can be explained in various ways. Cultivar Elles may show less stimulation of hormone production in response to nematode penetration, or simply be less sensitive to the signal. There are no data to prove or disprove this. Alternatively, root death and associated reduction of tissue nutrient concentration may be less pronounced in cultivar Elles than in Darwina, although this is not confirmed by experimental data (Van Oijen et al. 1995b). Whichever explanation is true, and irrespective of the validity of our model, the model does already indicate hypotheses for investigation.

Concluding remarks

In this chapter we have concentrated on the use of simulation models for examining the physiological processes underlying nematode damage, and the differences among cultivars therein. A simple model was shown to be useful in identifying the major damage mechanisms, and in assessing the importance of various host characteristics for tolerance of nematodes. The simulation models can, of course, also be used to investigate the effects of the environment on the damage relations. The relation between pre-planting density of cyst nematodes and final tuber yield is subject to many interactions with environmental factors and is, therefore, very variable (Trudgill 1986). Barker and Noe (1987) suggested the use of simulation models of plant-nematode systems to explain the variation.

We have shown that there is no need for explicitly incorporating nutrient or hormone relations into a model for potato cyst nematodes, even though the underlying damage mechanisms affect those relations. If one wants to use our model to explain observed interactions between, e.g. effects of nematode density and fertilization level (Trudgill 1980; 1986), it seems prudent to reparameterize the modeled relation between root length and rate of leaf senescence rather than to increase model complexity. It is our intention to apply the model in that manner.

Acknowledgements

We thank N. van Dijk for technical assistance in the preparation of this manuscript.

A growth model for plants attacked by nematodes

C.H. SCHOMAKER[1], T.H. BEEN[1] and J.W. SEINHORST[2]

[1] *DLO Research Institute for Plant Protection (IPO-DLO), P O Box 9060, 6700 GW Wageningen, The Netherlands*

[2] *Middelberglaan 6, 6711 JH Ede, The Netherlands*

Abstract. The relation between small and medium initial population densities and the relative total plant weight was derived as cross sections at right angles to the time axis of a growth model with three dimensions time after planting t, relative total plant weight Y and relative growth rate r_p/r_0 The relative growth rate is the (constant) ratio between the growth rate r_p of plants of a certain weight at a nematode density P and the growth rate r_0 of (younger) plants of the same weight without nematodes Therefore, the ratio between the time after planting that plants need to reach a certain weight in the absence of nematodes and at nematode density P, $t_0/t_p = r_p/r_o$ (2)

The relative growth rate $r_p/r_o = k + (1 - k)0\ 95^{P/T-1}$ for $P > T$ and $= 1$ for $P \leq T$ (3) Formally, k is the minimum relative growth rate as $P \rightarrow \infty$ As a result the arbitrary equation $y = m + (1 - m)0\ 95^{P/T-1}$ for $P > T$ and $= 1$ for $P \leq T$ (6) also applies to the relation between small and medium initial population densities and relative total plant weight T is the tolerance limit, below which growth and yield are not reduced by nematodes, m is the relative minimum yield

The relations between small and medium initial population densities of potato cyst nematodes and relative tuber weight of potatoes can be derived from the growth model in an analogous way However, there is one complication tuber initiation does not start at the same haulm weight in plants with and without nematodes, but at the smaller haulm weight the larger the nematode density As a consequence, tuber weights of plants with a certain total weight at nematode density P are not equal to those of plants with the same total weight without nematodes, but $r_p\Delta t$ units of weight larger, Δt being the difference between the actual time of tuber initiation and the time total plant weight becomes the same as that of plants without nematodes at the initiation of tuber formation

Relative total and tuber weights of plants with 'early senescence' and at large nematode densities are smaller than estimated by the model and Eq 2 This indicates that at large initial population densities growth reducing mechanism(s) become active that were not operating at smaller densities

Introduction

> *I shall take the simpleminded view that a theory is just a model of the universe or a restricted part of it, and a set of rules that relate quantities in the model to observations that we make. It exists only in our minds and does not have any other reality (whatever that may mean). A theory is good theory if it satisfies two requirements: It must accurately describe a large class of observations on the basis of a model that contains only few arbitrary elements, and it must make definite predictions about the results of future observations. For example,*

A J Haverkort and D K L MacKerron (eds), Potato Ecology and Modelling of Crops under Conditions Limiting Growth, 197–214
© 1995 *Kluwer Academic Publishers*

Aristotle's theory that everything was made out of four elements ( ) was simple enough to qualify, but did not make any definite predictions. On the other hand, Newton's theory of gravity was based on an even simpler model ( ). Yet it predicts the motions of the sun, the moon and the planets to a high degree of accuracy. (Stephen Hawking 1988, A brief History of Time).

In many biological and nematological studies one is not very particular about these requirements. Several nematologists or biologists have formulated mathematical relations between pre-plant nematode densities and crop yield that were indeed based on observations – although mostly limited ones – but that lacked a theoretical base and predictive value. Examples are the linear and log-linear functions (Lownsberry and Peters 1956; Hesling 1957; Seinhorst 1960; Hoestra and Oostenbrink 1962; Brown 1969), the quadratic curve (Peters 1961); the inverse linear function (Elston et al. 1991) and an exponential function (Van Haren et al. 1993). These equations do not represent a vision of the mechanisms involved in nematode-plant interactions. Calculations of limits of variables or parameters in some of them produce bizarre results.

Others (e.g. Evans and Haydock 1990; Trudgill 1992) tried to build theories on the effect of nematode attack on plants without using a model. This approach makes it difficult to study a (biological) problem methodically or to detect flaws in one's chain of reasoning. Moreover, there is no efficient transfer of these theories to fellow-nematologists for further development or criticism.

An approach which at least partly satisfies one of Hawking's requirements is the 'comprehensive simulation model'. This type of model tries to describe an often-limited research area with detailed information in a large number of sub-models and parameters. They were first applied in nematology by Ward et al. (1985) in a preliminary model on the population dynamics of potato cyst nematodes and were later extended by Schans (1993), who tried to describe the population dynamics of potato cyst nematodes and the associated damage to potatoes with information about some plant physiological processes and their changes with environmental conditions. Only because of the arbitrary nature of the sub-models and the multitude of parameters, which, although treated as constants, are more often variable than not, these models did not contribute much to theorem building and have a limited predictive value. In the model of Schans this was the more so as description of plant physiology was confined to short-term measurements of photosynthesis, respiration and transpiration on one or a few leaves, which did not explain differences in yield, at a limited range of nematode densities which, moreover, were not always constant during plant growth. Besides, as external conditions during crop growth are indispensable data in these models, they are unsuitable as a base for recommendations about control measures for nematode pests, which, unlike those for fungi or insects, must be taken, at the latest, at the time of sowing or planting of the crop to be protected (nematostatics), but generally much earlier (nematicides, crop rotation, resistant cultivars). Therefore, these simulation models would not qualify as good models.

In nematology Seinhorst's models, especially those for potato cyst nematodes (Seinhorst 1986a,b; 1993) come closest to the classical approach recommended by Hawking. The models are based on a theory which describes the principal mechanisms of nematode/nematode and nematode/plant interactions, they don't conflict with a large number of observations in different countries on different nematode species over a large period of time. The integrated, simplified, stochastic versions of the models (Been et al. 1994) deduced from more complicated ones that suit mainly scientific purposes, contain only a few parameters with known distribution functions and allow predictions to be made within sufficiently narrow limits, albeit with some constraints. The predictive value of the models, which makes them suitable for an advisory system for the control of potato cyst nematodes on farmer's fields, will be demonstrated by Been et al. (1994), elsewhere in this book; the theorem of one of Seinhorst's models: the effect of nematodes on growth and yield of plants, is explained in this article.

Causes of yield reduction by nematodes

Crop returns are reduced by nematode attack as a result of reduction of crop weight per unit area (which is mostly equivalent to average weight of marketable product per plant) and reduction of the value of the product per unit weight. For example, carrots attacked by rootknot nematodes (*Meloidogyne*) may be worthless because of branching and deformation of the tap root, although they have the same weight per unit area as carrots without nematodes. Onions of normal weight but infested with few stem nematodes (*Ditylenchus dipsaci*) at harvest will, nevertheless, be lost in storage. Attack of crop plants by potato cyst nematodes (*Globodera rostochiensis, G. pallida*) does not only reduce potato tuber weight, but may also reduce tuber size. However, potato cyst nematodes attacking potatoes and almost all root infesting nematodes attacking crop plants of which the above ground parts are harvested, hardly ever affect the value per unit weight of harvested product, unless crop weight is reduced so much by large initial population densities that growing it was uneconomical from the start. Therefore, prediction of crop reduction by these nematodes can, in general, be based on models of the relation between nematode density at sowing or planting and average weight of single plants at harvest. In the following the term 'yield' will be avoided. The yield in the agronomic sense must be derived from individual plant weights, taking the restrictions mentioned above into account.

To make a model of the relation between population density before planting (P) and the proportion (y) of uninfested plants (onions, flower bulbs) or relative plant weight (y) after a certain period of growth of the plants (the ratio between the yield of plants at density P and the yield of plants without nematodes, grown under the same conditions) a theory was developed about the mechanisms involved. The theory was tested by comparing it with the patterns of observed relations between nematode density and plant growth and plant weight at a certain time after sowing or planting. This also allows the estimation of values of system parameters under various experimental conditions.

Seinhorst's (1965) model for the relation between the population densities of stem nematodes (*Ditylenchus dipsaci*) at sowing and the proportion of infested onion plants is identical to that of Nicholson (1933) for the infection of caterpillar pupae by parasite wasps. The chain of reasoning is as follows: Only infested and non-infested onions are distinguished. The degree of infestation of single plants is irrelevant as only nematode-free onions are marketable. Only three assumptions (which in a slightly different formulation also apply to the root infesting species) were needed to formulate the model:

1. The average nematode is the same at all densities. This means that population density does not affect the average size or activity of the nematodes.
2. Nematodes do not affect each other's behaviour. They do not attract or repel each other directly or indirectly.
3. Nematodes are distributed randomly over the plants in a certain area.

If now a density of one nematode per unit area or weight of soil attacks a proportion d of the onion plants in a certain area (and leaves a proportion $1 - d$ uninfested), then a density of two nematodes per unit area or weight of soil will attack, on the average, a proportion d of already damaged plants (which has no additional effect on yield reduction), plus a proportion d of the still uninfested proportion $(1 - d)$, leaving a proportion $1 - d - d(1 - d) = (1 - d)^2$ of the plants uninfested. Generalized: a nematode density P leaves a proportion

$$y = (1 - d)^P = z^P \qquad (1)$$

of the onions uninfested. In Figure 1 values of z^P are plotted, not against P, but against log P. This has not only the advantage that the shape of the curves is the same for all z ($d(\log P)/dy$ is only determined by y), but also that, if P is estimated by counting nematodes in a soil sample, the variance of log P depends less on true P than the variances of P. The values of z are determined by conditions that affect the efficiency of the nematodes in finding and penetrating plants. In patchy infestations of stem nematodes these conditions for attack appeared to be more favourable in the centre of the patch than towards the borders, resulting in an increase of z with increase of the distance from this centre. This results in persistency of the patchiness. The model also applies when nematodes spread from randomly distributed infested plants to neighbouring ones, which results in overlapping circular patches of infested plants.

Patchy infestations of nematodes do not always develop this way. Infestation foci with potato cyst nematodes, for instance, occur by introduction of cysts in fields: small numbers of cysts are mostly transmitted by seed potatoes and larger numbers of cysts by agricultural machinery. After multiplication on potatoes they are spread by tillage and harvesting machines. As a result the patches have the same shape in all potato growing areas in the Netherlands (Seinhorst 1982; 1988; Schomaker and Been 1992). If good host plants are grown in short rotations, patchy infestations can become larger until the whole field is infested more or less uniformly. Examples of such infestations are found in south-east Groningen and east Drenthe, where starch potatoes are grown every other year.

Reduction of plant growth

Tylenchid root infesting nematodes (for instance cyst nematodes, root knot nematodes and *Pratylenchus* species) generally reduce crop yield in a less direct way than stem nematodes. The rate of growth and development of attacked plants is reduced, resulting in smaller weights than of plants without nematodes at given times after sowing or planting or, in exceptional cases, in reaching the same final weight as of plants without nematodes later. In general such a delay of the ripening of the crop is prevented by the external conditions at the end of the growing season of the plants.

Seinhorst (1979b; 1986b) based a growth model on two simple concepts: one of the nature of the plant; an element that increases in weight in the course of time, and one of the nature of plant parasitic nematodes: elements that reduce the rate of increase of this weight and, in principle, more so the larger the population density. Further the following principles are applied in addition to those mentioned above for the model on the relation between stem nematode density and proportion of onions attacked:

4. Root infesting nematodes are distributed randomly in the soil.
5. Nematodes enter the roots of plants randomly in space and time. Therefore the average number of nematodes entering per quantity of root and time is constant. This number is proportional to the nematode density P (number of nematodes per unit weight or volume of soil).
6. If the growth rate of plants at a given time t after sowing or planting is the increase in total weight (Y) per unit time (dY/dt), represented by r_0 for plants without nematodes and r_P for plants at nematode density P, then the ratio r_p/r_0 for plants of the same total weight (and, therefore, of different age) without nematodes and at nematode density P, is constant throughout the growing period. Further, according to Figure 2, $r_0 = \tan\alpha = \Delta Y/\Delta t_0$ and $r_P = \tan\beta = \Delta Y/\Delta t_p$. Therefore,

$$r_P/r_0 = t_0/t_P \qquad (2)$$

The relation between population density of the nematodes and its total effect on the growth rate of the plants is also considered to be according to Nicholson's competition model (Eq. 1, Figure 1). Eq. 1 is a continuous function for $0 \leq P \leq \infty$. However, all sufficiently accurate observations on the relation between the population density P of various nematode species, including potato cyst nematodes, on the weight of various plant species demonstrate the existence of a minimum density T, below which the nematodes do not reduce plant weight. It may be concluded that at densities $< T$ they also do not reduce the growth rate of plants. This is corroborated by the results of the few experiments

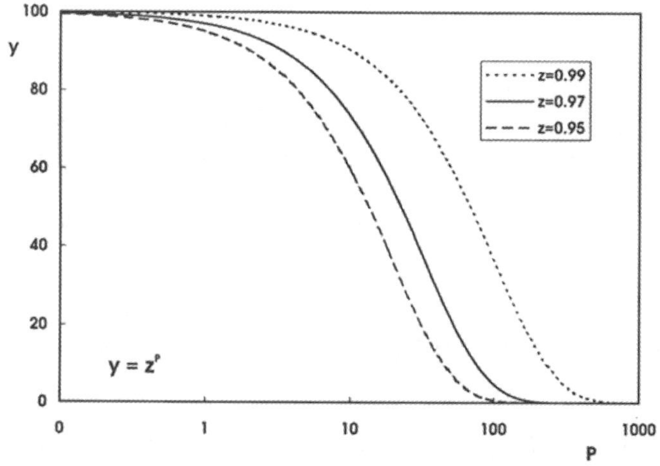

Figure 1. The relation $y = z^P$ (Eq. 1) between the initial population density (P) of stem nematodes and the proportion y of the onions that is not attacked.

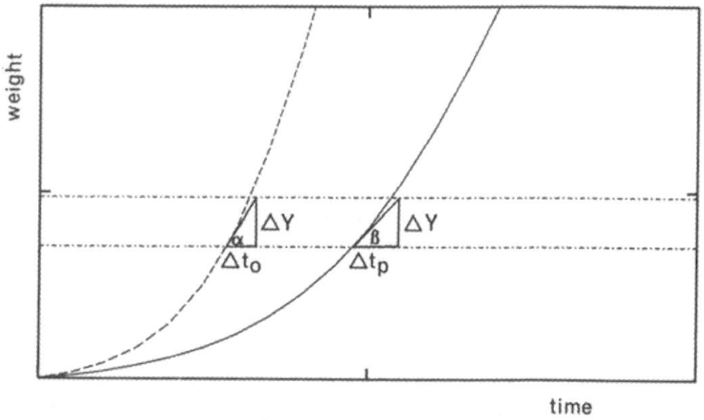

Figure 2. Growth curves of plants without nematodes and at nematode density P. Y = total plant weight; t = time after sowing or planting; t_0 and t_p are the times plants without and with nematodes need to reach the same total weight Y; r_0 (= $\tan\alpha = \Delta Y/\Delta t_0$) and r_p (= $\tan\beta = \Delta Y/\Delta t_p$) are the growth rates of plants without nematodes and at nematode density P respectively.

in which growth rates were actually determined. Moreover, at large nematode densities plant weight approached zero in only few of these experiments and growth rates never did. Therefore, Eq. 1 is adapted by replacing P by $P-T$ and introducing the minimum relative growth rate $r_P/r_0 = k$ for $P \rightarrow \infty$. The second equation constituting the model then becomes:

$$r_P/r_0 = k + (1-k)z^{P-T} \text{ for } P > T$$
$$\text{and} \quad r_P/r_0 = 1 \text{ for } P \leq T \qquad (3)$$

in which z is a constant smaller than 1. The value of k is independent of nematode density and time after sowing or planting, but may vary between experiments. Growth curves of plants for different nematode densities can be derived from a growth curve of plants without nematodes with the help of the Eqs. 2 and 3. These curves may vary in shape, as long as they are continuous and the growth rate decreases monotonously from shortly after sowing or planting. Figure 3 gives an impression of the three dimensional model with axes for total plant weight *Y*, relative nematode density *P/T* and time after planting *t* for a given value of *k*.

A simple model for the relation between nematode density and relative plant weight

The primary results of experiments are almost always weights of plants attacked by known nematode densities at a given time after sowing or planting. To investigate whether these relations are in accordance with the model, they must be compared with cross sections orthogonal to the time axis of the model, through growth curves of plants for ranges of densities *P/T* and different values of *k*. These cross sections were in close accordance with the now arbitrary equation:

$$y = m + (1 - m)z^{P-T} \text{ for } P > T \qquad (4)$$

in which *m* is the minimum relative plant weight, *z* is a constant < 1 with the same or a slightly smaller value than in Eq. 3 and *T* is the tolerance limit with the same value as in Eq. 3.

Seinhorst (1986b) gives the relations in pot experiments between *P/T* of different tylenchid nematode species, including *Heterodera avenae* (Seinhorst 1981), *G. rostochiensis* and *G. pallida* (Greco et al. 1988) and the relative dry plant weight (*y*) several months after sowing or planting. Values of $y' = (y - m)/(1 - m) = z^{P-T}$ (5) in thirteen pot experiments are plotted against *P/T* (Figure 4). The relation between average *y'* and *P/T* is in close accordance with $y' = z^{P-T}$ which suggests that the deviations of individual observations in the different experiments from values according to this relation were due to experimental error. Average relative plant and tuber weights of different resistant and susceptible potato cultivars at different times after planting were in close agreement with those according to Eq. 5 with *T* = 1.8 eggs/g soil. As Eqs. 4 and 5 are identical, we may conclude that the results of Figure 4 and the other experiments are well described by the model, which apparently applies to tylenchid nematodes that feed and multiply in as different ways as cyst nematodes, *Pratylenchus penetrans* and *Tylenchorhynchus dubius*. It also indicates that the general relation between nematode density of tylenchids and relative plant weight of attacked plants some time after sowing or planting is independent of the shape of the growth curve of plants without nematodes and of external conditions. Although values of *z* and *T* differed between nematode species and plant species and *m* also between experiments with the same

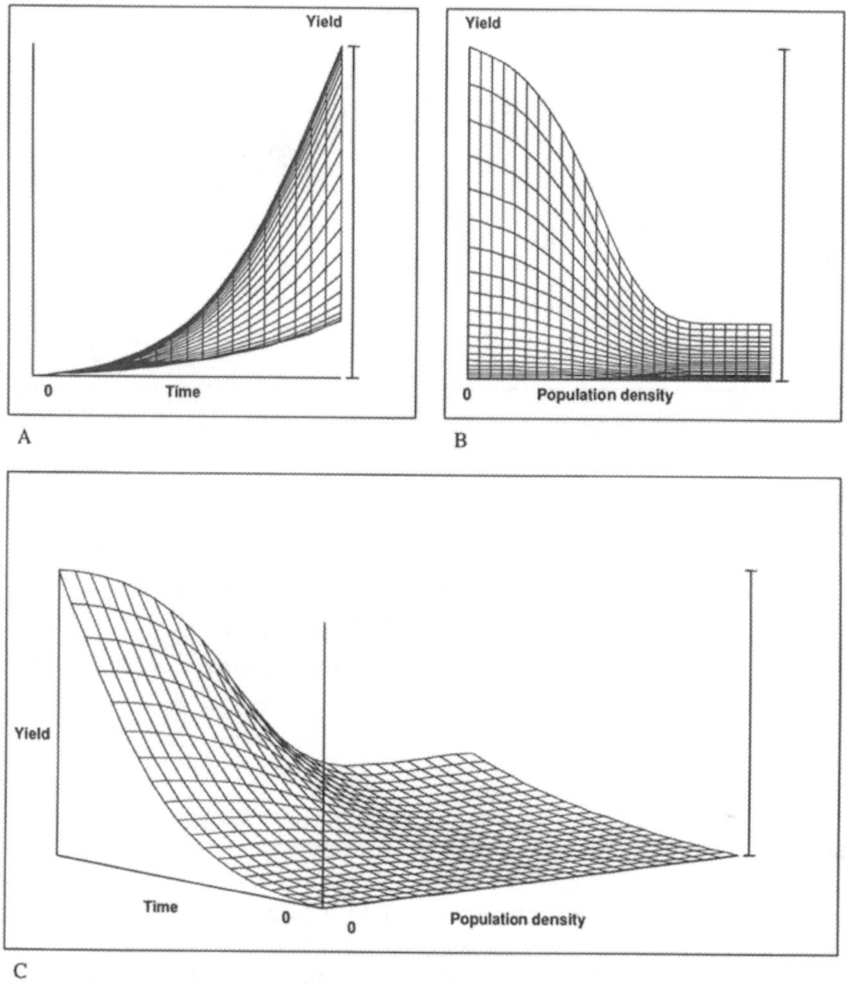

Figure 3. Surface plots of the three-dimensional model which represents the relation between total weight (Y) and nematode density P as cross sections at right angles with the time axis (t) of growth functions of plants at different nematode densities, $Y(r_p,t)$, and that without nematodes, $Y(r_0,t)$:
a at 0° rotation, which shows the relation between Y and t at different nematode densities.
b at 270° rotation, which exposes the relation between Y and population density P.
c at 230° rotation, which illustrates the relation between (Y,P) and (Y,t).
All rotations are clockwise. The growth rates of plants of the same weight without nematodes (r_0) and at nematode densities P (r_p) are related by Eqs. 1 and 2.

combination of nematode and plant species, z^T differed too little from 0.95 between experiments to distinguish the variation from that caused by experimental error. Therefore, for fitting curves according to Eq. 3 to experimental data, z^T is generally assumed to be 0.95, which transforms Eq. 4 to:

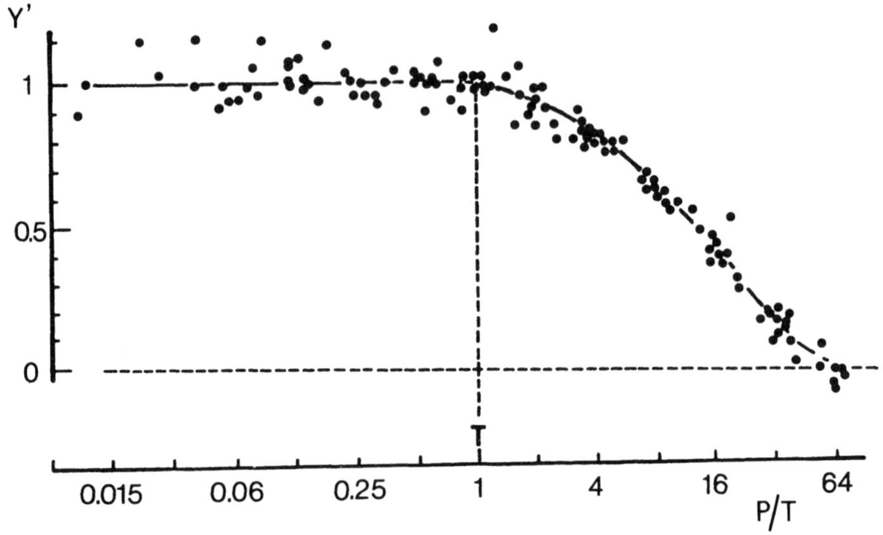

Figure 4. The relation between nematode density *P/T* at sowing or planting and the relative weight *y'* of plants after a certain growing time in thirteen experiments with several tylenchid nematode species and several plant species. $y' = (y - m_x)/(1 - m_x) = 0.95^{P/T-1}$, in which $y = m_x + (1 - m)0.95^{P/Tx-1}$ for the *x*-th combination of nematode species and plant species and all values T_x are coinciding.

$$y = m + (1 - m)0.95^{P/T-1} \tag{5}$$

The value of *T* depends on nematode species and pathotype and on the plant species, but seems in general not to be affected by external conditions. For potatoes with potato cyst nematodes, planted in spring, *T* was about 1.8 eggs/g soil. An exception is the sensitivity to daylength of the value of *T* for potato cyst nematodes on potatoes (Been and Schomaker 1986; Greco and Moreno 1992). In contrast, *m* for a given nematode species varies between cultivars or plant species within experiments and for the same cultivar between experiments under the influence of so far unknown external conditions. It is not established whether there are consistent differences between cultivars or even plant species. Therefore, *T* is the main measure of the degree of tolerance of a plant species or cultivar to a certain nematode species or pathotype, whereas the importance of *m* remains to be investigated. The shortcoming of most discussions on differences in tolerance between cultivars or crop plants is, that they do not refer to these two parameters.

Implications of the model

The similarity between very different nematode species in their reaction upon nematode attack is not restricted to its effect on the size of the plant. Nematode densities up to at least 16*T*, but generally more than 32*T*, affect neither water consumption per unit weight of plant and per unit duration of time, nor dry matter content (Seinhorst 1981). There are indications that this also applies to shoot to root weight ratios (Seinhorst 1979a; Been and Schomaker 1986). The only difference between plants without nematodes and those at small and medium nematode densities of the same weight stated so far is that the latter may be slightly taller.

The model implies comparison of growth rates, as affected by nematode attack, of plants of the same weight and not of the same age, because the latter also differ in developmental stage. The importance of this consequence of the model is evident from the investigation on the effect of *H. avenae* on the shoot to root ratio of young oat plants (Seinhorst 1979a) and on the relation between weight of potato plants with and without *G. pallida* and nitrate and potassium content (Been and Schomaker 1986).

Seinhorst (1986b) rated the mechanisms by which nematodes should reduce growth rates of plants (extraction of nutrients, mechanical damage to root tissue resulting in a hampered uptake of water and minerals, obstruction of plant vessels causing wilting, extraction of food), suggested in the literature, as myths and fairy tales. A more probable mechanism is the production of a growth reducing substance only during the penetration of the nematodes in the roots but not any more when they have settled. Because of the constant number of nematodes penetrating per unit quantity of root and per unit duration of time, the growth reducing stimulus will then remain constant per unit weight of plant. It remains to be investigated how the reduction of the efficiency per nematode with the increase of population density comes about. Further major problems to be investigated are the way growth reduction is prevented up to a certain nematode density and what determines the ratio *k*, keeping it independent of *T*.

Strictly taken, the model only applies to nematode species with a single generation per growing season, as of potato cyst nematodes, oat cyst nematodes and *Meloidogyne naasi*. However, Figure 3 demonstrates that it also applies at least to nematodes with small rates of reproduction (e.g. ten to twenty fold in a growing season). Seinhorst (in press) shows that the reason why may not be simple.

The effect of potato cyst nematode attack on tuber growth

After the start of tuber initiation the rate of weight increase of the haulms decreases strongly to change finally to a decrease. In the model increase of haulm weight is assumed to stop shortly after the initiation of tuber growth, after which increase of total weight is due entirely to that of tuber weight.

If the effect of nematode attack on potato plants would only be retardation of development, tuber initiation of plants without and with nematodes would start at the same haulm (= total) weight of plants, irrespective of the retardation of the development. If t_{s0} is the time tuber initiation of plants without nematodes starts, then tuber initiation at nematode density *P* would start at a time

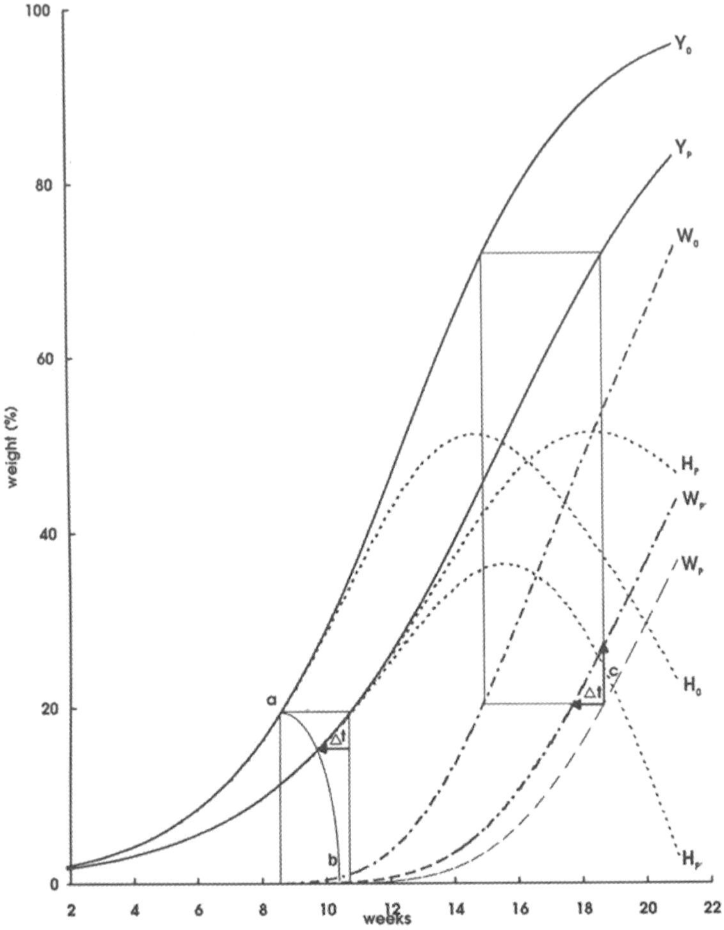

Figure 5. Growth curves for total, haulm and tuber weight of potatoes with and without nematodes. The growth rates for plants of the same total weight or the same tuber weight (measured some time after tuber initiation) with and without nematodes are related by Eq. 2: $r_p/r_0 = t_0/t_p$. The curves for tuber weights of plants of the same total weight with and without nematodes are related by $W'_P = W_0 + r_p\Delta t$. In the figure, the curve ab represents the relation between haulm weight and the time tuber growth starts; c = $r_p\Delta t$; Y_0 and Y_p represent total weight of plants without nematodes and at density *P* respectively; W_0 and W_p are tuber weights of plants without nematodes and at nematode density *P* if tuber initiation had started at the same haulm weight as in plants without nematodes; H_0 *and* H_P are haulm weights of plants without nematodes and at nematode density *P*, if tuber intiation had started at the same haulm weight as in plants without nematodes. H_P represents the actual haulm weight at nematode density *P*.

$$t_{sP} = r_0\, t_{s0}/r_P \qquad (6)$$

However, the ratio between tuber weights of plants of different age but of the same total weight at nematode density *P* and without nematodes (W_P/W_0) increases most probably from a density *T*, and there are indications that the initiation of tuber growth is delayed less than according to Eq. 7 (Seinhorst et al. submitted) and therefore starts at a time

$$t'_{sP} = (r_0\ t_{s0}/r_P)\ -\ \Delta t \qquad (7)$$

Δt increases with r_P/r_0 in such a way that t'_{sP} approaches a maximum of t_{s0} + about 2 weeks, possibly because the external conditions inducing the initiation of tuber growth become stronger than the delaying effect of nematode attack (relation between r_P/r_0 and t'_{sP} according to line *ab* in Figure 5). As a result tuber weights of plants at nematode density P (W_P') are $r_p\ \Delta t$ units larger than tuber weights of plants without nematodes of the same total weight.

The relation at given times t between nematode density and relative tuber weights in a model, constructed as described above, are again according to Eq. 6 up to $P/T = 50$ eggs/g soil and with $k > 0.4$ ($m > 0.05$) from the time tuber to total dry weight ratios larger than 0.5 are reached and with the same value of T but a smaller one of m than for total plant weight. Within these ranges of tuber to total dry weight ratio's and values of P/T and k actual relations between nematode density and relative tuber weights were according to those in the model, if not affected by growth reduction by other mechanisms than that described by the model.

More than one mechanism of growth reduction

According to Seinhorst (1981) nematodes cause two kinds of growth reduction: the 'first mechanism of growth reduction' operating at all population densities, and a 'second mechanism of growth reduction' additional to that of the 'first mechanism', with a noticeable effect only at medium to large nematode densities. The model only applies to growth reduction caused by the 'first mechanism' that retards growth of plants and, occasionally, also increases the length of haulms compared to those of plants of the same weight without nematodes. As long as only the first mechanism is active, water consumption during short periods is proportional to plant weight and, therefore, relative water consumption at different nematode densities and times after sowing or planting is a measure of relative plant weight. Actual plant weights are these relative weights times the actual weights of plants of the same age without nematodes, determined at the same time (Seinhorst 1981).

The 'second mechanism' reduces water consumption per unit plant weight and the (active) uptake or excretion of K^+ and Na^+ and increases the (passive) uptake of Ca^{2+} and dry matter content of plants (Seinhorst 1981; Been and Schomaker 1986). There probably is a negative correlation between age of the plant and nematode density at which the effect of the 'second mechanism' becomes noticeable. For potato cyst nematodes this density rarely is as small as $16T$, but more commonly, also for other nematode species on other plant species $> 32T$. Contrary to the 'first mechanism' it tends to advance the initiation of tuber growth.

A 'third mechanism of growth reduction' is 'early senescence' of potato plants attacked by *G. pallida*: a sudden ending of the increase of haulm length and weight. The time after planting at which it occurs may be negatively correlated with nematode density. The earliest occurrence was nine weeks after planting in the early cultivar Ehud and the smallest nematode density 25*T* in cultivar Darwina. Not all cultivars are equally sensitive (Seinhorst et al. submitted). The cause of 'early senescence' is unknown.

The estimation of *t* and *m*

There are strong indications that the value of *T* for potatoes planted in spring is not affected by differences in external conditions and can, therefore, be determined in pot experiments in both greenhouse and (much more laborious) field experiments. The only requirement of greenhouse tests is that large enough pots are used to guarantee about the same root density in the soil as in the field and to prevent the plants from becoming pot bound, which affects the relation between nematode density and plant weight and obscures the true value of *T* (Seinhorst and Kozlowska 1977). According to the model and the results of pot experiments, *T* for total weight, a short time after planting, suffices as an estimate of *T* for final tuber weight, which allows the use of smaller pots. Experiments should be done with ranges of nematode (egg) densities with ratios not larger than 1 to 2 between successive densities and a sufficient number of densities $< T$ and pots to provide an accurate estimate of plant weight at $P < T$. The largest density in the range should be about 30*T*. The accuracy of the estimates is mainly a matter of uniformity of plant material and growth conditions (light, water content) and carefully filling and handling of the pots.

At best *m* is characteristic of a combination of nematode and plant species with a very large variability between experiments. Therefore, a small number of tests in greenhouses, as will suffice to obtain a reliable estimate of *T*, will not produce such an estimate for a combination of a potato cultivar and a potato cyst nematode pathotype. However, it may be possible to derive differences, if they exist, between host varieties from the results of greenhouse experiments. In addition, of at least some host varieties per nematode pathotype, a sufficient number of values of *m* must be estimated in field experiments to establish a distribution function.

As growing potatoes at large potato cyst nematode densities is uneconomical, the relation between nematode density and growth and yield reduction by the 'second mechanism' can be ignored in a model for advisory purposes.

Estimation of *T* and *m* in field experiments is much more laborious than estimation in pot experiments. The same range of densities is needed and nematode density must be the only variable. Ranges of nematode densities in patches of potato cyst nematode infestation (but not necessarily of other species) come closest to this requirement, if free of other causes of variation of tuber yield with intractable spatial distributions. Ranges of nematode densities cannot be created by applying different dosages of nematicide or other biocide, as has often been done, as this has unpredictable effects on crop yield, other than by killing nematodes. The ranges of nematode densities, that are of interest for the estimation of the parameters in question, must be determined in sufficiently large samples from each field plot to guarantee a coefficient of variation of egg counts not larger than 15% (density differences 1:2 just distinguishable). According to Seinhorst (1988) soil samples of 4 kg per plot are then needed to estimate population densities of 1 egg per g of soil, given a coefficient of variation of the number of eggs per cyst of 16% and a negative binomial distribution of egg densities in samples from small plots with a coefficient of aggregation of 50 for a kg soil. As the coefficient of variation per unit weight of soil is negatively correlated with nematode density, required sample size also is. For instance, to estimate densities of 0.5 or 0.25 eggs/g soil with the same accuracy, eggs from soil samples of 10 and 20 kg must be counted respectively. Another requirement is that plots must on the one hand be small (e.g. 1 m^2) to reduce the effect of medium scale density variation, whereas on the other hand a large enough area per small density interval must be available to guarantee a small variability of tuber weight per unit area. Again, there must be a sufficient number of plots at densities smaller than T to estimate the maximum yield accurately.

Therefore, to avoid unnecessary handling of large soil samples, T and m should, as much as possible, be estimated from greenhouse experiments. Field trials are most efficiently used to confirm of falsify these estimates or ratios of estimates for different combinations of pathotypes and cultivars under more natural external conditions.

Final considerations

As a result of the striking conformity between the effects of all root infesting nematodes investigated upon the growth of attacked plants, whatever the host status of the plants, the ways the nematodes attack, and the reactions of the affected tissues, the growth model discussed above applies to small and medium population densities of all tylenchids. It also results in the same formal relation during the year of sowing or planting between nematode density at sowing or planting and relative total plant weight and, in potatoes, also relative tuber weight, from some time after planting, as long as later generations of species with large rates of reproduction (e.g. *Meloidogyne* species) do not cause additional growth reduction. This is corroborted by a large number of pot experiments and the few field experiments, that were suitable for the purpose. It allows a characterization of the sensitivity to growth and yield reduction (the tolerance) of a crop plant (annuals and perennials during the first year after sowing or planting) by small and medium population densities at sowing or planting of a particular nematode species by the values of two parameters, which is preferable to the (customary) single characterization without reference to nematode density. The most important of the two parameters, the tolerance limit *T*, most probably is insenitive to variation in external conditions that normally occur during the period the crop is grown. That of potatoes to potato cyst nematode attack depends on change of day length during the plant growth. The values of these parameters, especially that of the tolerance limit, are key factors in the calculation of combinations of cost of reduction by control measures of a given nematode population density, expected at the time of sowing or planting of a crop to be protected, and the cost of crop loss, caused by the surviving nematodes, resulting in maximum net returns. These calculations are again formally the same for all tylenchid nematodes, annual crops and control measures.

On the other hand the model gives strong indications of the nature of the mechanisms resulting in the growth reduction by nematode attack and is, therefore, a base for investigations on its biochemistry, the counteraction of this reduction, resulting in the existence of tolerance limits, and the adding up of the effects of very large numbers of nematodes to a maximum reduction of the growth rate of the attacked plant by an amount which, in general, is smaller than the actual growth rate of plants without nematodes. So far, the physiological mechanism(s) resulting in growth reduction by nematode attack have hardly been investigated seriously. Guided by this growth model, this type of research, which may open new ways for the management of these nematodes, could make a better progress.

An advisory system for the management of potato cyst nematodes (*Globodera* spp)

T.H. BEEN[1], C.H. SCHOMAKER[1] and J.W. SEINHORST[2]
[1] *DLO Research Institute for Plant Protection (IPO-DLO), P O Box 9060, 6700 GW Wageningen, The Netherlands*
[2] *Middelberglaan 6, 6711 JH Ede, The Netherlands*

Abstract. An advisory system is presented for the management of potato cyst nematodes (*Globodera pallida*) It emphasizes the use of partially resistant potato cultivars, which provide the possibility of keeping population densities of potato cyst nematodes at a low level in short fixed rotations Using stochastic models based on the population dynamics of potato cyst nematodes and the relation between pre-plant nematode densities and relative yield it is possible to calculate the probabilities of population development and the reductions in yield caused by these population densities A simulation model is developed which integrates both models, using the frequency distributions of some of the most variable parameters relevant to a particular combination of potato cultivar and nematode population Also, the natural decline in population density when non-hosts are grown is incorporated in the model The model makes it possible to calculate the probability of a certain yield reduction, given a certain potato cultivar, nematode population and rotation Therefore, it becomes feasible for a farmer to evaluate risks and the costs of different control measures in fixed rotations The application of this model in the starch potato growing areas could lead to significant improvements in financial returns and a major reduction of the use of nematicides

Introduction

Potatoes are among the most profitable agricultural crops in arable farming in the Netherlands. Therefore, they are grown as frequently as possible, especially in those areas, where farmers have almost no choice of other profitable crops. This frequency is limited by build up of potato cyst nematodes which, without control leads to considerable crop losses. Possible control measures other than crop rotation (growing susceptible potato cultivars in rotation with five to seven years of non-hosts) are growing highly- or partially-resistant potato cultivars and chemical control. Chemical control not only has a poor cost-to-benefit ratio, its use is also increasingly restricted by legislation. A 50% reduction by 1995 and a 80% reduction by the year 2000 in the use of the so-called fumigants are main objectives in the Multi-Year Crop Protection Plan of the Dutch government. Cultivars that are highly resistant to the pathotypes Pa 2 and Pa 3 of *Globodera pallida* are rare (industrial processing) or not available (human consumption). As crop loss is strongly associated with nematode density in the field at the time of planting, control should aim at preventing nematode

A J Haverkort and D K L MacKerron (eds) Potato Ecology and Modelling of Crops under Conditions Limiting Growth, 305–322
© 1995 *Kluwer Academic Publishers*

densities from becoming too large. Once the population density in the field is reduced to an acceptable level a proper combination of relative susceptibility of a potato cultivar and crop rotation can keep it small. It is in the interest of potato growers and the environment to integrate the use of partially-resistant potato cultivars and other control measures in farming practices in such a way that a maximum return is obtained by minimizing the sum of cost of control and cost of yield reduction. This requires an advisory system based on prediction of population development and yield reduction in given rotations which emphasizes the use of partially resistant potato cultivars. As treatments with nematicides are expensive (about 1200 Dutch guilders/ha) and their effectiveness in reducing population densities is overrated, net returns will be considerably larger than with chemical control, when partially resistant potato cultivars are grown in the proper short rotation and, therefore with only minor crop losses due to the resulting small nematode densities. As a result the greater part of the reduction of the use of chemicals required in the Multi-Year Crop Protection Plan can be achieved without adverse economical consequences, if advices according to the advisory system are followed.

An advisory system has been developed based on a stochastic simulation model which can include a number of sub-models ranging from those of spatial distribution of population densities (resulting in new soil sampling methods), to economics for calculation of maximum returns. This paper is restricted to a discussion of the possibilities of using partially-resistant potato cultivars in certain crop rotations to minimize yield reductions and the cost of control. Four sub-models in the integrated simulation model will be discussed:

1. The simplified version of population dynamics of potato cyst nematodes applying at small to medium nematode densities at planting;
2. The concept of relative susceptibility ('susceptibility' = 'host status');
3. The relation between pre-plant nematode densities and relative yield (yield as fraction of the yield when potatoes are grown without nematodes);
4. Population decrease as a result of growing non-host crops.

Population dynamics

The relation between population density at planting or sowing (*Pi*) and the population density of the new generation at harvest (*Pf*) for nematode species with only one generation per season (potato-, oat-, white clover cyst nematodes, *Meloidogyne naasi*) is described by Seinhorst (1967; 1970; 1986a; 1993) (Figure 1). Seinhorst's (1993) most extended equation for this relation contains ten parameters of which the most important ones are the maximum rate of reproduction (occurring at very small initial population densities) and a theoretical maximum density, the number of eggs that would have been produced per unit weight of soil at very large initial nematode densities if the size of the plant would not have been reduced by the nematodes. Because of the reduction of the size of the plants by nematodes the actual maximum nematode

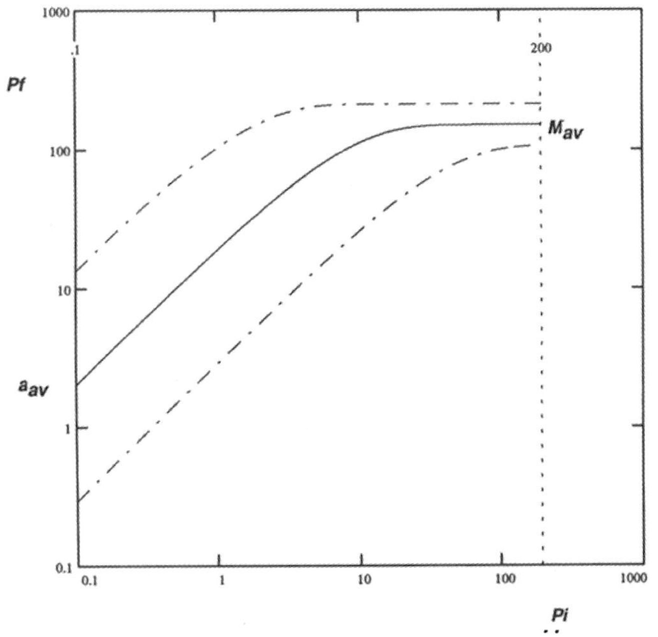

Figure 1. The relation between initial and final egg densities according to Eq. 1. Dotted lines = 95% confidence interval of the log-normal distribution of *a* and *M* ($0{,}145a < a_{av} < 6{,}9a$ and $0{,}71M < Mg_{av} < 1{,}41M$). The scale of *Pf* (eggs/g soil) applies to *G. pallida* ($a_{av} = 20$ and $M_{av} = 150$ eggs/g soil on susceptible cultivars).

density after a potato crop in the field occurs at medium initial densities and also includes non hatched eggs of the parent generation (Seinhorst 1967; 1984; 1986a).

Seinhorst's (1993) equation for this relation, if extended to incorporate the part of the parent population surviving in soil that was not exploited by roots (Seinhorst 1986a), contains eight parameters; too many to be useful for the prediction of population densities and their frequency distributions after growing a cultivar with a certain degree of resistance. However, it can be used as a basis for the formulation of the constraints on the simpler equation:

$$Pf = M \times (1 - e^{-aPi/M}) \qquad (1)$$

in which: *Pi* – initial egg densities (before planting) in eggs/g soil; *Pf* – final egg densities (after harvest) in eggs/g soil; *a* – the maximum rate of reproduction; *M* – a theoretical maximum final population density.

It fits to the relation between initial and final nematode densities up to the *Pi* value where the maximum *Pf* value is reached according to the extended equation. Taking this maximum as an estimate of *M* did not result in appreciable differences between *Pf* values according to Eq. 1 and observed ones in field experiments (e.g. Seinhorst 1986a), where a considerable plot-to-plot variation existed.

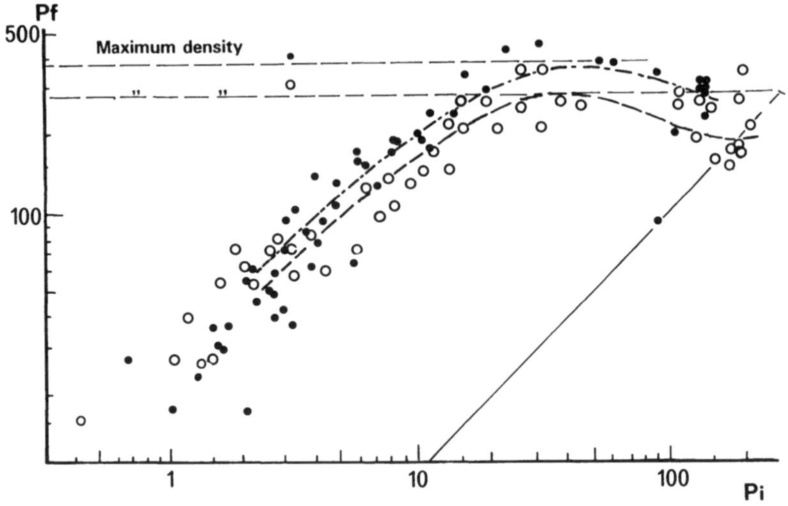

Figure 2. The relation between *Pi* and *Pf* of pathotype Ro 1 of *G. rostochiensis* on cultivar Bintje on two parts of a farmer's field (Seinhorst 1986a). *Pi* and *Pf* in eggs/g soil. The solid line indicates *Pf/Pi* = 1.

As the values of *a* and *M* are not only determined by the potato cultivar but also by external conditions, the final population density at one initial population density can vary strongly between fields and years. Both parameters not only differ between years and fields; variation can occur within a single field (Figure 2). Therefore, it is impossible to predict the development of population densities in individual fields using only average values of *a* and *M*, let alone from *a* only, as is common practice in contemporary advisory systems and legislation. A better approach is to establish the frequency distributions of *a* and *M* and to calculate the probability of all possible combinations of *a* and *M* and of the resulting densities *Pf* and their probabilities. According to observations on seventeen farmers' fields on several soil types in several years *a* for *G. rostochiensis* varied between 3 and 157 with a geometric mean of 25 and *M* varied between 200 and 400 eggs/g soil with a mean of 300 eggs/g soil (Seinhorst 1986c). A detailed analysis of these data will be presented elsewhere. In further calculations, the frequency distribution of both parameters is assumed to be log-normal. There are no indications that these distributions are different for partially-resistant cultivars. Therefore, the calculations for these cultivars are based on smaller average values (depending on their degree of resistance) with the same variability as for fully-susceptible cultivars.

There are not enough observations from field experiments to estimate mean values and their variances of *a* and *M* for *G. pallida*. A value of 0.8 for the ratio between *a* for *Globodera rostochiensis* pathotype Ro 1 and for *G. pallida* pathotype Pa 3 and a value of 0.5 for that between *M* for these pathotypes could be deduced from values on the susceptible cultivar Irene grown in pot experiments (Den Ouden 1974a; Seinhorst and Oostrom 1984; Seinhorst 1986b).

Therefore, it is assumed that a_{av} and M_{av} for *G. pallida* on susceptible cultivars in the field are 20 and 150 eggs/g soil respectively. These assumptions must be verified against observations from field experiments when these are available.

Measures of partial resistance and their relation

When partially-resistant potato cultivars are grown fewer females will mature than on susceptible cultivars; also the number of eggs/cyst may be smaller. Therefore, nematodes multiply less strongly on these cultivars than on susceptible ones and sustain a smaller maximum population density. However, both *a* and *M* are too variable to be suitable as a measure for partial-resistance. Therefore, the concept 'relative susceptibility', *rs*, was introduced, based on the population dynamics of the potato cyst nematode. The relative susceptibility is the ratio between the maximum multiplication rate *a* of the nematode population on the tested cultivar and on a susceptible reference cultivar or the equivalent ratio of the maximum population density *M* on these cultivars. These present two measures of partial resistance or relative susceptibility, provided that the tested cultivar and the susceptible reference are grown under the same conditions in the same experiment. Relative susceptibility is independent of external conditions which influence both *a* and *M* and contribute to their large variability. Figure 3 shows the relation between *Pi* and *Pf* of pathotype Pa 3 of *G. pallida* on the

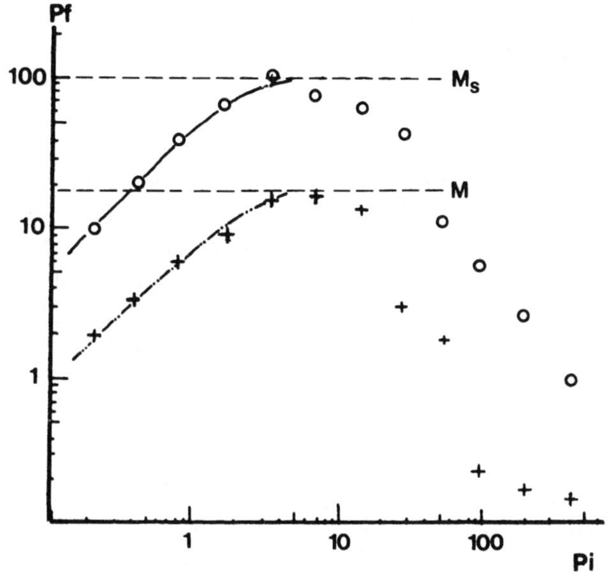

Figure 3. The relation between initial and final population density *Pi* and *Pf* of *Globodera pallida*, pathotype Pa 3 on cvs. Irene (o) and Darwina (+), (after Seinhorst and Oostrom 1984). *Pi* and *Pf* in eggs/g soil. Lines according to Eq. 1.

partially-resistant cultivar Darwina and on the susceptible cultivar Irene according to Eq. 1.

Jones et al. (1981) and Phillips (1984) assumed that the ratios $(a_{partially\ resistant}/a_{susceptible}) \times 100\%$ and $(M_{partially\ resistant}/M_{susceptible}) \times 100\%$ are numerically equal. According to Seinhorst (1984), Seinhorst and Oostrom (1984) and Seinhorst et al. (in press) this also applied to 9 out of 11 cultivars tested with pathotype Pa 3. However, rs_M was smaller than rs_a with cv. Activa and breeding line Karna 77/281 (Seinhorst et al. in press) and cv. Ehud (Seinhorst 1984). Seinhorst and Oostrom's 1984 data also indicate that, despite considerable differences in rates of reproduction at small initial egg densities of the same pathotype on the same cultivar in different experiments, the variation of the relative susceptibility was largely if not entirely due to experimental error, resulting from limited numbers of pots and insufficient number of cysts and eggs counted per replication.

The great similarity between rankings of cultivars according to degrees of susceptibility in pot and field experiments observed by Forrest and Holliday (1979), Phillips and Trudgill (1983; 1985), Phillips et al. (1987) supports the conclusion that relative susceptibilities do not depend on external conditions. The methods to determine the relative susceptibility of a potato cultivar with a high degree of accuracy are described in detail by Seinhorst et al. 1993.

Table 1 presents the relative susceptibilities of sixteen potato cultivars to different pathotypes of *G. pallida*, measured in 10 l pots at the IPO-DLO. The pathotypes include one of the most virulent populations found so far: Pa 3 (1) (the 'Rookmaker' population). A wide and continuous range of relative susceptibilities is apparently available in the cultivars tested.

Relation between nematode density and yield

Another essential part of the integrated simulation model is the relation between population density of eggs at the time of planting and tuber yield at harvest expressed as a proportion of the yield in the absence of nematodes. As large yield losses must be avoided, large *Pi* values must always be reduced by extra control measures to acceptable *Pi* values before a crop is grown. Therefore, the relation at small to medium nematode densities (Seinhorst 1986b) suffices:

$$y = m + (1-m) * 0.95^{(P-T)/T} \quad \text{for} \quad P > T$$

$$\text{and} \quad y = 1 \quad \text{for} \quad P \leq T \qquad (2)$$

in which: y – the yield at egg density P at planting as a proportion of that at $P \leq T$; m – the minimum relative yield (therefore a constant < 1); T – the tolerance limit, the density P up to which no yield reduction is caused.

T for susceptible cultivars varied little between experiments about an average of 2 eggs/g soil in field experiments with pathotype Ro 1 (Seinhorst 1982a; 1986b), microplots with pathotypes Ro 1 and Pa 3 (Greco et al. 1982) and pot

Table 1. Relative susceptibilities (a/a_s) of sixteen potato cultivars for pathotype Pa 1, Pa 2 and Pa 3 of *Globodera pallida* expressed in percentages of the susceptibility of the susceptible cultivar Irene for these pathotypes. Pa 3 (1), also known as the 'Rookmaker' population, is highly virulent, Pa 3b is a population with average virulence

Cultivar	Pa 1	Pa 2	Pa 3 (1)	Pa 3 (2)
Irene	100	100	100	100
Amalfi	–	3	–	39
Amera	76	63	–	98
Producent	15	3	40	34
Multa	–	12	43	40
Pansta	–	6	32	36
Promesse	–	4.5	35	21
Proton	–	1	31	18
Darwina	4.7	0.3	12	5
Santé	–	1	18	8
Atrela	–	0.7	20	9
Karna 77/270	–	–	28	–
Karna 77/281	–	–	12	–
Activa	–	–	25	–
Ellen	–	–	17	–
Seresta	–	–	2	–

– = relative susceptibility not measured. Pa 1 does not occur in the Netherlands.

experiments with pathotypes Ro 1, Ro 3 and Pa 3 (Seinhorst 1982b). Values of *m* varied more; 0.2 – 0.6 according to Seinhorst (1982a). But using an average of $m = 0.4$ in predictions does not result in unacceptable deviations between actual and calculated losses at densities $\leq$ 15 eggs/g soil. Over- or under-estimations do not exeed 5% (Figure 4). Predicting relative yield losses at larger nematode densities has no practical value as yield reductions at these densities are too high to be tolerated. Farmers will not grow a potato crop at such nematode densities without first taking control measures. In the Netherlands this implies the application of a nematicide to reduce *Pi* ánd also, as fumigation is mostly not sufficiant to attain acceptible nematode densities, the use of a systemic nematicide just before planting. The latter provide a temporal protection by delaying nematode invasion, thereby increasing the minimum yield *m*.

Little is known about the tolerance limit of partially-resistant cultivars except that *T* for pathotype Pa 3 on cv. Darwina is, also, 2 eggs per g soil and that results of field tests do not contradict such a value of *T* for other cultivars. The similarity of ratios M_r/M_s to a_r/a_s for the cultivars including cv Darwina, partially-resistant to pathotype Pa 3 (Seinhorst et al. in press), suggest that *T* for pathotype Pa 3 on these cultivars is not a multiple of the 2 eggs/g soil for cv. Darwina. An accurate assessment of the tolerance limit is desirable for two reasons. It is required for the calculation of losses to be expected at given egg population densities at the time of planting and, as the maximum population density tends to become proportional to the size of the plant when *Pf* approaches *M*, this

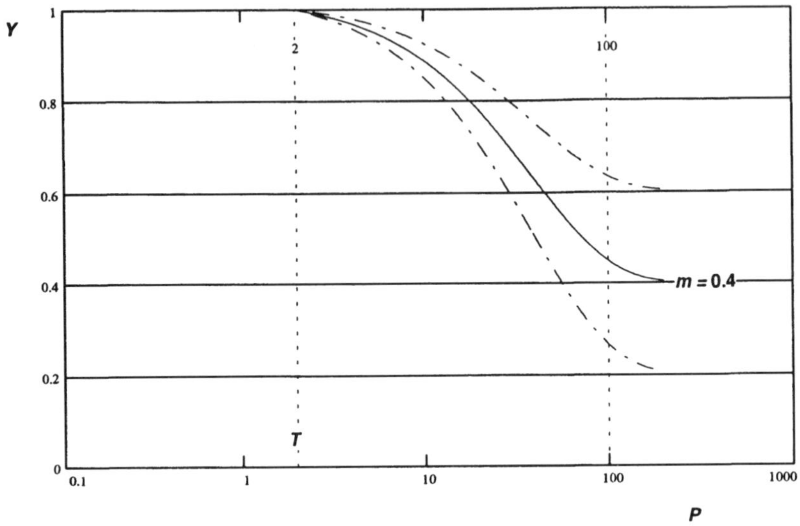

Figure 4. The relation between the initial population density (*P*) and the relative yield *y* according to the equation: $y = m + (1 - m) \times 0.95^{(P-T)/T}$ with $T = 2$ eggs/g soil and $m = 0.4$ (the mean) and $m = 0.2$ and 0.6 (the boundaries of the 95% confidence interval).

maximum will be correlated positively with the tolerance limit. However, small differences in tolerance (e.g. a ratio of tolerance limits of 1 to 2) would not have a detectable effect on M.

Population decline in the absence of a host

The nematode density at planting of potatoes is the density left by the previous potato crop multiplied by the survival rate over the period during which no potatoes were grown. This rate can be considered to be independent of nematode population density. Reduction of the population in the absence of a host crop is largely due to spontaneous hatching during a short period in spring of an apparently fixed proportion of the eggs. According to Huijsman (1961) the survival rates over a period of six years was 65% per year and this rate was (largely) independent of the age of the population as was the viability of the surviving population (Den Ouden 1963). Cole and Howard (1962) found a survival of 80 % per year during three years, and Den Ouden (1970) reported an average of 79% over 14 fields in 1969 but only 51% over six fields in 1973 (Den Ouden 1974b). However, the magnitude of field-to-field and year-to-year variation in the survival rate is unknown as, generally, sampling error was either ignored or was so large that differences could not be distinguished. However, there are indications that the survival rate during the first year after a potato crop is lower (Den Ouden 1960; Cole and Howard 1962; Andersson 1987; 1989) than in subsequent years. Therefore, in the calculations of the simulation model, a lower survival rate is used the first year after the potato crop than during later years (about 50% changing to 65%). The survival rate is used to calculate a factor for crop rotation, *c*, having a value of 0.5 for a 1:2 rotation; $0.5 \times 0.65 = 0.32$ for a 1:3 rotation; $0.5 \times 0.65^2 = 0.21$ for a 1:4 rotation and 0.5×0.65^3 for 1:5 rotation. But more information on survival rates in the absence of potatoes is required to adjust the model if necessary.

Simulation model

Basics

To make computations the frequency distributions of *a* and *M* were assumed to be log-normal with means 1.30103 and 2.17609 and standard deviations 0.419 and 0.07525 respectively for *Globodera pallida* pathotype Pa 3 on susceptible cultivars. Both frequency distributions were divided into 24 classes with a width of 0.25 times the standard deviation (s). Hence, a range of classes between (log *a* or *M*) + 3*s* and (log *a* or *M*) − 3*s* were used, comprising 99.7 of the frequency distribution. Each class of *a* and *M* is presented by the antilog of its class mid-mark and divided by a_{av} and M_{av} respectively and a relative frequency. Values of log *a* and *M* beyond this range are ignored. All relative frequencies are multiplied by 0.997^{-1} to obtain a total of 1 for the cumulative probability of the range considered.

Pf values following a given *Pi* for a potato cultivar with a given *rs*, are calculated according to Eq. 1 with every class mid-mark of the distribution of *a* times $rs \times a_{av}$ combined with all class mid-marks of the distribution of *M* times $rs \times M_{av}$. The relative frequency of each *Pf* is the product of those of the *a* and *M* used. Calculating *Pf* values for the next crop rotation as indicated above with each of these 24^2 values as *Pi* would require an unnecessarily excessive amount of work. Therefore, the *Pf* values, also, are divided into 24 classes and new mid-marks are calculated. The relative frequency of each of these class mid-marks is the sum of relative frequencies of *Pf* values belonging to that class. Now, there are 24 *Pi* values to start with, resulting in $24^3 = 13824$ *Pf* values after the second potato crop. This procedure is repeated until the required number of potato crops has been simulated.

As this paper is limited to the use of partial resistance to keep nematode densities small actual population densities as determined by soil sampling are not used as input. Instead, the almost stable frequency distribution of population densities established after five years of growing partially-resistant potatoes of the same relative susceptibility is used. These are practically independent of population densities in the first year of cropping. A more extensive description of this part of the simulation model will be presented in another paper.

Crop losses in rotations with potato cultivars of given relative susceptibility

For the assessment of relative crop losses $(1 - y)$ the relation between nematode density and relative tuber yield according to Eq. 2 is used to calculate y for the mid-marks of classes of Pf values. For the tolerance limit T a value of 2 eggs/g soil is used and for the minimum yield m 0.4. Estimates of percentages of fields with nematode densities and crop losses exceeding certain limits occurring in a given year or percentages of years in which these occur in a given field when a potato cultivar with a certain rs is grown in a certain rotation are given in Fig. 5. The probability of suffering a crop loss larger or equal than a certain value (on the x-axis) can be read off from the y-axis. From these data the average yield reduction can be calculated. When crop rotation (and the same applies to other control measuers) is used to reduce the population density between two potato cultivars with a factor c than the Pi at the next potato crop is cPf of the population density after the last potato crop. The development of population densities than depend on ca_{gem} and cM_{gem}. This implies that the frequency distributions of expected crop losses in Figure 5, which are obtained when cropping potatoes with a certain

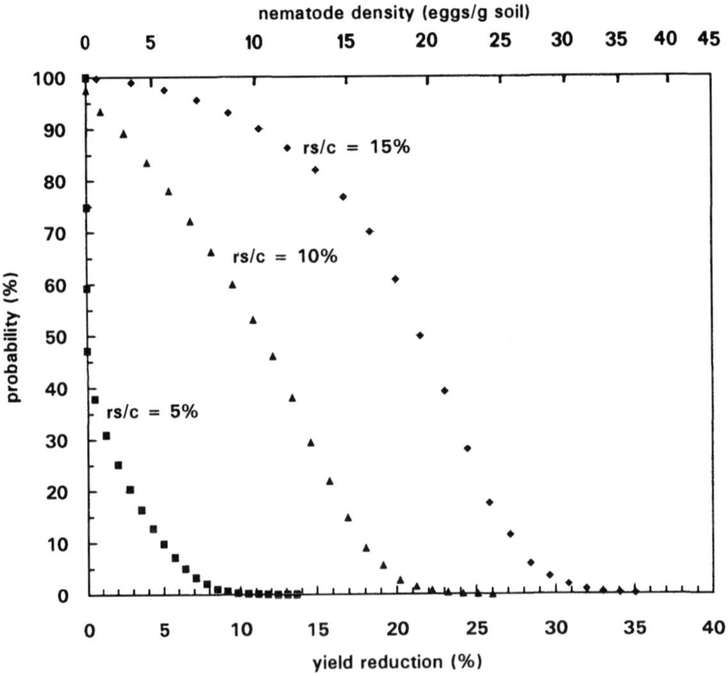

Figure 5. Probabilities of nematode densities (top x-axis) and resulting relative yield reduction ($(1 - y) \times 100\%$) (bottom x-axis) after five croppings of a partially-resistant potato cultivar with rs of $5/c$ (■); $10/c$ (▲); and $15/c$ (♦) respectively at an initial density of 5 eggs/g soil in the first year ($c = 0.5$ at 1:2 cropping frequency; 0.32 at 1:3 frequency; 0.21 at 1:4 frequency and 0.14 at 1:5 frequency). The relative yield (y) according to Eq. 2 and Figure 4 with $T = 2$ eggs/g soil and $m = 0.4$. $M_{av} = 150$ and $a_{av} = 20$ eggs/g soil.

Table 2. Relation between relative susceptibility, cropping frequency, average yield reduction and probabilities of a certain yield reductions after five croppings of a partial resistant potato cultivars at a *Pi* of 5 eggs/g soil in the first year

Relative susceptibilities (%) at cropping frequencies of					Average yield reduction (%)	Prob (%) of yield reduction larger than					Percentage yield reduction in 95% of the fields smaller than
1:1	1:2	1:3	1:4	1:5		0%	5%	10%	12,5%	15%	
1	2	3	5	7	0,0	0,0	0,0	0,0	0,0	0,0	0,0
2	4	6	9	14	0,0	0,1	0,0	0,0	0,0	0,0	0,0
3	6	9	14	20	0,1	3,7	0,0	0,0	0,0	0,0	0,0
4	8	12	18	27	0,4	18,0	1,2	0,0	0,0	0,0	2,6
5	10	15	23	34	1,2	37,8	7,4	0,0	0,0	0,0	5,8
6	12	18	27	41	2,5	60,0	20,0	1,8	0,0	0,0	8,6
7	14	21	33	47	4,2	70,9	35,4	8,9	2,1	0,0	11,1
8	16	24	36	54	6,1	86,4	51,1	21,9	9,2	2,3	13,7
9	18	27	41	61	8,1	90,5	63,8	37,0	21,4	7,9	16,1
10	20	30	45	67	10,1	93,3	74,0	51,0	35,4	19,5	18,2
11	22	33	50	74	12,2	95,7	81,4	62,7	49,7	32,9	20,2
12	24	36	54	81	14,1	96,8	86,6	72,3	61,6	47,0	22,2
13	26	39	59	88	16,1	100	91,0	79,3	70,8	59,5	24,2
14	28	42	63	94	17,9	100	93,8	84,4	77,8	69,3	25,7
15	30	45	68	100	19,7	100	95,6	88,4	83,5	76,6	27,5

rs continously, are the same when a potato cultivar with *rs/c* is cropped in the corresponding rotation cycle. For a better overview concerning a range of relative susceptibilities Table 2 shows the different average yield reduction, the probabilities of more than 0%, 5%, 10%, 12.5% and 15% yield reduction and the maximum yield reduction in 95% of the fields in 5 different crop rotations and with different values of *rs*. From these figures it can be concluded that cultivars with partial resistance of ≤ 8/*c*% always control potato cyst nematodes sufficiently without the need for additional control measures. However, the level of *rs* that is still useful depends on the costs of control and the extra yield as a result of applying that control. It also depends on the financial situation of the potato grower. If he has a healthy financial reserve he can choose relative susceptibility on the basis of average yield reductions predicted. However, if no financial reserves are available and crop losses may not exceed a certain amount, risks of certain yield reductions have to be considered and the choice of relative susceptibilities becomes more limited. The percentages in Figure 5 and Table 2 can also be interpreted as chances that such densities and losses will occur in any field and year.

By applying a certain amount of control on all fields, after cropping a susceptible cultivar, calculated to prevent unacceptable yield losses even in the most heavily infested fields, too much control will be applied on a large number of fields. If the amount of control is adapted to obtain maximum average returns, yield losses in a certain proportion of the fields will be too high, while in the other fields still too much control is applied. Therefore, it would be justifiable to adjust the amount of control, after growing susceptible potatoes, to actual population densities, estimated by taking soil samples (Seinhorst 1982b; Schomaker and Been 1992). Sampling methods are then needed which provide estimates of densities of eggs and larvae with only a small error.

However, the need for soil sampling becomes obsolete when potato cultivars with high partial resistance are grown as then the small maximum population density M on these cultivars limits population increase sufficiently to densities where the cost of additional control exceeds the increase of net return obtained, whereas the required cropping frequency for potatoes is determined by other factors than crop loss caused by potato cyst nematodes. Soil sampling could then be restricted to once during a couple of rotation cycles to check for the presence of a more virulent nematode pathotype.

Sensitivity analysis

The values of ca_{av} and cM_{av} which are supposed to apply to a certain partially-resistant cultivar grown in a certain rotation in a field with a given nematode pathotype depend on several factors: The susceptibility (*rs*) of the cultivar, the values of a_{av} and M_{av} of the pathotype involved on the susceptible reference cultivar (of which a_{av} is the hardest to estimate in field trials), and the survival rate c of the nematode pathotype during the years without potatoes are all subject to experimental error. For practical application the indirectly derived values of a_{av} and M_{av} for pathotype Pa 3 on susceptible cultivars should be replaced, as soon as possible, by accurate averages and estimated frequency distributions derived from sufficient numbers of directly observed values of a and M from several fields and years.

Average reduction in yield (%) and the probabilities of reductions in yield > 10% were calculated using the best estimate of a_{av} of a susceptible control (20) and using the value of 25 (presuming an underestimation of the best estimate by 20%). The values for several cultivars with different *rs* are presented in Table 3. The underestimation of a_{av} resulted in only slightly larger calculated average crop losses than those using the best estimate of a_{av}. However, the underestimation of a_{av} resulted in a considerable underestimation of the chances of crop losses ≥ 10% in single fields. Presuming an overestimation of the best estimate of a_{av} by 20% (a_{av} then is 16) resulted in equivalent changes; calculated average crop losses were only slightly smaller, but crop losses ≥ 10% have considerable smaller probabilities than originally estimated.

Figure 6 shows the sensitivity of the relative yield to the value taken for the relative susceptibility of a cultivar. It can be seen that an over- or under-estimation of *rs* has a far greater effect on the average yield reduction than the same error in the estimation of a_{av}. Up to an rs/c of 4 the average yield reduction is negligible. Between $5\% < rs \times c < 15\%$ the relation becomes linear and average yield reduction increases by 2% per unit $rs \times c$. Therefore, an error of 5 pecentage points in the estimation of *rs* causes a deviation of $10\% \times c$ in the

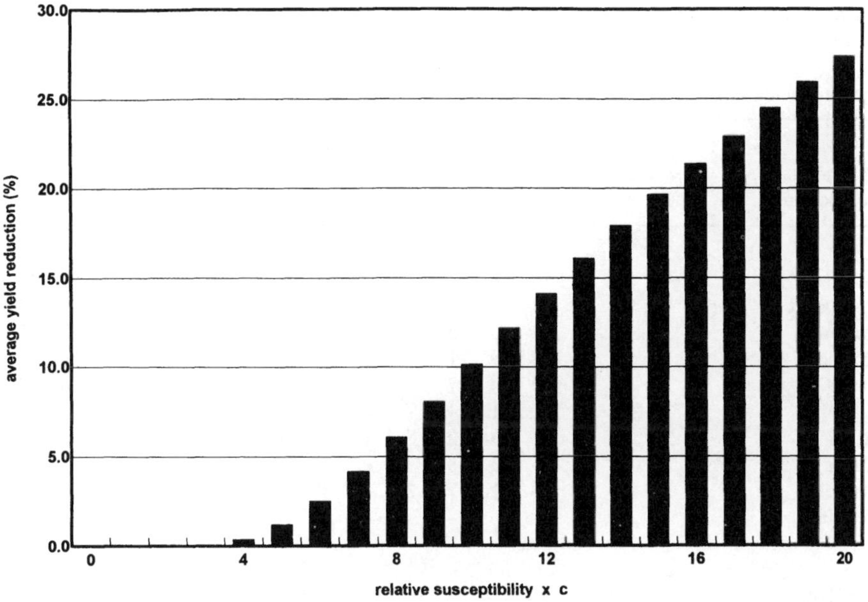

Figure 6. Sensitivity analysis of the effect of variation of estimated $rs \times c$ on the average yield reduction expressed as percentage of the expected yield when no nematodes are present after 5 croppings of a partially-resistant cultivar with the same ca_{av} and cM_{av} and Pi = 5 eggs/g soil in the first year.

predicted average yield reduction. It demonstrates the necessity of minimizing testing errors by conducting potato cultivar testing for relative susceptibility with the utmost care.

Which relative susceptibilities for which rotation schemes

The yield of potato cultivars, with the same characteristics grown in fixed rotations in fields infested with potato cyst nematodes, is negatively correlated with their relative susceptibility but there will be no difference in cost of nematode control. Therefore, the least susceptible of otherwise similar cultivars is always the best choice either because losses caused by the nematodes are smaller or because a shorter and more profitable rotation cycle can be practiced. To determine the upper limit of susceptibility that still is acceptable in a given rotation not only the average loss but also the probability of more than a certain percentage loss of a single crop is important, as, once a large nematode density has built up, the chance of larger than average population densities will be increased during the following years. Table 2 provides general information.

Table 3 presents a list of cultivars with relative susceptibilities for two different pathotypes of Pa 3 ranging from 2 to 44% with average yield losses and probabilities of more than 10% yield loss.

A farmer can decide to use this information in different ways. Assume he

Table 3. Relative susceptibilities of different potato cultivars on two populations of pathotype Pa 3, average yield reduction and probabilities of a yield reduction larger than 10% at two values of M_{av}/a_{av}. Pi = 5 eggs/g soil; – = < 0.5%. The relative susceptibilities are averages of all measurements up to now and can differ from previous publications

		Average yield reduction (%)								Prob. (%) of yield reduction > 10%							
Pa 3		*a* = 20				*a* = 25				*a* = 20				*a* = 25			
population 1	% *rs*	1:1	1:2	1:3	1:4	1:1	1:2	1:3	1:4	1:1	1:2	1:3	1:4	1:1	1:2	1:3	1:4
Irene	100	58	50	41	30	58	50	42	31	100	100	100	99	100	100	100	100
Astarte	44	48	30	19	10	48	31	21	12	100	99	87	48	100	100	94	63
Producent	40	45	27	17	8	46	29	19	10	100	98	81	35	100	99	90	50
Proton	31	39	21	11	4	40	22	13	6	100	90	55	8	100	95	70	15
Karnico	30	38	20	10	4	39	21	12	5	100	88	51	6	100	94	66	11
Santé	25	33	15	7	2	34	17	9	3	100	76	27	1	100	86	40	2
Ellen	22	30	12	5	1	31	14	7	2	98	63	13	–	100	75	22	–
Atrela	20	27	10	4	–	29	12	5	1	97	51	6	–	99	65	11	–
Karna 77/281	16	21	6	2	–	23	8	3	–	92	22	1	–	96	34	1	–
Darwina	15	20	5	1	–	21	7	2	–	88	15	–	–	94	24	–	–
Seresta	2	–	–	–	–	–	–	–	–	–	–	–	–	–	–	–	–
Pa 3		*a* = 20				*a* = 25				*a* = 20				*a* = 25			
population 2	% *rs*	1:1	1:2	1:3	1:4	1:1	1:2	1:3	1:4	1:1	1:2	1:3	1:4	1:1	1:2	1:3	1:4
Irene	100	58	50	41	30	58	50	42	31	100	100	100	99	100	100	100	100
Astarte	40	45	27	17	8	46	29	19	10	100	97	81	35	100	99	90	50
Proton	18	25	8	3	–	26	10	4	1	96	37	2	–	98	51	4	–
Atrela	9	8	1	–	–	10	1	–	–	37	–	–	–	51	–	–	–
Santé	8	6	–	–	–	8	1	–	–	22	–	–	–	34	–	–	–
Darwina	5	1	–	–	–	2	–	–	–	–	–	–	–	–	–	–	–
Seresta	2	–	–	–	–	–	–	–	–	–	–	–	–	–	–	–	–

wants to grow potatoes in a fixed rotation, for instance once in two years. He chooses which risk of yield reduction he is prepared to accept, taking into account the amount of money he could save by applying no other control measure (for instance the use of a nematicide), and chooses a cultivar with the required or better relative susceptibility that promises the largest net return because of other cultural characteristics. Another possibility is to determine in which crop rotation his favourate potato cultivar can be grown with the largest net return with or without additional control.

An example: Let's investigate whether a cultivar with a relative susceptibility of 18% yields a better net return in a 1:2 crop rotation than in a 1:3 crop rotation over a period of six years. In both cases no nematicides will be used. It is assumed that the net return of potatoes is 47% and that of a certain, extra, non-host crop is 27% of the gross return of potatoes (Kwantitatieve informatie 1990–1991, IKC-agv & PAGV). According to Table 2 the potato crop suffers an average yield reduction of 8.1% in a 1:2 rotation and of 2.5% in a 1:3 rotation. The only difference between the two rotations is that one potato crop in the 1:2 rotation is exchanged for the extra non-host crop. Therefore, the other three non-host crops can be disregarded in the calculation. Then the average net return in the remaining three years will be 3 × (47% − 8.1%)/3 = 38.9% using a 1:2 rotation and (2 × (47% − 2.5%) + 27%)/3 = 38.7% using a 1:3 rotation. So, there is still a slight advantage in growing potatoes once in two years with a cultivar with 18% relative susceptibility. However, at larger susceptibilities the balance will tip in favour of a 1:3 crop rotation.

Conclusion

By combining Eqs. 1 and 2 and using the frequency distributions of the relevant parameters, probabilities of different relative yield reductions can be calculated. As input for Eq. 1 sampling results can be used provided that these estimates of population density give a good approximation of the real density within the sampled area. When highly partially-resistant potato cultivars are grown in the proper rotation, sampling data become obsolete, as at the maximum population density M no such increase of yield can be obtained, that can balance the cost of sampling and control. The calculated frequency distribution of a and M for *G. pallida* and the rate of decrease of population density in the absence of potatoes are now verified in field experiments by the IPO-DLO on 20 farmers' fields during several years.

As the tolerance limit T is 2 eggs/g soil for most combinations of potato variety and nematode pathotype, the two important variables for Eq. 2 are $P = Pf$ of the previous potato crop (the output from Eq. 1) times c (crop rotation factor) and m. When using a value of 0.4 for m in the Netherlands, reliable predictions of yield loss can be made at economically interesting population densities. With the results of these calculations the farmer can evaluate the risks associated with the cropping of potato cultivars with a known relative susceptibility in a certain cropping frequency and choose combinations with the greatest probability of a maximum financial return based on net returns from potatoes and alternative crops in his fields.

Sensitivity analysis demonstrates that the accurate estimation of relative susceptibility of a cultivar is more important than an equally precise estimation of a and M (Eq. 1). Emphasis should be put on stabilizing experimental error when screening potatoes for partial resistance. Presently, the CPRO-DLO is testing more than 40 cultivars for their relative susceptibility against a number of populations of *G. pallida*, pathotype Pa 3, ranging in virulence from moderate to high. Field tests with some of these cultivars are being performed by the Research Station for Arable Farming and Field Production of Vegetables (PAGV).

The use of an advisory system as described above requires a mental reorientation by farmers, who still are inclined to aim at attaining maximum yields and, as a consequence, tend to opt for maximum security as actual yield losses are considered to be unpredictable. Nematicide treatments are, therefore, seen as a necessary insurance. However, in the Netherlands the use of nematicides is, at present limited to once in four years by statutory measures.

Moreover, the frequent application of nematicides causes adaptation of the microflora in soils resulting in accelerated breakdown of the fumigant (Smelt et al. 1989a,b). Therefore, chemical control cannot be used any further as an 'insurance' against losses by nematode attack in the areas producing potatoes for industrial processing, where a 1:2 cropping frequency is prevalent.

Farmers should strive to optimize returns instead of yields, not only to decrease the use of nematicides, but also to make more profit. An advisory system would provide the necessary information to apply a more profitable method of control, but whether it will be used depends on acceptance of the advice by the potato growers. The primary impulse to use the information generated by this advisory system will be the need to prevent yield reductions in those cropping years when potatoes are grown but nematicide application is not allowed.

Factors involved in the development of potato late blight disease (*Phytophthora infestans*)

J.G. HARRISON
Scottish Crop Research Institute, Invergowrie, Dundee DD2 5DA, Scotland, UK

Abstract. Late blight, caused by infection by the fungus *Phytophthora infestans*, is the most serious disease of potatoes and can completely destroy a crop. Blight is controlled by frequent applications of fungicides to the foliage, but there is increasing concern over the liberal use of agrochemicals. Loss of tuber yield depends on the amount of green foliage destroyed by the disease which, in turn, depends on the timing of the onset of blight and the rate of increase in blighted foliage. Modelling potato blight attempts to describe disease onset and progress using as input the many factors that affect blight.

Sexual reproduction of *P. infestans*, resulting in the formation of oospores, is probably important in the survival of the fungus in the field and in increasing the genotypic diversity of *P. infestans*. However, most of the build-up of disease results from the asexual reproductive cycle that consists of spore germination, infection of foliage, colonisation of foliage, sporulation, and spore dispersal and survival. The different stages in the asexual cycle of reproduction are affected by environmental factors that are largely determined by weather conditions and genotypic and physiological features of the pathogen (*P. infestans*) and the host. Most factors affecting blight interact with each other in complex ways, and their effects are difficult to quantify. Furthermore, many factors change continually and also vary horizontally and vertically within a crop, so the development of a realistic model for potato late blight is not easy.

Introduction

Late blight (blight) is the most serious disease of potatoes and can completely destroy a crop. Blight first appeared in Europe in Belgium in June 1845 and spread rapidly. Within a few months it had affected the potato crop throughout Ireland with devastating effects (Bourke 1965). Most people depended on potatoes as their main food and millions died from starvation or emigrated to escape the famine. Blight is caused by a phytopathogenic fungus, *Phytophthora infestans*, that infects and kills the leaves, stems and tubers of the potato plant. Fungicides have been used to control blight since the nineteenth century and today control depends on their frequent application, although genetic resistance of the potato plant to blight is increasingly important. The liberal application of agrochemicals has been questioned in recent years and there is concern over their potentially harmful effects on both man and the environment. Effective control of blight with a minimum of fungicides is therefore desirable. Fungicides

A.J. Haverkort and D.K.L. MacKerron (eds.), Potato Ecology and Modelling of Crops under Conditions Limiting Growth, 215–236.
© 1995 *Kluwer Academic Publishers.*

are usually sprayed on to the haulm of a potato crop every 7–14 days to control blight, so there may be up to ten applications during the growing season (Krause et al. 1975). However, it has long been known that the date of blight first appearing in a crop and its subsequent progress depend, *inter alia*, on the weather. Blight is associated with cool, moist conditions and leaf infection occurs only if the foliage is wet (Rotem et al. 1971). The destructive nature of blight and the realisation that its progress depends on the weather provided the stimulus for the development of blight prediction systems that could be used to schedule the application of fungicides. Van Everdingen (1926) produced the first system for forecasting the onset of the disease, based on temperature, rain, cloudiness and the duration of dew. Subsequently there have been many attempts to formulate new systems and to improve existing ones (Harrison 1992) with the consequence that the forecasting of blight is now more advanced than that of any other plant disease.

Forecasting systems led to the development of computer programmes e.g. BLITECAST (Krause et al. 1975) from which computer-based simulations describing disease development were later developed (Bruhn and Fry 1981). With ever increasing sophistication, some systems now consider residual fungicide activity and tolerance of *P. infestans* to fungicides (Levy et al. 1991), and others use weather forecasts (Raposo et al. 1993). Computer-based models describing the behaviour of blight have recently been linked with those describing crop growth to simulate host-pathogen interactions and provide estimates of both disease progress and crop production (Van Oijen 1992). Computer simulations can be used to assess the consequences of altering one or more factors on blight severity and its effects on crop growth and yield. Although computer simulation models produce results much more quickly and with a lower labour input than do field experiments the simulations do depend for their accuracy on a realistic assessment of the factors they include, and on the inclusion of all factors that affect the parameter(s) under investigation. Because the factors affecting blight are not fully understood the simulation models all contain approximations, assumptions and guesses, so their potential for error is great. This paper considers some of the many interacting factors that should be taken into account in modelling blight development, in a synthesis of existing knowledge.

Life cycle of *P. infestans*

Knowledge of both the life cycle of *P. infestans* and the related cycle of the disease it causes are essential in order to understand how blight is affected by the various factors that influence its development. The life cycle of *P. infestans* is shown in Figure 1 which is largely self-explanatory. The asexual part of the cycle, i.e. the production of sporangia and the release and then germination of zoospores to produce hyphae, predominates in the build-up of disease. The sexual stage with the production of oospores that may survive for long periods

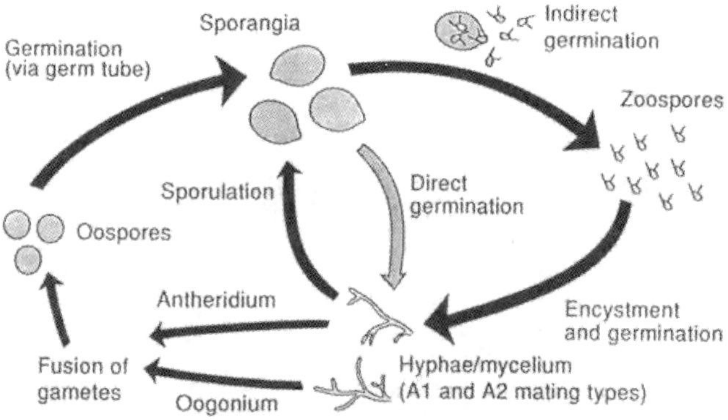

Figure 1. Life cycle of *Phytophthora infestans.*

in field soil normally requires the presence of both the A1 and A2 mating types. Its importance outside Central America has been uncertain until recently, but there is now compelling evidence that sexual reproduction frequently occurs, in the field, in Europe (Drenth 1994). Genetic recombination occurring during sexual reproduction provides a mechanism to increase the genotypic diversity of *P. infestans* so that it can adapt more readily to adversity, for example, by becoming tolerant of fungicides or overcoming host resistance. Genetic recombination may occur by parasexual mechanisms (Leach and Rich 1969), but these probably play a minor role in increasing the diversity of *P. infestans*.

Disease cycle and epidemic development

The disease cycle (Figure 2) reflects the life cycle of the pathogen as it occurs on the host, with leaf infection, lesion expansion, sporulation, and spore dispersal and survival largely determining disease development.

Sporangia arising on shoots growing from infected tubers, often in cull piles (Van der Zaag 1956), are blown to a new crop and form the primary inoculum. Blight is often initiated by a single sporangium or several sporangia each of which produces a leaf lesion. The lesions expand and sporangia are formed from a few days after infection (Figure 3) with eventually up to about 10^6 sporangia present on a single lesion (Harrison and Lowe 1989). Sporangia are dispersed by rain and wind (Fitt and Shaw 1989) and infect more leaves, so the unrestricted progress of blight follows a compound interest pattern (Van der Plank 1963) and the rate of increase in the amount of blighted foliage is explosive (Figure 4). Cycles of infection and lesion expansion are shown in Figure 4 as irregularities in the approximately sigmoidal disease progress curve. Mathematical models and computer simulations of blight progress attempt to describe this curve but its form is determined by a multitude of factors, some favouring blight and others not (Figure 5). Spores washed into the soil can infect tubers that act as an inoculum source for subsequent crops.

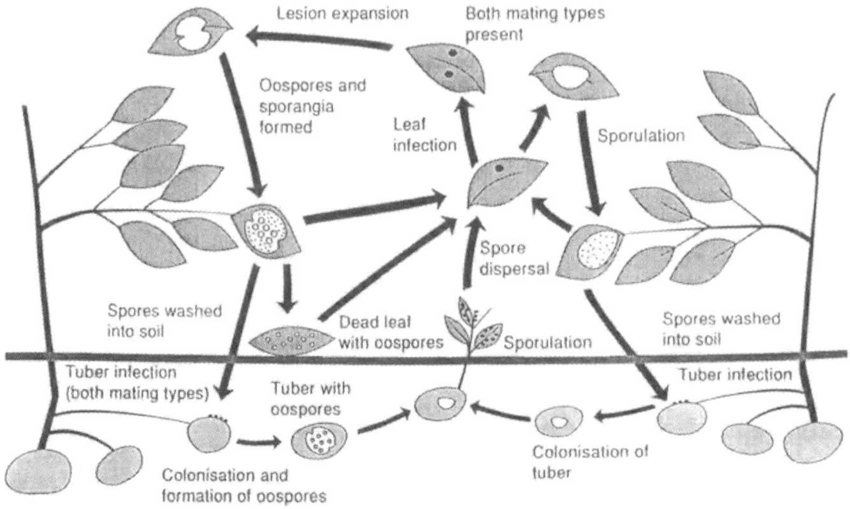

Figure 2. Disease cycle of potato late blight.

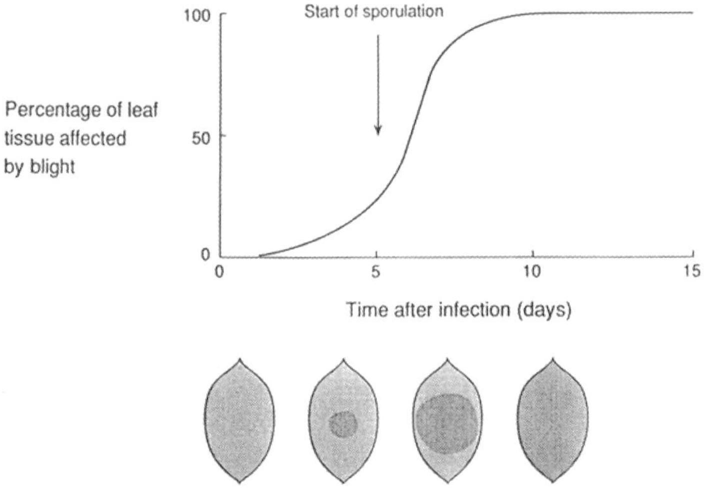

Figure 3. Increase in area of single blight lesion on a potato leaf under constant conditions.

Effects of blight on crop growth and yield

Effects of blight on crop growth and tuber yield depend on the disease progress curve (Figure 6). In the absence of senescence, the potential for photosynthesis of uninfected tissue is not significantly affected by the pathogen in other parts of the plant (Van Oijen 1990) and the loss of harvested yield is almost entirely due to a reduction in leaf area duration caused by the disease (Haverkort and

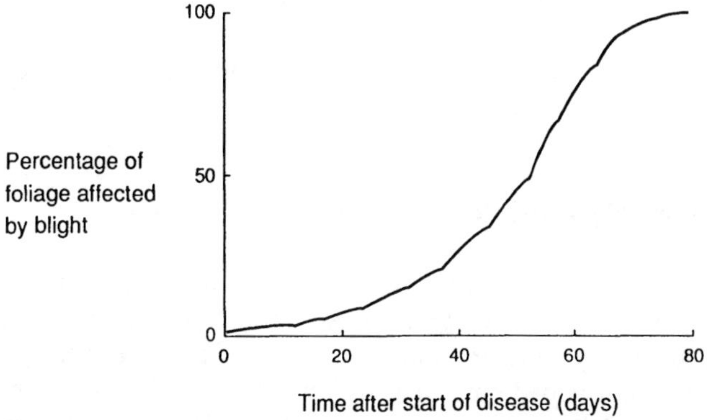

Figure 4. Hypothetical increase in blight lesion area on potato foliage of whole crop.

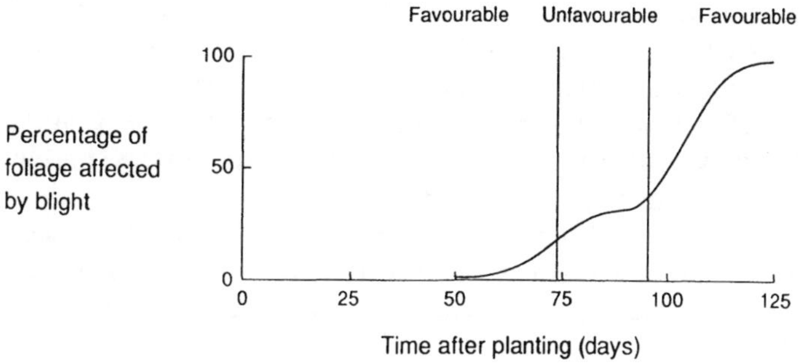

Figure 5. Disease progress curve during favourable and unfavourable periods for development of potato blight.

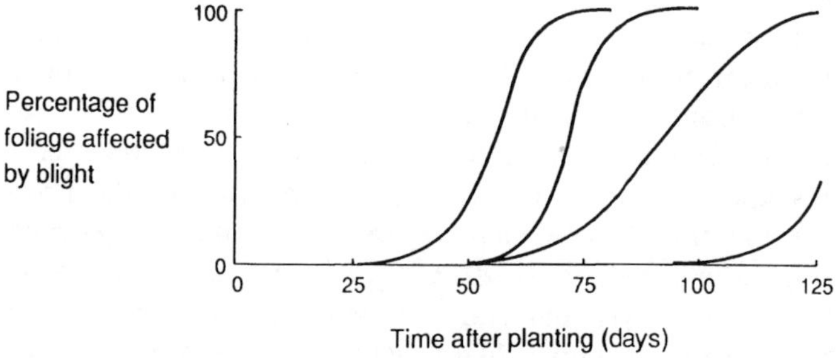

Figure 6. Disease progress curves showing effects of a) time of initiation of potato blight. b) rate of blight progress.

Bicamumpaka 1986; Van Oijen 1990; 1991a,b). Van Oijen (1991b) considered that the area under the disease progress curve was closely and inversely related to yield. Clearly, if blight develops only slowly, or if it starts late in the season after the bulk of the potential tuber yield has been formed, then losses at harvest will be relatively small (Figure 6). Infected tubers, however, frequently rot during storage (Dowson and Rudd Jones 1951) so blight can also cause losses after harvest.

Factors affecting blight and the asexual cycle of *P. infestans*

The factors that affect the timing of the onset of blight and its subsequent progress can be divided into three broad categories: environmental factors, determined mainly by weather conditions; biotic factors; and fungicides. Biotic factors can be sub-divided conveniently into those of the host, i.e. the potato plant, and those of the pathogen, i.e. *P. infestans*. However, it cannot be over-emphasised that effects on blight are the result of complex *interactions* between factors. In order to understand how each factor affects blight, it is necessary to explore the interactions, although the resulting study may sometimes be related only tenuously to field conditions. The dynamic nature of the field environment further complicates matters. Weather changes continuously, as do the physiology of the host and pathogen, and often also the genotype of the pathogen. It may be difficult to relate results from experiments conducted under controlled conditions to a crop in a field.

Because the asexual life cycle dominates disease development, and is usually its sole determinant, I will discuss the effects of various factors on infection and colonisation of foliage by *P. infestans*, sporulation, and spore dispersal and survival.

Environmental factors

Temperature. Sporangia can germinate either directly by the formation of a germ tube, or indirectly by releasing zoospores but leaves are not infected from direct germination (Schöber and Ullrich 1985). The optimum temperature for indirect germination appears to be between 12 and 16 °C (Crosier 1934; Yamamoto and Tanino 1961) although zoospores are released slowly at temperatures approaching 0 °C. At higher temperatures the proportion of sporangia germinating directly increases, although indirect germination does occur at temperatures well above 20 °C. The length of time zoospores remain motile before they encyst and start to germinate is highly dependent on temperature. They swim for up to 24 h at temperatures close to 0 °C, but for only 15–20 min at 24–25 °C. Zoospores are extremely sensitive to desiccation and die almost immediately unless they are in water. They depend for their survival on the plant surface remaining wet until they have infected so, temperature and duration of surface water together determine whether or not infection occurs. Crosier (1934) and Rotem et al. (1971) observed germination of zoospores between about 3 and 28 °C. It was fastest at 21–24 °C. Germ tube elongation was most rapid at about 21 °C and infection could occur within 2–3 h of inoculation.

It is unrealistic to separate germ tube growth from colonisation of plant tissues by hyphal growth and, indeed, they have similar cardinal temperatures, with hyphae growing slowly at 2–4 °C and not at all above about 28–30 °C (Zan 1962). Lesions readily visible to the naked eye do not normally appear until at least 2–3 days after inoculation, even at temperatures that favour hyphal growth. The production of sporangia (sporulation), a process that can be considered to be modified hyphal growth, has a similar minimum temperature to germination and colonisation of tissues, but the maximum temperature of 25--26 °C for sporulation is a few degrees below that for the other processes. The optimum temperature for both the speed of formation of sporangia and their final abundance has been variously reported to be 16–22.5 °C. Fresh sporangia can be formed under optimum conditions a few hours after the removal of old spores from a lesion.

Temperature probably has little direct effect on spore dispersal but it is a major factor in the survival of *P. infestans*. The low temperatures required to kill the fungus would completely destroy a potato crop, so they are irrelevant to disease development during the growing season. Effects of temperature on survival are therefore of concern only at high temperatures. Survival of sporangia in air depends on an interaction of factors, chiefly temperature, humidity, irradiation and the physiological state of spores, determined by factors such as their age and water content. For example, a change in temperature, caused by a change in irradiation, affects the relative humidity and the rate of water loss from sporangia and so their sensitivity to the potentially damaging effects of high temperatures. Sporangia die rapidly in air above about 25 °C (Smith 1915), and Wallin (1953) reported that temperatures as low as 20 °C reduced their viability. However, survival at different temperatures is highly dependent on humidity. Rotem and Cohen (1974) found that sporangia deposited on leaf surfaces, where the humidity is usually high due to transpiration, survived adverse conditions well. There is little information on direct effects of temperature on zoospore survival but the drying of plant surfaces, resulting in desiccation of zoospores, is itself affected by temperature. Once established in plant tissues, *P. infestans* is remarkably tolerant of high temperatures. While it may not grow at temperatures above 30 °C, it can survive much higher temperatures, even those above 40 °C for short periods and recommence growth and development when the weather becomes more favourable (Rotem and Cohen 1974), although its capacity to produce sporangia may be impaired by exposure to high temperatures (Kable and MacKenzie 1980). It is important to appreciate that leaf and stem temperatures can be considerably higher than air temperatures, especially when foliage is exposed to direct sunlight (Ansari and Loomis 1959).

Infection of tubers is encouraged by moderate soil temperatures and is markedly decreased above 18 °C (Sato 1979).

In the interaction between temperatre and other factors such as duration of surface moisture and humidity, temperature limits are broad when other factors are favourable and narrow when they are unfavourable (Rotem et al. 1971).

Temperature also has an indirect effect on asexual development of *P. infestans* through its effects on the potato plant, considered later in this chapter, and loss of fungicide activity.

Humidity. Humidity is a major determinant of blight development, yet it is impossible to control humidity precisely, or to measure it in the zone of air where infection and sporulation occur as no instrument is small enough to measure humidity adjacent to plant surfaces. Transpiration humidifies the air adjacent to leaves so, although the air above a crop may be relatively dry, it is close to saturation next to an unwilted leaf. The gradient of decreasing humidity away from a leaf surface depends on the rate at which transpired water vapour is swept away which in turn depends on air speed. Further, leaves heated by sunlight warm the adjacent air and lower its relative humidity (RH), further complicating the estimation of humidity next to a leaf. The net result of these effects is that we do not know precisely how humidity affects microorganisms on plant surfaces. However, sporangia and zoospores germinate only in the presence of liquid water so humidity influences infection indirectly through its effects on the rate of drying of surface moisture.

Once *P. infestans* has become established within the potato plant, RH between 80 and 100% has no effect on colonisation (Harrison and Lowe 1989). Lower humidities may cause desiccation of severely infected tissues, limiting the availability of water to the pathogen and reducing its growth rate.

There are numerous reports that sporangia are formed only at high humidities (e.g. De Weille 1963) but the precise requirements close to a plant surface are unknown. Harrison and Lowe (1989) demonstrated an interaction between ambient humidity and air speed in determining numbers of sporangia formed on leaflets. Abundant sporangia were formed at low air speeds when the ambient humidity was 90–100% RH, but not at 85 or 80% RH. At higher air speeds numerous sporangia were formed at 100% RH, but not at 95–80% RH. These results suggest that sporangia are formed only if the humidity immediately surrounding the pathogen is at or very close to saturation. Harrison and Lowe also investigated sporulation when the humidity alternated between 80 and 100% RH. Numbers of sporangia formed were intermediate between those at each humidity continuously and higher than the number formed at a constant 90% RH.

Hirst (1958) thought that variations in humidity aid dispersal of sporangia, although the changes in RH that he observed were largely the result of temperature changes. Rotem and Cohen (1974) considered that low humidities enhance dispersal.

Humidity has a large effect on survival of spores. Sporangia of *P. infestans* can survive for several days in saturated air (Rogoshin and Filippov 1983) and survive well on leaf surfaces, but viability is lost rapidly in drier air (Rotem and Cohen 1974) and more so at high than at low temperatures. Warren and Colhoun (1975) considered that survival of sporangia depended on a surrounding film of water, without which sporangia lost their viability after only 5 min in air at 95% RH. Minogue and Fry (1981) presented evidence that the germination ability of partially desiccated sporangia depended on the rate of subsequent rehydration. A low rate favoured survival, while only a few spores germinated after rapid rehydration. Survival of sporangia in relation to environmental conditions almost certainly depends on spore age (considered later in this chapter).

Zoospores die rapidly in the absence of liquid water (Lapwood 1968) and there are no quantitative data on their survival at different humidities.

Precipitation and dew. Rain, drizzle, mist, overhead irrigation and dew formation result in wet foliage, a prerequisite for infection. Dew usually forms at night as temperatures fall. The rate at which it forms on leaves is related to temperature, ambient humidity and wind speed (Collins and Taylor 1961) and, conversely, the rate at which surface water, however deposited, evaporates is also related to temperature (determined partly by irradiation intensity), ambient humidity and wind speed. These meteorological variables therefore interact to determine the duration of surface moisture and the likelihood of sporangia releasing zoospores and of zoospores germinating and infecting the host. Duration of surface moisture itself interacts with inoculum density and temperature to determine whether or not germination and infection occur. Longer periods of wetness are required for germination as the temperature deviates from the optimum, and high spore loads inhibit infection when the wet period is short (Rotem et al. 1971). A short period during which the leaves are dry, initiated within the first 3 h after inoculation, substantially reduces the number of lesions that develop, but temporary surface dryness occurring later generally has less effect, presumably because infection is better established (Hartill et al. 1990). Subsequently, surface water probably has little or no effect on colonisation of plant tissues.

Although formation of sporangia occurs after rain or dew, liquid water is not required (Harrison and Lowe 1989). Sporulation is undoubtedly encouraged by the high humidities close to foliage associated with surface moisture, rather than by the liquid water itself. Indeed, abundant water on foliage may suppress sporulation of *P. infestans* (Rotem et al. 1978).

Precipitation and wind are the main factors controlling the dispersal of sporangia of *P. infestans*. Rain, drizzle and overhead irrigation wash spores from leaves and stems into the soil where they can remain infective for several weeks (Lacey 1965) and may infect tubers (Lapwood 1977) or may be splashed on to leaves to initiate lesions. Intensity and duration of precipitation determine the extent to which these occur. Sporangia are readily incorporated into water falling on them. Splash droplets containing spores, produced when water droplets hit sporangia on a leaf or stem are responsible for transporting the pathogen over short distances within a crop (MacKenzie et al. 1983). Fitt et al. (1986) found that fungal spores were dispersed in splash droplets in still air as far as 1 m from infected leaves. Numbers of spores declined exponentially with distance from the source and decreased rapidly with increasing height. Fewer than 10% of spores were splashed to a height of more than 300 mm. Large raindrops dispersed more spores than did smaller drops. Duration and intensity of precipitation and the size distribution of water droplets are therefore important in determining the extent of dispersal in this manner. Hirst (1958) thought that about half the foliage blight lesions in a crop originated from rainwater dripping from one leaf to another. A small proportion of droplets splashed from foliage and containing spores may be launched into the prevailing air flow. The role of wind in the dispersal of *P. infestans* is dealt with in the following section. Rain is important in removing sporangia from the air (Hirst 1953).

Wind. Wind is not known to affect colonisation of tissues by *P. infestans* but modifies infection of aerial plant parts indirectly through drying surface moisture. Wind affects sporulation through its effect on humidity close to foliage depending *inter alia* on the amount of haulm and the geometry and orientation of individual leaves. In the more sheltered, lower parts of the canopy wind speed is less than at the top of the haulm so there is usually a gradient of decreasing humidity with increasing height from the soil surface, reflected in a tendency for fewer spores to be produced in the upper parts of the canopy. However, turbulence increases with wind speed, reducing gradients of both humidity and temperature, so that the environment within the canopy becomes more uniform at higher wind speeds. Wind speed just above a crop also varies with time – wind is usually gusty, and in space depending on the topography of the land and shelter from trees and buildings. This dynamic interaction of factors that determines air movement within a crop and thus the duration of surface moisture and conditions suitable for sporulation and germination, makes it difficult to quantify from weather data the sporulation and infection that are likely to have occurred. Conditions may favour sporulation and infection in some parts of a field and not in others.

The principal effect of wind is on dispersal of sporangia but few sporangia are detached by wind alone, even at speeds of up to 15 m/sec (Stepanov 1935; Hirst 1958). Wind can spread rain-splash droplets containing spores within the crop or to adjacent fields (Hirst and Stedman 1960) often resulting in a fan-shaped pattern of disease from spreading a locus of infection. Being blown by the wind is the sole means by which spores can travel long distances. Once they have entered the air above a crop, droplets containing sporangia can be carried many kilometres (Robinson 1976) and to heights up to 1 km, particularly in a strong wind. The importance of wind in long distance spread of blight, however, depends on sporangia remaining infective, favoured by cool, moist, dull conditions.

Irradiation. The spectrum of solar radiation has both direct and indirect effects on *P. infestans* and is an important factor reducing the survival of sporangia (Rotem et al. 1970). The indirect effects are caused by the radiation heating plants, soil and the air adjacent to them, reducing RH, and increasing the rate of evaporation (see the earlier discussion). Bashi et al. (1982) thought that the death rate was aggravated by high temperatures and fully turgid sporangia may be less damaged by exposure to sunlight than partially desiccated spores.

Information on direct effects of solar radiation on the asexual development of *P. infestans* is sparse. At low intensities, the UV radiation may increase the proportion of sporangia that are able to germinate and infect a potato plant, but it can kill spores at high intensities (De Weille 1963; 1964). There are conflicting reports of effects of visible radiation on production of sporangia and its effects on colonisation may be due solely to changes in the physiology of the host which are considered later.

Biotic factors – the potato plant

Genetic resistance. There are two forms of genetic resistance of the potato plant to blight, controlled by major and minor genes respectively, although the distinction between them is sometimes unclear. Major and minor genes are often present together in a potato plant. Major gene resistance, also known as race-specific resistance, usually acts at the infection stage and can effectively result in total resistance to blight. So far, at least eleven major resistance genes (R1 to R11) have been identified each of which can lead to immunity to blight. However, for every host R-gene, *P. infestans* can possess a corresponding virulence gene (virulence genes or factors 1 to 11) that can completely overcome the complementary R-gene and lead to disease susceptibility. A strain, or isolate, of *P. infestans* with virulence gene 1 (i.e. that can overcome resistance gene R1 in the potato plant) is referred to as race 1, etc. A plant may not possess any R-genes (RO) so that it is susceptible to all races of *P. infestans* or it may possess one or more R-genes. Similarly *P. infestans* may have no virulence genes (race 0), or one or more may be present. No blight will develop in the field unless there is a race of *P. infestans* compatible with the crop genotype. However, *P. infestans* readily adapts to overcome R-gene resistance (Malcolmson 1969), which is therefore short-lived, so the incorporation of R-genes into potato genotypes has largely been discounted for blight control, although many potato cultivars do possess R-genes. There is evidence that not all R-genes are expressed to the same extent. In particular R10 is sometimes incompletely expressed (Wastie et al. 1993) so that *P. infestans* without virulence factor 10 can infect and colonise the tissues albeit at a reduced rate. The expression of major gene resistance may be partially dependent on the physiology of the potato plant which itself may depend on the environment and plant age (Stewart 1990). Although the resistance of both foliage and tubers is controlled by R-genes, there is often only a poor correlation between foliage resistance and tuber resistance (Stewart et al. 1992). For example, genes R2 and R3 provide leaf immunity to races of *P. infestans* without virulence factors 2 and 3, but the same isolates may grow in the tubers. In contrast, R1 appears to provide comparable immunity in leaves and tubers (Roer and Toxopeus 1961). These differences may be due to physiological/biochemical factors, e.g. only tubers can produc sesquiterpenoid phytoalexins (Rohwer et al. 1987), or to the involvement of other genetic factors.

Minor-gene resistance is thought to be controlled by many genes whose effect is additive. It can reduce the proportion of zoospores that infect the foliage, but its main effect is in reducing the rate of colonisation (Colon et al. 1992) so it does not confer immunity. However, the relative importance of the different resistance components depends on cultivar. Being under multi-genic control, minor gene resistance is much more durable than R-gene resistance and offers the best potential for long-term blight control using a minimum of fungicides (Shtienberg et al. 1994). However, it is sometimes difficult to separate the effects of minor and major gene resistance (Gees and Hohl 1988). There is also evidence for synergism between the two forms of resistance (Darsow et al. 1987). It has become increasingly apparent in recent years that expression of minor gene resistance of some cultivars is strongly influenced by the environment (Simmonds and Wastie 1987). This genotype-x-environment interaction was investigated by Harrison et al. (1994) who identified three environmental components (photoperiod, light intensity and temperature) each of which could effectively switch on or off minor gene resistance in leaves. Depending on the cultivar, resistance could be switched by one or more environmental factors, or be unaffected by any of them. The possibility that R-gene resistance may be similarly influenced by the environment cannot be excluded. Minor genes can also determine the resistance of tubers to *P. infestans* (Bjor and Mulelid 1991). Again, foliage and tuber resistance may not be closely correlated.

Neither major- nor minor-gene resistance appears to have a direct effect on sporulation of *P. infestans* once infection has occurred.

The structure of the leaf canopy, which is largely genetically determined, can also affect blight susceptibility by, for example, impeding air circulation resulting in slow drying of foliage and high humidities within the canopy (Bonde et al. 1940).

Physiology. The physiological state of a potato plant has a large influence on blight susceptibility. Although the mechanisms involved are poorly understood, certain factors have been identified that affect infection and colonisation. Effects on sporulation are usually indirect and simply reflect the extent of colonisation.

The physiological age of a plant, itself determined by growing conditions, particularly temperature and the length of the growing period, affects the susceptibility of leaves to infection. Leaves of young plants are highly susceptible to infection. As plants age the leaves first become more resistant to infection and later become more susceptible again (Stewart 1990). Carnegie and Colhoun (1982) reported the self-evident effect that the most blight resistant cultivars showed the greatest changes in susceptibility. Most reports relating differences in blight susceptibility to plant age do not distinguish between effects on infection, colonisation and sporulation (e.g. Darsow et al. 1988).

Leaf position, which is generally confounded with leaf age, also affects susceptibility to infection. Fry and Apple (1986) thought that the oldest leaves were most susceptible but Warren et al. (1971) found that, in mature plants, leaves in the middle zone of a stem were the most resistant, and that leaf position had no effect in young plants. Populer (1978) reported that effects of leaf position on susceptibility to infection depended on cultivar.

Rate of colonisation of leaf tissues similarly depends on leaf position and age. Carnegie and Colhoun (1982) reported that the rate of lesion expansion generally increased linearly from the apex to the base of a plant, but that on young plants these differences were absent.

As with foliage, the age of tubers affects their susceptibility to blight (Pathak and Clarke 1987).

The maturity class of a potato cultivar is another factor that modifies the incidence of blight with differences probably related to foliage growth (Van Oijen 1991b). Early varieties are generally less attacked by blight than later ones (Kolbe 1982), possibly because growth is often completed and the crop harvested before inoculum levels become high enough to cause extensive disease. Resistance to infection, colonisation of tissues or sporulation are probably unaffected by maturity class.

Mineral nutrition of the host affects its susceptibility to blight and Rotem and Sari (1983) suggested that infection of leaves, colonisation and sporulation are all affected by host nutrition. Plants grown with high levels of nitrogen are particularly susceptible to both foliage and tuber blight and have higher rates of lesion expansion (Carnegie and Colhoun 1983; Kurzawinska 1989) with genotypic differences (Main and Gallegly 1964). Awan and Struchtemeyer (1957) reported that large applications of phosphorus or potassium decreased the size of leaf lesions, with P having the greater effect. Nutrient levels in leaves can change lesion diameter by a factor of 2–3 which, presumably, reflects on the potential number of sporangia that can be formed. Borys (1964) found that mineral nutrition, especially chloride and sulphate ions, strongly affected the susceptibility of leaves to infection, the effects depending on leaf age and cultivar. Some of the effects of mineral nutrition, especially N, on blight susceptibility may be indirect, by encouraging luxurious haulm growth or by delaying leaf senescence.

The carbohydrate status of potato plants has been implicated as a factor affecting blight susceptibility (Grainger 1979) but its importance remains unclear.

Carnegie and Colhoun (1980) reported that potato plants under water stress had increased resistance to infection by *P. infestans*, but they offered no explanation of this observation.

Virus infection. Infection of potato plants with any of several viruses tested increased the resistance to blight of leaves (Pietkiewicz 1974) and tubers (Fernandez de Cubillos and Thurston 1975). The presence of virus increases resistance to both infection and colonisation by *P. infestans*, delays the onset of sporulation and reduces the number of sporangia formed. However,

Richardson and Doling (1957) reported that the rolled leaf surfaces of plants infected with potato leaf roll virus remained wet for longer than the flat surfaces of uninfected leaves, encouraging infection, so that more blight lesions developed in the field, although in laboratory tests, small leaf pieces from virus-infected plants were less susceptible to infection than pieces of virus-free leaves.

Surface microflora.
Naturally-occurring bacteria that colonise the surface of tubers have recently been shown to be antagonistic to *P. infestans* (S.A. Clulow, personal communication) but there is no known effect, either as promotors or inhibitors, of microorganisms that occur naturally on the phylloplane.

Biotic factors – P. infestans

Genotype. The importance of virulence factors in overcoming R-gene resistance to blight in potato foliage and tubers has been explained. *P. infestans* can acquire new virulence factors either through mutation (Denward 1970) or through somatic variation (Caten and Jinks 1968). Even in the absence of sexual reproduction, new races can be produced when colonies of two different races are grown together (Leach and Rich 1969; Malcolmson 1970) or when a single isolate is passaged through hosts with different R-genes (Graham et al. 1961). There is a strong selection pressure in the field favouring races of *P. infestans* that are pathogenic on the crop, so the population structure of the pathogen may change during the growing season (Caten 1970; Shattock et al. 1977).

The aggressiveness of *P. infestans*, a characteristic that is independent of virulence, depends on the isolate of the pathogen and has a major effect in determining rates of infection and colonisation of leaves and tubers, and generation time (period between inoculation and sporulation) on leaves (Latin et al. 1981). Aggressiveness, like most physiological parameters, is difficult to quantify. It is also rather unstable and can increase or decrease over many generations of an isolate (Jinks and Grindle 1963). Caten (1970) suggested that directional selection for rapid growth and profuse sporulation operates in the field. He provided evidence for adaptation to specific cultivars, supporting the report of Latin et al. (1981) that minor gene resistance and isolate (aggressiveness) interact to determine blight severity. The results of Bjor and Mulelid (1991) using tubers without R-genes also suggested that adaptation of *P. infestans* to specific cultivars may occur and that minor gene resistance to blight may eventually become eroded.

Spielman et al. (1992) related individual components of pathogen fitness (aggressiveness) to disease progress. They concluded that sporulation capacity was most closely related and infection frequency was least related.

P. infestans can become adapted to the climate of a region. For example, the optimum temperature for sporulation of isolates from Egypt is several degrees higher than that of British isolates (Shaw 1987). Such differences, presumably genetically-determined, may account for apparent discrepancies in reports on the sensitivity of *P. infestans* to temperature.

There are numerous reports of fungicide tolerance in *P. infestans* (e.g. Holmes and Channon 1984). Populations of *P. infestans* in the field are normally heterogeneous, particularly late in the growing season, and consist of both fungicide-susceptible and fungicide-tolerant types in a dynamic equilibrium. The use of a fungicide shifts the balance between the two in favour of strains that tolerate the fungicide (Levy et al. 1983). In the absence of fungicide, metalaxyl-resistant isolates of *P. infestans* showed strong competitive ability and their proportion increased from 10 to 100% after 8–10 sporulation cycles (Kadish and Cohen 1988). This strong competitive ability was attributed to a rapid rate of colonisation of leaves, rather than to differences in infection or sporulation capacity. However, metalaxyl-resistant isolates may not survive in tubers as well as sensitive isolates (Kadish and Cohen 1992).

Physiology. There is a paucity of information on possible effects of the physiological state of *P. infestans* on blight development, despite the fact that effects of the environment clearly have a physiological basis. Moreover, the genotype of *P. infestans* has a strong influence on its physiology, so environmental effects, genetics and physiology are all inter-related.

Physiological and biochemical changes associated with ageing may render all stages of the pathogen more vulnerable to adverse conditions, so that, for example, old sporangia may lose their viability and die more rapidly than freshly-produced spores. The author has observed that zoospore release is slower from old than from young sporangia. It therefore follows that old sporangia would require a prolonged period of surface wetness for infection. Bashi et al. (1982) reported that sporangia dispersed late in the day were more infectious than those that had been dispersed earlier. De Weille (1963) considered that newly-formed sporangia needed a period of maturation before they would germinate. As mature spores age their environmental requirements for germination may become more rigorous so that zoospores can be released only within a narrow temperature range. Effects of ageing or nutritional status on colonisation of the host and on sporulation are speculative.

Inoculum density. The amount of inoculum, effectively number of sporangia, obviously has a major bearing on disease development. In the absence of inoculum, no disease develops at all. Under a given set of conditions the amount of blight that develops on foliage is directly proportional to inoculum quantity, but it also depends on the amount of healthy foliage available for infection and colonisation, which at any particular time, itself depends on preceding rates of disease progress and haulm growth. In the initial stages of blight, the inoculum depends on number and proximity of infected shoots, often in cull piles or refuse heaps, but sometimes in the crop itself or on volunteer plants in other fields (Van der Zaag 1956). The inoculum also depends on the suitability for sporulation and spore dispersal of weather conditions and on the direction of the prevailing wind. As blight progresses, sporangia formed within the crop dominate, although spores from sources outside the crop remain important in the introduction of new races, strains or a second mating type of *P. infestans*.

There is a marked diurnal periodicity in the concentration of airborne sporangia, with a maximum usually occurring during late morning (Hirst 1953), which, together with the suitability of conditions for infection, will influence the number of new lesions initiated.

The extent of tuber infection clearly depends on numbers of sporangia formed on the foliage.

Fungicides

Although the scope of this paper is restricted to the 'natural' ecology of blight, the almost universal application of fungicides for its control means that factors affecting their efficacy in slowing down blight development should be considered briefly. Again, many factors interact, this time to determine the effect of fungicides. The most important are fungitoxicity, itself dependent on fungicide tolerance by *P. infestans*, rate and frequency of application of fungicide and its distribution within the crop canopy, type of fungicide (e.g. systemic or contact), its redistribution and the rate of decline of fungicidal activity (determined mainly by precipitation and temperature, and the properties of the fungicide), and the rate of growth of new foliage. Again, some of these factors, such as the distribution and dynamics of fungicides, are difficult to quantify.

Factors affecting the sexual cycle of *P. infestans*

The potential importance of sexual reproduction of *P. infestans* in the epidemiology of late blight has already been stressed, but knowledge of factors that affect the production and germination of oospores is scant. While oospores are produced freely when hyphae of both mating types grow together, they can also be formed in the laboratory as a result of self fertilization of isolates of either mating type (Skidmore et al. 1984; Campbell et al. 1985). The importance of self-fertilization in the field, if it occurs at all, is not known. Chang and Ko (1990) claimed that *P. infestans* isolates of mating type A1 could change into type A2 when grown in the presence of metalaxyl. As the author knows of no other reports of a change of mating type, that report may be anomalous.

Harrison (1992) and Drenth (1994) found that oospores could be formed at temperatures from 5 to 25 °C. Numbers of oospores formed depended on host cultivar, with fewer in leaf tissues that were highly susceptible to blight and that rotted quickly, than in tissues that were moderately resistant (Drenth 1994). Harrison considered that the rate of oospore formation was closely related to the rate of hyphal growth. Smoot et al. (1958) observed germination of oospores from 12 to 25 °C. There is evidence that continuous illumination inhibits oospore formation (Romero and Gallegly 1963), and of a light requirement for germination (Shattock et al. 1986). Drenth (1994) hypothesized that oospores at or near the soil surface released zoospores during rainy conditions and that these could infect plant parts in contact with the soil, or be splashed to initiate infections higher in the canopy. Oospores readily survive temperatures as low as −80 °C and up to 35 but not 40 °C (Drenth 1994), but the conditions favouring survival and length of time they can survive in the field are not known.

Conclusions

This paper has attempted to explain how development of potato late blight is controlled by many inter-dependent and constantly-changing factors. There may be others that remain to be discovered, particularly concerning the physiology and biochemistry of the host-pathogen interaction.

Modelling blight progress requires quantification of factors that affect it. While quantification of some environmental factors may not be straightforward, there are much greater problems in attempting to quantify biotic factors, particularly those that determine the degree of susceptibility to blight. A further major difficulty in attempting to develop a realistic model to describe blight progress is the heterogeneity within a field, with disease developing at different rates depending on location. Addressing these two problems offers the greatest potential for improving the modelling of blight. Systems already exist for the precise quantification of infection, colonisation, sporulation and, perhaps less precisely, spore dispersal and survival. A systematic study is required of effects of the biotic factors identified earlier, including interactions between biotic factors and between them and environmental factors, on these components of blight development. Similarly, effects of fluctuating environmental conditions on these same parameters need to be investigated, paying particular attention to possible lag phases resulting from carry-over effects of previous environments. It should be stressed, however, that each of these would be a major undertaking best tackled by large co-ordinated groups of workers. Although no mathematical model will ever be totally accurate, there is still considerable scope for improving existing systems.

Simulation models of potato late blight

M. VAN OIJEN
Research Institute for Agrobiology and Soil Fertility (AB-DLO), P.O. Box 14, 6700 AA Wageningen, The Netherlands
Present address: Wageningen Agricultural University, Dept. of Theoretical Production Ecology, P.O. Box 430, 6700 AK Wageningen, The Netherlands

Abstract. Plant disease epidemiology has its roots in the study of infectious diseases in man. The oldest mathematical model of a human disease stems from the 18th century (Bernoulli 1760), but it was not till two centuries later that mathematical analysis of plant diseases was initiated (Van der Plank 1963). Potato late blight featured prominently in Van der Plank's analyses, and by now there is a rich literature of blight models. Before the mid 1980s blight models focused on the life cycle of the fungus, and on effects of the environment on the various stages in that life cycle. In recent years more emphasis has been given to the role of host growth in epidemic development and loss of yield. Sophisticated pathogen-crop combination models are now available.

Potato late blight modelling has served two purposes mainly. Firstly, blight models were used to evaluate strategies for disease control, especially the scheduling of fungicide application. Secondly, blight models have been used to explain differences in loss of yield among cultivars, and among various temporal and spatial patterns of disease development.

Many models have been made and their behaviour differs strongly. The sensitivity of the models to changes in parameters or inputs depends largely on the structure of the model. The models may be put to better use if more attention is paid to correct initialization and parameterization, and if comprehensive sensitivity analyses are carried out.

Introduction

Potato late blight, caused by the fungus *Phytophthora infestans*, is a major infectious disease of potatoes and other *Solanaceae spp*. The life cycle of the fungus, in which leaf infection is followed by a latent period before lesion growth and sporulation start and spore dispersal leads to new leaf infections, has been described in some detail by Harrison (1995). The life cycle strongly resembles that of many infectious diseases of humans. Present-day models of potato late blight, therefore, still show many similarities to much older models developed for medical purposes. In this chapter we will briefly describe the medical origins of plant disease epidemiological models before discussing recent developments and prospects of blight modelling. We will not strictly follow a chronological order, nor restrict ourselves completely to potato late blight, but the key blight modelling studies that have appeared since 1963 are listed chronologically in Table 1.

A.J. Haverkort and D.K.L. MacKerron (eds.), Potato Ecology and Modelling of Crops under Conditions Limiting Growth, 237–250.
© 1995 *Kluwer Academic Publishers.*

Table 1. Key simulation model studies of *P. infestans*. References are listed chronologically, with indication of new features of potato late blight modeled in the particular study, or areas of improvement on older models. The list is not claimed to be exhaustive

Reference	Keywords
Van der Plank (1963)	Mathematical analysis of epidemics
Waggoner (1968)	Computer simulation of fungal life cycle, Meteorology, EPDEM
Sparks (1980)	Stem lesions
Stephan and Gutsche (1980)	Forecasting, SIMPHYT
Bruhn and Fry (1981)	Tuber yield, Contact fungicides, Cultivar resistance
Minogue and Fry (1983)	Spatial distribution of disease
Paysour and Fry (1983)	Interplot interference in experiments
Fohner et al. (1984)	Fungicide scheduling
Michaelides (1985)	Sensitivity to weather
Milgroom and Fry (1988)	Buildup of fungal resistance to metalaxyl
Ferrandino (1989)	Host growth, Spore distribution, Spatial aggregation of disease
Van Oijen (1989)	Lesion growth, Resistance components
Shtienberg et al. (1989)	Fungicide scheduling against early and late blight simultaneously
Kluge and Gutsche (1990)	Validation in field studies
Levy et al. (1991)	Fungicide mixtures
Michaelides (1991)	Plant density, Overhead sprinkling irrigation
Van Oijen (1991b)	Upward spread of disease through leaf layers, Induced resistance
Van Oijen (1992a)	Host maturity class, Tolerance components
Van Oijen (1992b)	Initialization, Parameterization, Multi-parameter changes
Raposo et al. (1993)	Incorporation of weather forecasts
Shtienberg et al. (1994)	Fungicide scheduling

Human disease epidemiology: the development of SEIR-models

The oldest mathematical model of an infectious disease was devised by Bernoulli (1760). The purpose of his study still sounds very modern although the name of his subject disease happily does not: Bernoulli used his model to evaluate vaccination strategies against smallpox.

Almost two centuries later the first model appeared that, in its structure, resembled the life cycle of infectious disease agents, and therefore may be called the first disease *simulation* model (Kermack and McKendrick 1927). This model, consisting of two linked differential equations, was applied to infectious diseases in human populations of fixed size. In the model *susceptible*, *infectious* and *removed* individuals were distinguished. Susceptible individuals became infectious at a rate directly proportional to the product of the densities of susceptible and infectious individuals. The proportionality constant, later termed *daily multiplication factor* (*DMFR*) in the botanical literature, thus was a measure of rate of production, dispersal and infectivity of infectious propagules. Infectious individuals were removed at a constant relative rate accounting for loss of infectiousness, isolation or death. The reciprocal of the removal rate was the average *infectious period* of diseased individuals. The infectious period thus was implicitly assumed to be exponentially distributed.

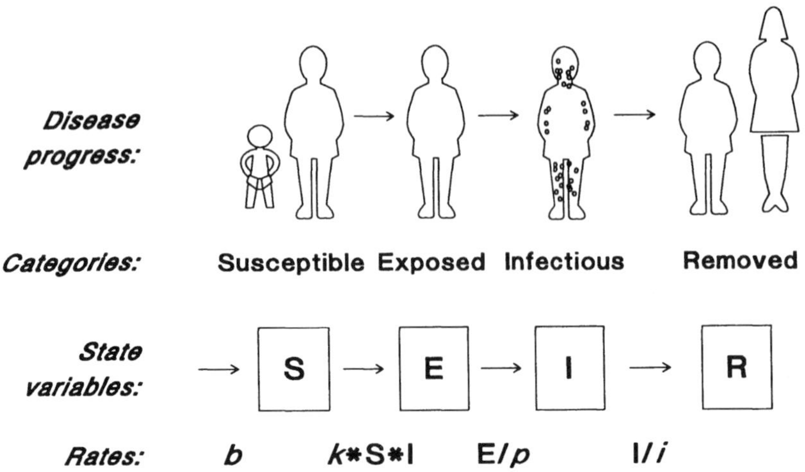

Figure 1. General structure of SEIR-models. Disease progresses by the transition of susceptible individuals (S) to subsequent categories of exposed (E), infectious (I) and recovered or removed (R) individuals. The picture shows terms often used for the transition rates, and their dependence on birth rate (*b*), contact frequency (*k*), latent period (*p*) and infectious period (*i*).

The model of Kermack and McKendrick has been termed the first 'SIR-model', after the initial letters of the three categories of individuals distinguished in the population (see Hethcote 1976). A fourth category was later distinguished: individuals that had already been *exposed* to the disease, but were not yet infectious (SEIR-models, Anderson and May 1982). The period before the start of infectiousness was called the *latent period.* A schematic representation of the structure of SEIR-models, including the most common transition rate terms, is given in Figure 1.

SEIR-models in plant disease epidemiology

Van der Plank (1963) introduced mathematical analysis into plant disease epidemiology. He applied SIR- and SEIR-models to plant diseases, including potato late blight. The latent and infectious periods in his SEIR-model, however, were not exponentially distributed, but constant for all disease units, i.e. lesions. His model is therefore formulated as a time-delayed differential-difference equation, later termed the 'paralogistic' equation (Zadoks and Schein 1979).

Preliminary analysis of the paralogistic equation was carried out by Van der Plank (1963) himself. The analysis was hampered by the fact that the paralogistic equation is analytically less tractable than the original SEIR-model from human epidemiology, even though its dynamics are not dissimilar (Jeger 1986). Some analytical results can be derived if the removal term is left out, which reduces the paralogistic to a 'para-exponential' SEI-model, describing

Table 2. Definitions of the five components of resistance to infectious diseases of plants, as used in this chapter. Note: the daily multiplication factor (*DMFR*), not listed here, is a higher-level parameter that represents the product of sporulation intensity, efficiency of spore dispersal and subsequent infection

Resistance component		Epidemiological meaning
Infection Efficiency	(e: %)	Fraction of sporangia, landed on plant tissue, that successfully initiate a new lesion
Latent Period	(p: d)	Interval between infection and start of lesion expansion and sporulation
Lesion Growth rate	(g: m d^{-1})	Radial rate of expansion of lesions
Infectious Period	(i: d)	Duration of sporulation of lesion area
Sporulation Intensity	(s: m^{-2} d^{-1})	Rate of production of sporangia per unit area of sporulating lesion

only the initial phase of epidemics. Oort (1968) presented a sensitivity analysis of the exponential rate in the initial phase by varying the latent and infectious periods and *DMFR*. He found the strongest sensitivity for changes in the latent period. The parameters tested, and some others, are generally referred to as 'resistance components', for together they determine the level of host resistance to the pathogen (Table 2). Oort's analysis, in fact, constituted the first use in plant disease epidemiology of mathematical models for testing the sensitivity of disease progress to the levels of resistance components. Unfortunately, the initial settings of the model parameters did not correspond to values appropriate for real epidemics, but more realistic and extensive numerical analyses were given by Zadoks (1971; Rabbinge et al. 1989), pointing to a major importance of the latent period in the analyses of 1971, and of *DMFR* in 1989.

Waggoner (1968) presented the first computer simulation model for potato late blight, in which the effect of environmental conditions on the resistance components was extensively modeled. He showed that a wealth of information on late blight could be integrated in one simulation model, but he did not apply his model to any practical or scientific problems. Nevertheless, Waggoner's intricate model served as a paradigm for later plant disease modelers, so that crop disease simulation models, generally, are still rich in parameters and can only be analyzed numerically.

Alternative distributions for the latent and infectious periods

As indicated earlier, latent and infectious periods were assumed to be exponentially distributed in the human disease SEIR-models, while in the first plant disease SEIR-models both periods were supposed to have zero variance among lesions. Alternative distributions were later introduced. Berger and Jones (1985) used distributed delays for the latent period in their model, but kept the infectious period fixed. The method of distributed delays offers a range of distributions, from a step function to almost normal distributions. Berger and Jones (1985) used four delay intervals and thus approached a normal distribution quite closely. This seems to be a realistic distribution for the latent period of fungal leaf diseases (Shaner 1980). Knudsen et al. (1987) used distributed delays for both latent and infectious period.

It is still unclear how important the distributions of latent and infectious period are in epidemic models. Van der Plank (1963) thought that the distribution had little effect on calculated disease progress, while Berger and Jones (1985) had the opposite view.

Stem lesions of potato late blight represent a special category of lesions with extremely long infectious periods, albeit often accompanied by low sporulation intensity. Their special epidemiological role as survival mechanisms under adverse weather conditions was incorporated in the forecasting model of Sparks (1980).

Modelling lesion growth

Zadoks (1977) mentioned five components that should be included in an analysis of the epidemiological importance of the various stages in the fungal life cycle, viz. the resistance components: infection efficiency, latent and infectious period, sporulation intensity and *lesion growth rate* (Table 2). However, a common feature of the models discussed so far is the absence of lesion growth rate as a separate resistance component. This may have historical reasons: no analogon for lesion growth was present in the earlier models for human diseases. Lesions are assumed to occupy a fixed lesion area from the beginning of their latent period. For example, a fixed lesion area of 0.3 cm^2 was used by Michaelides (1985) in his simulation model of *P. infestans*. Michaelides treated sporangium dispersal in great detail, thus complicating the experimental determination of parameter values for his model, while losing realism by oversimplifying lesion growth.

Incorporating lesion growth rate in epidemiological models might be expected to be very important in diseases with indeterminate lesion growth, i.e. without fixed final lesion sizes, as in potato late blight. Berger and Jones (1985) incorporated a constant relative growth rate of the total diseased leaf area in their model. This neglects both the increasing limitation of susceptible host area when the disease progresses, and the fact that even individual lesions generally have decreasing relative area growth rates. This also applies to potato late blight: blight lesions have a constant radial growth rate (Gees and Hohl 1988; Van Oijen 1989) so that lesion areas increase as a quadratic function of time instead of exponentially. A submodel of lesion expansion, with a dynamically changing lesion size distribution, was incorporated in a model of potato late blight by Van Oijen (1989), in which circular lesions were Poisson-distributed over leaflets, while their diameters changed by a radial growth rate that was proportional to the fraction of free leaf area on infected leaflets. Lesions in most pathosystems are not distributed randomly, so the Poisson-distribution might better be replaced by a negative binomial distribution (Waggoner and Rich 1981). Lapwood (1961) reported that even within potato leaves and leaflets lesions of late blight are not randomly distributed. He found relatively many lesions on the distal leaflet and on tips and edges of leaflets. Waggoner and Rich (1981) further suggested abandoning the direct proportionality between lesion formation rate and the product of densities of susceptible and infectious sites. Such nonlinear incidence rates are investigated intensively in human epidemiology (Liu et al. 1987).

Spatial heterogeneity

Epidemiological models usually simulate 'general epidemics', defined by Zadoks and Schein (1979) as epidemics that develop homogeneously in space from homogeneously distributed initial disease. Gradually, however, more models now incorporate the spatial aspect of development of epidemics. Minogue and Fry (1983) modeled the spatial dynamics of potato late blight, and found that blight foci expand with a constant velocity, determined by host resistance and fungicide use. Paysour and Fry (1983) used a model to calculate the level of interplot interference in experiments with potato late blight. They examined the effect of plot shapes, sizes and distances, and showed that a square plot shape minimizes interplot interference. Ferrandino (1989) also made a spatial model of potato late blight, which he used to examine the relation between spatial distribution of disease and yield loss. From his model study he concluded that tuber yield loss is maximal if the disease shows a highly aggregated pattern.

The vertical distribution of disease in the crop may be of epidemiological importance. Potato late blight usually starts in the lower leaf layers and gradually spreads to the top of the canopy (Lapwood 1961; Van Oijen 1991c). It is not yet clear whether this is caused by higher susceptibility of older leaves, better microclimatic conditions for infection in the lower canopy or greater deposition of sporangia low in the canopy. Björling and Sellgren (1955) found 2–4 times as many sporangia deposited on middle and bottom leaves as on top leaves, both in incipient and well-developed epidemics. A preliminary model of the gradual upward spread of blight infection was discussed by Van Oijen (1991b). He showed that resistant cultivars are characterized more by retarded upward spread of the pathogen than by reduced rates of spread within leaf layers.

Host growth

Until a few years ago, only few blight models simulated host growth in a realistic manner. Many models simply considered host leaf area as a forcing function of time (e.g. Bruhn and Fry 1981; Michaelides 1985). Slightly more sophisticated is the approach in which leaf area grows towards an asymptote according to a logistic function of the uninfected leaf area (Berger and Jones 1985; Van Oijen 1989).

Recent years have witnessed a trend to include host growth by means of a more realistic submodel. Several authors have based their host growth submodel on the principle, outlined by Monteith (1977), of a linear relation between crop growth rate and the amount of light intercepted by green leaf area. Ferrandino (1989), Michaelides (1991) and Van Oijen (1992a) all built models of potato late blight according to this principle. The slope of the relation, i.e. the light-use efficiency, is unaffected by late blight (Haverkort and Bicamumpaka 1986; Van Oijen 1991a). So, the fungus does not impair host photosynthesis (Van Oijen 1990), and a host crop growth model need not include sophisticated subroutines for calculating effects of infection on leaf photosynthesis parameters (Rossing et al. 1992).

Waggoner (1990) gave equations for interception of light by green leaf area, when the disease is heterogeneously distributed horizontally or vertically. These equations may conveniently be incorporated in the models based on light interception and utilization.

An important object of study for potato late blight models is the relation between host size and disease escape. Most models show that whenever the susceptible leaf area decreases below a certain threshold, the area of infectious tissue can no longer increase, and that some susceptible tissue will remain uninfected when the epidemic has ended (Kermack and McKendrick 1927; Van der Plank 1963). The size of this fraction of the host area escaping the disease depends on a nonlinear function of all resistance components that influence the epidemic rate, thus pointing to a close link between host characteristics that determine escape and those that determine resistance (Van Oijen 1989).

Stochasticity

The early models of human infectious disease, e.g. the model of Kermack and McKendrick (1927), were deterministic (Bailey 1975). Gradually, however, stochasticity has been incorporated and is now featured in most human epidemiological models (Becker 1979). In contrast, most models of plant disease epidemics are still deterministic (Gilligan 1985). The results of these models may differ from those of stochastic models, in spite of the large number of disease units involved, because of the non-linearity of pathosystems (Rouse 1991). Preliminary stochastic models have been presented by Teng et al. (1977) and Sall (1980), for barley leaf rust and grape powdery mildew, respectively. In these models, parameter values for a subset of the resistance components are drawn from uniform or normal probability distributions. In some recent models, e.g. the blight model of Ferrandino (1989), spore dispersal is treated stochastically to account for variable wind patterns not incorporated in the model.

The stochastic models provide estimates of the variation in rates of disease progress. Such estimates of variation are useful if the model is used in disease forecasting, because they set boundaries to the reliability of the forecasts. However, if the model is used for explaining system behaviour in terms of its components or to assess the importance of resistance components for breeding purposes, this variation is unwanted because it obscures the relation between the individual components and rate of disease progress, and thus complicates the identification of the major components.

Blight models applied in fungicide scheduling

Much effort, costs and environmental hazards are involved in the regular spraying of potato crops with fungicide against *P. infestans* (Harrison 1995). To assist in this activity, many mathematical models have been made to forecast the development of blight epidemics depending on past and expected weather conditions. With few exceptions (Kluge and Gutsche 1990), most of these models are of empirical nature, directly linking spray advice to weather variables, without explicitly modelling fungal population dynamics, let alone the relation with the host crop. Therefore, these models fall beyond the scope of this chapter, but recent reviews on this topic are available (Bajic 1988). However, simulation models have been applied in their own way to the question of fungicide scheduling. Rather than function as forecasters themselves, simulation models have repeatedly been used to evaluate the effectiveness of empirical strategies for fungicide scheduling (Fohner et al. 1984; Shtienberg et al. 1994). Unfortunately, the general conclusion from these simulation studies was that little economic gain may be expected from refraining from spraying regularly, weekly or biweekly. Apparently, late blight is too explosive a disease to warrant taking the risks of skipping sprays while weather forecasts are still unreliable.

A special role of simulation models in guiding use of fungicides has been to assess the risk for buildup of resistance against systemic fungicides like metalaxyl (Milgroom and Fry 1988; Levy et al. 1991). Levy and coworkers (1991) showed in a modelling study that the use of fungicide mixtures may delay the buildup of resistance in the pathogen population.

Blight models applied in resistance breeding

Straightforward multiplication of the resistance components can be considered as the simplest model for assessing their combined effect on the initial exponential development of epidemics (Zadoks 1977). Van der Zaag (1959) indeed found that the ranking of potato cultivars for partial resistance to *P. infestans* closely followed the ranking of their multiplied resistance components. However, the analyses of Van der Plank (1963) and Oort (1968) have shown that

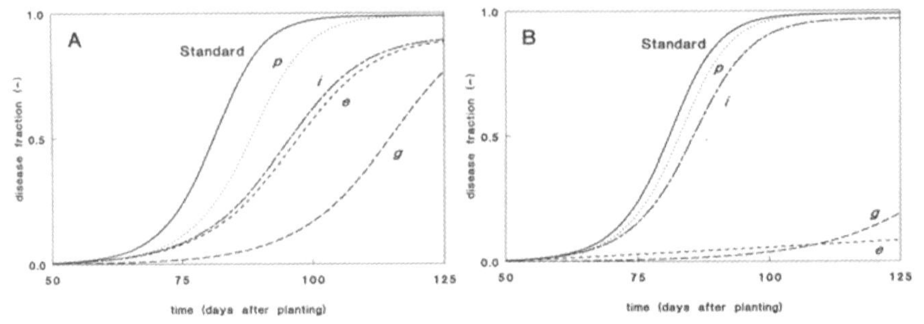

Figure 2. Simulated disease progress curves of potato late blight in various host cultivars. (A) A standard susceptible cultivar (cv. Bintje) and hypothetical partially resistant genotypes with half the standard value for infectious period (*i*), infection efficiency (*e*), or lesion growth rate (*g*), or a doubled value for latent period (*p*); (B) the same standard susceptible cultivar, and genotypes in which *i*, *e*, *g* or *p* is set to the most resistant value observed among current cultivars. (Source: Van Oijen 1992b.)

this method unjustifiably considers all components to have the same effect on total resistance, and the method has not been applied frequently.

Van Oijen (1992a,b) used his model to identify the resistance components that should be improved to breed more resistant cultivars. A sensitivity analysis of the model showed that disease progress rate was most sensitive to changes in lesion growth rate, followed by infection efficiency and infectious period, and, finally, the latent period (Figure 2A). However, this result in itself did not warrant advising breeders to concentrate solely on decreasing lesion growth rate. Breeding perspectives depend not only on the intrinsic importance of plant characters, but also on the scope for changing the characters, i.e. the available genetic variation. Therefore a second sensitivity analysis was done, in which the genetic variation in the different resistance components was taken into account. This more realistic analysis pointed to both lesion growth rate and infection efficiency as prime targets for improving the level of partial resistance to potato late blight in breeding programmes (Figure 2B).

Problems with use of blight models: initialization and parameterization

When the performance of epidemiological models is tested, the status of the initial inoculum is rarely considered. However, the way a model is initialized affects its behaviour in various ways. For example, model epidemics starting from equally-aged *latent* lesions differ strongly from epidemics started from *infectious* lesions, if the latent period is not treated as a constant (Jeger 1986). When the model is initialized with *too many* lesions, sensitivity of disease progress rate to changes in infection efficiency can be underestimated (Van Oijen 1992b). Other problems with initialization are found in models with time delays, as the paralogistic equation of Van der Plank (1963), where the level of disease must be defined for some period (equal to the maximum delay in the model) preceding the simulated period.

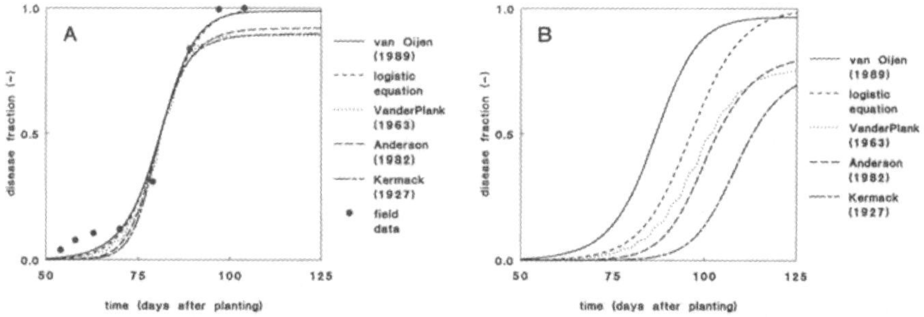

Figure 3. Simulated disease progress curves of potato late blight, generated by five different models. (A) Best fit of the models on disease progress data for the susceptible cultivar Bintje (field measurements 1988); (B) the effect of reducing the infection efficiency parameter by 25% in each of the five models. (Source: Van Oijen 1992b.)

Parameterizing epidemiological models can be difficult if the model parameters correspond poorly to quantities that are actually measured. Methods of measuring resistance components often aim at maximum discrimination between cultivars or treatments, or ease of measurement. For modelling purposes, however, components should be measured in a way that corresponds best to their function in the life cycle of the pathogen. Therefore, measurements of lesion *growth rate* are more useful than measurements of lesion *size* at an arbitrary time after inoculation. Determining the total sporangium production *per unit of area* of diseased leaf tissue is better than the more common practice of quantifying sporulation as number of sporangia washed off *per leaflet or lesion*, again at arbitrary times after inoculation. Only measurement of sporulation intensity on an area basis allows a clear separation of sporulation intensity from lesion size (i.e. integrated lesion growth rate) and infectious period in fungal diseases with expanding lesions.

Rate of disease progress is generally quantified differently in measurements and component models, thus complicating model validation. Total visibly diseased tissue is measured, while latent, infectious and removed tissue are modeled. Models with a fixed size for all lesions, including latent ones, give special problems here, since in actual epidemics latent lesions occupy much less leaf area than visible lesions. The paralogistic equation of Van der Plank (1963), for example, calculates the total infected (latent + infectious + removed) leaf area, whereas only infectious and removed leaf area can be observed. This is a further argument in favour of models that explicitly include lesion growth, where latent lesions can realistically have zero or negligible area.

The sensitivity of the models to changes in parameters is often used as a measure of the importance of those parameters for disease progress rate. However, one should use this approach with caution, because sensitivity to parameter changes strongly depends on model structure. The sensitivity, therefore, is a feature of the model and may not reflect true sensitivity of the epidemic. This was shown in a comparison of five different models of potato late blight (Van Oijen 1992b). In each model, the parameter that quantifies infection efficiency was reduced by 25%. When infection efficiency was thus altered, disease progress rate was reduced in all models, but to widely differing extent (Figure 3).

Concluding remarks

As described in the preceding paragraph, many modelling studies have suffered from inadequate parameterization and insufficient validation. Even well-validated models, however, may be put to better use. Too often a sensitivity analysis for resistance components is missing (Jeger and Groth 1985), although the usefulness of such analyses for guiding breeding efforts has been emphasized repeatedly (Zadoks 1971; 1977). Whenever such a sensitivity analysis has been performed, the analysis has usually been restricted to single-parameter changes. Simultaneous changes of more than one resistance component should be evaluated too (Van Oijen 1992b). These multi-parameter changes should take into account the fact that some parameters may not vary independently in the real pathosystem, if they are genetically or physiologically linked.

Lesion growth rate should be explicitly included in models of fungal leaf diseases with indefinitely expanding lesions. The importance of this component has been shown by the few models that do include it: rate of disease progress calculated by these models shows great sensitivity to changes in the lesion growth rate.

If a disease spans a long period of the host growing season, which applies to most fungal leaf diseases, especially when partially resistant cultivars are evaluated, the disease is incorrectly simulated by models that consider the host leaf area to be constant. For such diseases the effect of the pathogen on host growth should be explicitly modeled, primarily if the model is to be used for analysing effects of host characteristics on loss of yield due to the disease. In the case of potato late blight, simulation of tuber growth and infection may require more attention than so far.

Despite the cautionary remarks given above, it is still clear that simulation models of potato late blight have been used successfully in many respects. Blight models have been useful in epidemiological analysis, for the evaluation of fungicide application strategies, and for the identification of host characteristics that improve resistance. Probably these areas of modelling, aiming at explanation and evaluation rather than forecasting, will remain the proper territories for blight simulation models in the near future.

Life cycle and ecology of *Verticillium dahliae* in potato

L. MOL[1] and A.J. TERMORSHUIZEN[2]

[1] *Department of Agronomy, Wageningen Agricultural University, Haarweg 333, 6709 RZ Wageningen, The Netherlands*

[2] *Department of Phytopathology, Wageningen Agricultural University, P O Box 8025, 6700 EE Wageningen, The Netherlands*

Abstract. *Verticillium dahliae* is a serious pathogen in most countries where potato is grown. The density of microsclerotia of *V. dahliae* in soil mainly depends on the cropping history. Plant roots can be colonised if microsclerotia germinate in the vicinity of the root tip. Colonisation is followed by systemic infection of the vascular system of the plant. During colonisation of the root cortex and systemic infection of the plant interactions with many soil organisms. Cultural practices can lessen the colonisation of the roots and the severity of the disease.

In the vascular system of the plant, *V. dahliae* is dispersed by conidia and mycelial growth. Wilting symptoms appear after the reactions of the plant to the presence of the pathogen. Yield reduction is mainly caused by closure of the stomata and early senescence of the canopy after blockage of the vascular system of the haulm.

The fungus forms microsclerotia on dead plant tissue. External climatological factors and haulm killing practices have a large influence on the number of microsclerotia formed per unit haulm material. Since *V. dahliae* has a very broad host range, attention should be paid to the control of this pathogen in all crops in a rotation. In some crops host specificity has been found, but this is a gradual property.

Introduction

Diseases caused by *Verticillium dahliae* Kleb. are wide-spread throughout the world, wherever susceptible crops are grown, and are of economic importance in most countries (Pegg 1984). *Verticillium dahliae* is supposed to be the major component of the potato early dying complex that damages crops by causing early maturity of the crop by wilting. Hosts include all dicotylodonous plants, with the most important crops affected being potato, cotton, egg-plant, tomato, mint, and olive. *Verticillium dahliae* survives by means of microsclerotia (MS). The high survival potential of MS and the wide host range make *V. dahliae* endemic to many agricultural soils (Powelson 1970).

Verticillium dahliae is classified as a soil invading or root inhabiting fungus (Powelson 1970). These fungi are characterised by a parasitic phase on the living host plant (Figure 1: I), and by a saprophytic phase after the death of the host (Powelson 1970) (Figure 1: II). The two phases will be discussed separately.

A J Haverkort and D K L MacKerron (eds), Potato Ecology and Modelling of Crops under Conditions Limiting Growth, 251–264
© 1995 *Kluwer Academic Publishers*

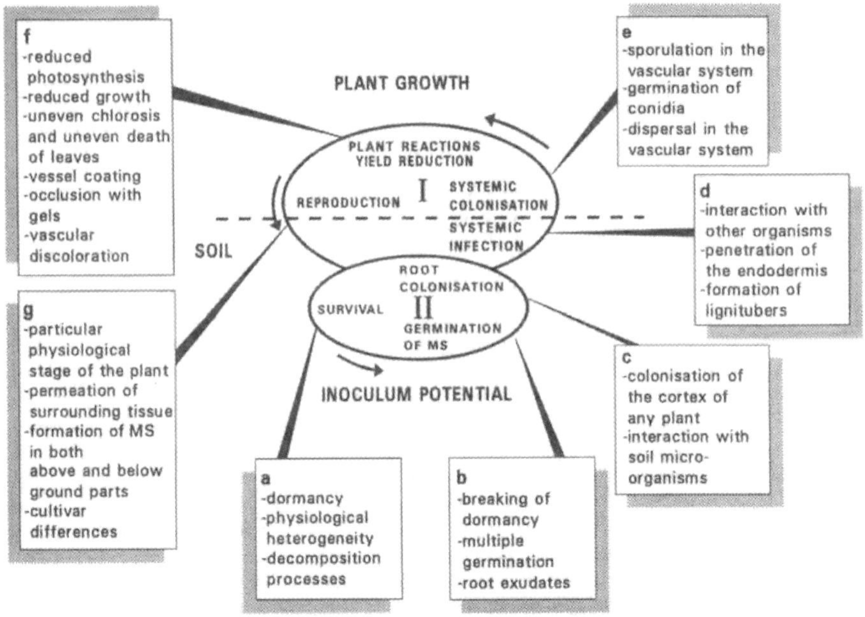

Figure 1. Schematic life cycle of *V. dahliae* showing a parasitic phase on a host (I), and a saprophytic phase after plant death and in soil (II). The scheme is explained in the text. MS = microsclerotia.

Dynamics of the soil population of *V. dahliae*

A schematic representation of processes concerning the MS population in the soil is shown in Figure 1: II. The population of MS in soil varies in composition and density depending on the cropping history of the field. After germination of a microsclerotium, the hypha may infect a plant root. Due to their low competitive saprophytic ability, the majority of the hyphae are not successful in reaching a plant root and they die. The next paragraphs will be focused on: survival of MS (Figure 1: a), germination of MS (Figure 1: b), and colonisation of plant roots (Figure 1: c).

Survival of MS (Figure 1: a)

Ben-Yephet and Szmulewich (1985) reported that MS of *V. dahliae* survive longer in the field than in the laboratory. After 5 years of storage of air-dried soil samples at 20–25 °C no *V. dahliae* could be detected, whereas 4% of the original population density remained viable after 7 years of crop rotation. Wilhelm (1955) found that *V. dahliae* persisted for 14 years in field soil with no hosts present. The long persistence of the fungus in the field is probably due to its ability to colonise and produce new MS on the root systems of nearly all plant species including monocots (Martinson and Horner 1962). The density of MS in soil was monitored by Itoh et al. (1989) following Chinese cabbage by using infested soil. A first order rate equation was fitted to the observed decrease in the number of MS which was more rapid at higher temperatures. A linear relation was observed between the logarithmic value of the half life and the temperature in the range of 5–31 °C.

Survival of MS of *V. dahliae* appears to be best under air-dry conditions (Coley-Smith and Cooke 1971). The mechanism for survival has not been elucidated. There is a clear relation between the presence of melanin in organisms and their persistence (Bloomfield and Alexander 1967). Temperature may exert an indirect influence on survival through direct effects on germination of MS (Coley-Smith and Cooke 1971) and decomposition of the plant material they are embedded in to achieve their release into the soil (Hancock and Benham 1980). The influences of the temperature and decomposition are entangled. Decomposition is the result of more than one process, and an increase in temperature, within a certain range, will accelerate most biological processes.

The release of MS from colonised plant debris lasts at least one year following its incorporation into the soil and consequently affects the inoculum density in soil. In two cultivated cotton-field soils studied by Evans et al. (1967), the density of MS in soil declined throughout the growing season, but increased again at harvest time when mechanical damage to colonised cotton plants caused the release of fresh MS. There was a further increase in the soil inoculum density when the dead cotton stalks were returned to the soil in preparation for the next crop. Huisman and Ashworth (1976) observed the soil inoculum density of *V. dahliae* in eight commercial fields with different cropping histories at about monthly intervals for 3.5–4.0 years. High inoculum densities in soil persisted under continuous cultivation of cotton. The inoculum density usually increased rapidly following one year of cotton, with a higher inoculum density occurring in the second year regardless of the susceptibility of the subsequent crop. The slow release of MS from decomposing plant debris probably causes a delay in the build-up of the inoculum density in soil. It is not known how the survival of MS embedded in plant material compares to the survival of released 'free' MS. Presumably, the death rate is higher in bound MS because of the higher microbial activity in and around the decomposing plant debris.

Germination of microsclerotia (Figure 1: b)

Microsclerotia of *V. dahliae* have no constitutive dormancy where development is delayed because of an endogenous property of the propagule (Pegg 1974), but they do have a requirement during germination for exogenous nutrients (Emmatty and Green 1969) from root exudates of either hosts or non-hosts (Schreiber and Green 1963; Mol 1994).

Concentrations of exogenous nutrients can be influenced by organic amendments to the soil, which influence general soil microbial activity and concentration of soluble compounds in the soil solution. Increased soil microbial activity has been reported to reduce the proportion of germination of microsclerotia and growth of germ-tubes of *V. dahliae* in the rhizosphere (Jordan et al. 1972). In greenhouse and field experiments, amendments with chopped barley or oat straw at several rates reduced the inoculum density of *V. dahliae* in soil and disease incidence in potato plants (Tolmsoff and Young 1959; Harrison 1976). The effect of organic amendments has not been elucidated, but it has been suggested to be an increase in production of antifungal substances by the enriched microflora of the rhizosphere (Curl and Truelove 1986). However, the interactions between the pathogen and the microflora in the rhizosphere are probably more important, since there is still doubt that antibiotic substances produced in soil can endure microbial degradation for long enough to act as effective inhibitors (Curl and Truelove 1986).

Microsclerotia have the ability to germinate more than once. Farley et al. (1971) showed that MS germinated and sporulated every time after re-moistening soil with a sucrose solution or water up to nine times. Germination percentage and the number of germ-tubes decreased with succeeding germinations. Where a susceptible crop is planted in a field in soil naturally infested with *V. dahliae*, the behaviour of the MS is likely to be different from that in experiments where artificial inoculum is applied (Menzies and Griebel 1967). Unless several years have elapsed since the last susceptible crop was grown, the inoculum may be mostly in the form of MS embedded in partly decomposed plant residues. Some of these MS may already have undergone a cycle of germination and be depleted, while others may be freshly exposed during cultivation operations and may go through a peak of germination as the host seedling roots are growing in their vicinity (Menzies and Griebel 1967).

Colonisation of plant roots by V. dahliae (Figure 1: c)

Germination hyphae of MS may infect susceptible hosts by penetration of the root cortex followed by systemic invasion of the xylem vessels (Powelson 1970).

Infectious hyphae of *V. dahliae* emerging from MS penetrate roots primarily near the root tip and in the root-hair zone (Fitzell et al. 1980; Gerik and Huisman 1988). The density of colonisation at a distance of more than 1 cm from the root apex appeared to be constant (Gerik and Huisman 1988). Most of the colonies are removed when the roots are surface-sterilised, indicating that *V. dahliae* is predominantly restricted to superficial sites in the root cortex (Evans and Gleeson 1973).

Host-pathogen interactions

A schematic representation of the host-pathogen interactions is shown in Figure 1: I. After colonisation of the plant root, the fungus crosses physical barriers in the plant to reach the vascular system. During systemic infection, colonisation of the plant may result in resistance reactions, yield reduction, and, finally, production of MS on the plant debris. The next paragraphs will be focused on systemic infection of plant roots (Figure 1: d), systemic colonisation of the plant (Figure 1: e), plant reactions and yield reductions (Figure 1: f), and reproduction of *V. dahliae* (Figure 1: g).

Systemic infection of plant roots (Figure 1: d)

In response to the initial invasion by the fungus, the inner tangential wall of the epidermis becomes swollen (Bell 1973). Occasionally a gum-like encapsulation zone may be present in the outer cortical cells of the plant roots around the hyphae of *V. dahliae*; these formations are referred to as lignitubers (Griffiths 1973). Lignitubers result from the extrusion of vesicles from the protoplast into the area between the cell wall and plasmalemma. When the fungus starts to penetrate the cell wall, the vesicles aggregate and lose their individual identity as they form the lignituber. Subsequently, the fungal cell wall becomes surrounded by the lignituber, while the host plasmalemma apparently remains intact. In most cases the invading hyphae lyse, but in a few cases successful penetration occurs and the fungus proceeds through the layer of cortical cells (Bell 1973).

Fungi that successfully penetrate the cortex encounter a second barrier at the endodermis (Bell 1973). Most of the penetrating hyphae that were not stopped by lignitubers are so at the endodermis; the few hyphae that penetrate this barrier progress into the vascular tissue and invade the vessels by penetrating through the pits. Alternatively, hyphae may reach the vascular system by colonising the young root tip, where the endodermis still has to be formed. The systemic infection of the plant root is the property determining the suitability of the plant as a host. For example, susceptible mint (*Mentha* spp.) roots showed more extensive systemic invasion, while the number of root cortex infections was similar to that of resistant mint species (Lacy and Horner 1966).

Infection of plant roots by *V. dahliae* may be influenced by other organisms. In most cases, potato early dying can be attributed to a complex of soil microorganisms including *V. dahliae*, *Globodera* spp., *Pratylenchus* spp., *Meloidogyne* spp., *Colletotrichum coccodes*, *Fusarium* spp. and *Rhizoctonia solani* (Scholte 1989; Scholte and s'Jacob 1989). Other pathogenic agents that have been associated with potato early dying include *Erwinia carotovora* and Potato Virus X (Rouse 1985). It has been established that potato early dying is frequently caused by *V. dahliae* in combination with endoparasitic nematodes. In microplot studies on potato cv. Russet Burbank, Kotcon et al. (1985) did not find a significant contribution by either *C. coccodes* nor *R. solani* to the disease syndrome either alone or in combination with *V. dahliae* or *P. penetrans*. However, in potato cv. Amethyst yield loss by *V. dahliae* was almost doubled in the presence of *C. coccodes* (Scholte et al. 1985), and Scholte and s'Jacob (1989) found a three-factor interaction on yield loss between *Meloidogyne* spp. or *P. neglectus* and *R. solani* and *V. dahliae* with several potato cultivars.

The mechanism of the interaction between *V. dahliae* and *P. penetrans* that results in the synergistic expression of symptoms and yield loss has been subject of speculation. Green (1981) considered that the mechanism could be wounding from nematode feeding, which gives the fungus easier access to the vascular system. Alternatively he proposed a physiological mechanism where the plant becomes more susceptible to the fungus because of a translocated substance. Wheeler et al. (1992) compared a model in which yield loss of potato is proportional to both *V. dahliae* and *P. penetrans* with a model in which yield loss is proportional only to the population density of *V. dahliae* and the presence of *P. penetrans* leads to a more severe yield loss function than in the absence of the nematode. However, their comparison was regarded not conclusive considering the variability in the expression of potato early dying and confounding effects of environmental conditions.

In evaluating his mechanistic model, Johnson (1992) described potato crop losses caused by multiple biotic stress factors, excluding nematodes. He concluded that crop loss by multiple stress factors was less than the sum of losses from each stress factor acting alone. This was illustrated by a competitive defoliation between *V. dahliae* and *Phytophthora infestans*.

Systemic colonisation of the plant (Figure 1: e)

The process of production of conidia in the vascular system is still unclear. A hypothesis put forward by several authors is that directly after *V. dahliae* has penetrated the vascular system, it starts to produce conidia (Howell 1973; Schnathorst 1981). Conidia are produced by simple conidiophores or by budding (Tolmsoff 1973) and are passively distributed through the vascular system throughout the plant. Conidia may germinate, and penetrate vessel walls (Garber 1973; Tolmsoff 1973). More research concerning the process of sporulation and distribution in the vascular system is needed. For example the influence of the nutrient concentration in the xylem and the influence of the vitality of the plant on these two processes are indicated as origins of differences in the number of *V. dahliae* particles in the stems of plants, but the effect has never been proven in experiments.

In susceptible and tolerant cotton cultivars, colonisation was equal up to the point where the pathogen passed the endodermis and reached the xylem (Garber 1973). The number of vessels invaded appears to be a good measure of the severity of wilt disease, because the number of invaded vessels was related to the number of systemic infections and to the number of hyphae that progressed through the cortex from the points of colonisation.

As long as petioles of potato plants are green, distribution of *V. dahliae* in the plant is limited to the xylem, but in plants with severe disease symptoms the pith, cambium, and cortex are invaded (Garber 1973). In the leaves, infection may be confined to a single pinna, or may involve the entire leaf (Garber 1973).

Plant reactions and yield reductions (Figure 1: f)

In potato, symptoms of *V. dahliae* are difficult to distinguish from normal senescence and may initially involve only reduced growth (Street and Cooper 1984; Haverkort et al. 1990). Early foliar symptoms may appear as unilateral chlorosis of lower leaves on a few plants. Later some wilting of whole leaflets or leaves may occur, but the unilateral death of lower leaves is more typical (Isaac and Harrison 1968).

The activities of the fungus stimulate the plant to produce a suberin-like coating, tyloses and gels inside the vascular elements. Tyloses are extensions from parenchyma cells into the xylem (Newcombe and Robb 1988). Gels arise from perforation plates, end-walls and pit membranes by a process of distension of primary wall and middle lamella constituents (Van der Molen et al. 1977). This can lead to occlusion of the vessels in the vascular system immediately above primary infection sites (Van der Molen et al. 1977; Harrison and Beckman 1982; Newcombe and Robb 1988). A light brown vascular discoloration is often visible at the stem base when sliced (Isaac and Harrison 1968). The accumulation of specific chemicals such as terpenoid aldehydes in the vessels accompanies or follows the presence of the pathogen and reduces the viability of the fungal propagules (Harrison and Beckman 1982). As colonisation of the xylem proceeds, the vessels may become plugged by hyphae (Garber 1973). In potato plants the plugging has been traced from the root tip up to the top of the stem.

Inoculated potato plants produced no symptoms until tuberisation commenced (Busch and Edgington 1967), suggesting that before symptom expression of Verticillium wilt becomes evident, the host must be in an advanced stage of development (Busch et al. 1978). By altering the photoperiod to prevent tuberisation, few or no symptoms developed (Busch and Edgington 1967). These observations are consistent with other observations of an association between lateness of cultivar and resistance to *V. dahliae* (Busch et al. 1978).

Kotcon et al. (1985) associated disease incidence of *V. dahliae* with reduced root growth, foliar weight, and tuber yield. Infected plants exhibited lower specific leaf areas (area produced/dry weight of leaf tissue), higher leaf weight ratios (dry weight of the leaf system/dry weight of the whole plant) and higher leaf area ratios (area of the leaves produced/dry weight of the whole plant), and, under dry conditions, lower relative growth rates and lower leaf growth rates (increase in dry weight/unit leaf area/week) (Harrison and Isaac 1969).

Bowden et al. (1990) and Haverkort et al. (1990) showed that the initial decrease in photosynthesis caused by *V. dahliae* was caused by stomatal closure. The low stomatal conductance was correlated with low leaf water potential (Bowden et al. 1990). In potato, *V. dahliae* caused a reduction in the light conversion efficiency, but a stronger reduction of stomatal conductance, resulting in decreased internal/external CO_2 ratios and in a higher net photosynthesis at similar values of stomatal conductance (Haverkort et al. 1990). The reduction of net photosynthesis does not seem to be responsible for more than 10% of the reduction of dry matter production. In areas where the disease causes an early and rapid senescence of the leaf canopy, reduction of intercepted radiation may be a more important component of damage than reduction of photosynthesis (Haverkort et al. 1990).

Although there are interconnections between the vascular bundles in the stem, these are absent in the petiole. As a consequence blockage of petiole bundles can be more damaging than a proportional blockage of stem bundles (Garber 1973).

A lesser root length of potato plants due to *V. dahliae* may decrease the water supply and cause the development of foliar symptoms (Kotcon et al. 1984). Also root surface areas and volumes were reported to be affected negatively by *V. dahliae*.

Reproduction of V. dahliae (Figure 1: g)

As infected plants become senescent, the fungus permeates the surrounding tissues and forms MS within dead tissue (Powelson 1970). There is no evidence of other than transient increases in inoculum from sources other than plant debris.

Potato stems colonised by *V. dahliae* will be filled with MS. A 1 cm segment of stem may contain 8,000–20,000 viable MS and populations up to 1,000 MS per gram of soil have been reported in fields repeatedly cropped to potato, which roughly equals 50 million MS in the soil volume occupied by the roots of one plant (Menzies 1970). Differences in production of MS from 7,000–9,000 propagules per gram of stem tissue have been reported in several potato cultivars (Slattery 1981).

In most plant species infected with *V. dahliae*, water is required for formation of MS (Powelson 1970). However, in the temperate zones, humidity is usually sufficiently high during the senescence of the potato haulms to ensure abundant MS formation without rain.

External factors can have a large influence on the formation of microsclerotia. Autumn-grown crops of potato in the Negev area of Israel had approximately 100 times higher microsclerotial production than the spring-grown crops (Ben-Yephet and Szmulewich 1985). Either the cool and moist conditions in the autumn-winter season enabled the plants to dry slowly, favouring formation of MS, or the cooler weather allowed better survival of the fungus in the plant tissue.

Ioannou et al. (1977) examined the formation of MS in tomato debris in soil subjected to different irrigation and flooding regimes under field conditions. Few, if any MS were produced during the flooding treatment. This inhibition was attributed to decreased O_2 and increased CO_2 concentrations in the flooded soil. Upon drainage, the concentrations of O_2 and CO_2 returned rapidly to normal atmospheric levels and formation of MS was resumed. The numbers of MS eventually produced following 10-, 20-, and 40-day flooding treatments were 90, 44, and 46% respectively, of the average numbers in the non-flooded treatments.

The major part of the new inoculum is produced in the aerial parts of the potato plant (Ben-Yephet and Szmulewich 1985; Mol 1994). A direct way for control of *V. dahliae* is to interfere with the formation and dispersal of MS. In practice this is accomplished by various sanitation measures such as removal or destruction of diseased plant material before it enters the soil or before it releases the inoculum otherwise. This process of field sanitation is not widely practised because of expense, lack of equipment, and through farmers' aversion to destroying organic matter. Another possibility for the control of the reproduction is the use of mechanical haulm killing techniques. Compared to haulm killing by the application of a herbicide, the reproduction in the aerial parts of the potato plant was reduced by up to 80% following mechanical haulm killing (Mol 1994).

Production of MS has been reported in plant roots without systemic infection. Microsclerotia were formed in large numbers in only few sections of wheat roots (Krikun and Bernier 1990). Taking into account total root length of wheat and the number of MS found (up to 100 per root fragment of 1.3 × 0.4 mm), the contribution to the inoculum in soil may be significant (Krikun and Bernier 1990).

Weed control is important to limit multiplication of *V. dahliae* (Woolliams 1966; Johnson et al. 1980). Host species include many common weeds and native plants. As in crop plants, symptoms are not always apparent in infected weed plants.

Host specificity

Adaptation of *V. dahliae*, leading to host specificity, may confuse the search for possible resistance or tolerance mechanisms. Zilberstein et al. (1983) reported that germination on agar and pathogenicity of MS of *V. dahliae* to egg-plant, potato and tomato was affected by growth medium and host origin. The virulence of an isolate of *V. dahliae* depends on the host species, and geographical origin, susceptibility of the host cultivar/genotype, and the organ on the plant (Zilberstein et al. 1983; Michail 1989). Isolates obtained from susceptible cultivars attacked these cultivars only. Isolates from resistant cultivars attacked both susceptible and moderately resistant cultivars (Michail 1989).

Production of MS variants might permit the fungus to adapt to new host species and varieties or to new environmental conditions (Tolmsoff 1973). Mint isolates were originally not pathogenic to tomato. After one passage through tomato, however, the isolates became more pathogenic to tomato and lost pathogenicity to mint. The isolates from a region of continuous cropping to one crop tend to be similar and all display high virulence against that particular crop but generally a weak virulence against other species which they can infect nevertheless (Vigoroux 1971). Thus a notion of 'preferential' and 'occasional' hosts is formed. It is known that great variability is possible in *Verticillium* species (Vigoroux 1971).

Puhalla and Hummel (1983) found 16 different vegetative compatibility groups in *V. dahliae*. Using another method, Joaquim and Rowe (1990; 1991) were able to reduce the number of vegetative compatibility groups to four. Isolates from two different vegetative compatibility groups showed significant differences in pathogenicity towards potato (Joaquim and Rowe 1991).

Thus, in soils, populations of individuals with different properties can be built up. It appears that the kind of crop constitutes a determinant factor for the quantitative and qualitative composition of the population of *V. dahliae* in cultivated soils (Vigoroux 1971; Tjamos 1981). So, the choice of cultivar, crop rotation, and cultural practices will not be sufficient to keep the crop from infection, but they will still be key factors in the controlling the severity of the disease.

Acknowledgements

We thank Dr. K. Scholte, Professor P.C. Struik, and Dr. J. Vos for critically reading the manuscript.

Use of a crop-growth model coupled to an epidemic model to forecast yield and virus infection in seed potatoes

THOMAS NEMECEK[1], JACQUES O. DERRON[1], ANDREAS FISCHLIN[2] and OLIVIER ROTH[3]

[1] *Federal Agricultural Research Station of Changins, 1260 Nyon, Switzerland*
[2] *Systems Ecology, Institute of Terrestrial Ecology, Swiss Federal Institute of Technology, Grabenstr 3, 8952 Schlieren, Switzerland*
[3] *Widenstr 3, 8302 Kloten, Switzerland*

Abstract. Pathosystems of vector-borne plant viruses consist of the elements viruses, vectors and plants, influenced by man and the natural environment Most virus epidemic models emphasize the importance of vectors and viruses, but do not consider the role of the plants

The model 'EPOVIR' is the first virus epidemic model coupled to a crop-growth submodel which, for its part, is coupled to a soil water-balance submodel The two submodels are used in three ways tuber yield and tuber size are calculated, the physiological age of the leaves and the drought stress are used to estimate the susceptibility to virus infections ('age resistance') and finally, the fraction of soil covered by the canopy is needed to calculate landing rates of the vectors in the potato field Since the rate of virus spread is a function of plant physiology and phenology as simulated by the crop submodel, the epidemic should react appropriately to changes in plant growth, caused by man or the environment As an example we show how planting density influences virus infection

The model EPOVIR is integrated in the decision-support system 'TuberPro', which forecasts tuber yields graded by size and the infection of the tubers by PVY and PLRV It supports optimization of haulm destruction dates in seed potato production The system calculates expected seed yield and the probability that virus infection remains below the tolerance limit used in seed certification The combination of the two factors gives the expected average certified seed yield for a production region, which can subsequently be optimized

Introduction

Pathosystems of vector-borne plant viruses consist of the elements virus, vectors and plants, all influenced by man and the natural environment (Robinson 1976). Most forecasting systems and virus epidemic models focus on vectors and viruses and their relations (e.g. Marcus and Raccah 1986; Kisimoto and Yamada 1986; Miyai et al. 1986; Madden et al. 1990; Kendall et al. 1992). The role of the plants in virus infection and translocation and their influence on the vector populations is not included, or at best, described by a few constant parameters or functions (e.g. Sigvald 1986; Ruesink and Irwin 1986; Van der Werf et al. 1989). However, to describe appropriately the effects of plant physiology and phenology on the epidemic process and the influence of the environment on these relations, a crop-growth submodel is necessary. Crop models have been used in plant pathology in combination with fungal epidemic models (Rouse 1988), but to our knowledge, crop models have never been coupled to virus epidemic models.

A J Haverkort and D K L MacKerron (eds), Potato Ecology and Modelling of Crops under Conditions Limiting Growth, 281–290
© 1995 *Kluwer Academic Publishers*

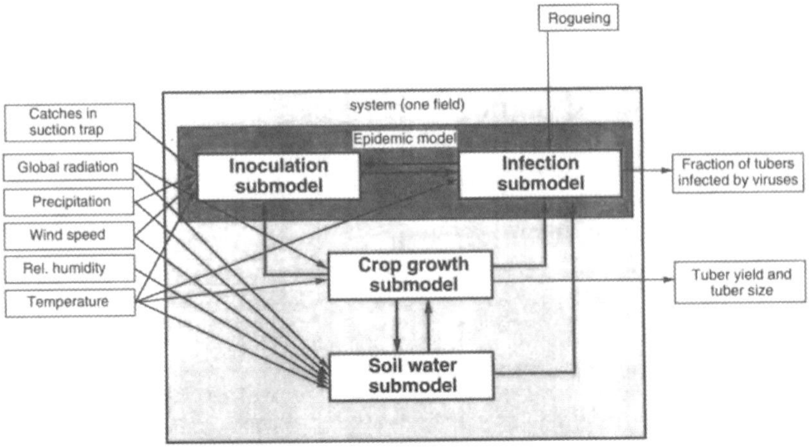

Figure 1 Structure of the model EPOVIR (= *ep*idemiology of *po*tato *vir*uses)

We present a project, where a potato crop submodel has been coupled to a virus epidemic model to predict yield and virus infection in seed potatoes. The crop submodel for its part is coupled to a soil water-balance submodel, which accounts for suboptimal water supply conditions. These two submodels influence the epidemic model, but are independent of it. The reason why we did not study the yield reduction due to viruses is that we are mainly interested in seed potato production, where the yield loss caused by virus infection is very small. Nevertheless, the economic impact of virus infection can be very high, since a seed lot can be rejected, leading to almost no revenue to the farmer. We have developed a decision-support system for seed potatoes that allows us to forecast virus infection of tubers and, particularly, the risk that virus infection will exceed the tolerance limits for a certain quality class. We also forecast tuber yields graded by size. The system, called 'TuberPro' (= *Solanum* **tuber***osum* **pro**gnosis), allows us to optimize haulm destruction dates in seed potato production (Nemecek et al. 1994).

The aim of this paper is to show, why and how the crop submodel is used in the virus epidemic model and within 'TuberPro'.

Model outline

The model 'EPOVIR' (= *e*pidemiology of *po*tato *vir*uses) forms the core of TuberPro. EPOVIR simulates crop growth and virus epidemic from crop emergence to haulm destruction in a single field. A detailed description of a first version is given by Nemecek (1993). It consists of four submodels (Figure 1):

- An *inoculation submodel*, which calculates vector intensity (after Irwin and Ruesink 1986) from vector abundance in the field, vector propensity and vector behaviour. Vector abundance is estimated from catches in a suction trap. Wingless aphid vectors are considered for PLRV transmission only. The submodel represents the role of the vectors and their relations with the viruses and the crop.

- An *infection submodel*, which determines the infection of plants and tubers by PVY and PLRV, respectively. Further, it estimates the fraction of plants serving as infection sources by accounting for the latent period and the age resistance of the plants to virus infections. The submodel represents mainly the role of the viruses and their relations with the crop.
- A *crop-growth submodel*, which calculates the dry mass of leaves, stems, roots, tubers and assimilates, the physiological state of the canopy and graded tuber yields.
- A *soil water-balance submodel*, which calculates the water content of the soil and the water stress of the potato plants from the ratio of actual and potential transpiration.

The crop model has been developed by Johnson et al. (1986; 1987). The original model used measurements of soil moisture, which were not available in our experiments so we replaced them by a soil water-balance submodel, based on work of Van Keulen and Wolf (1986). A complete description of the coupled crop-growth and soil water-balance submodels and validation results for cv. Maris Piper are given by Roth et al. (1995). The crop submodel was adapted to varieties cultivated in Switzerland and extended by a tuber-size submodel. These changes and the corresponding validation results are given by Nemecek and Derron (1994). The crop model was capable of reproducing experimental data collected under wet and dry conditions with sufficient reliability.

EPOVIR is implemented in the programming language Modula-2 (Wirth 1985) by using the simulation environment ModelWorks (Fischlin et al. 1994) on Apple™ Macintosh™ computers.

Use of the crop-growth submodel

Following our definition of a plant-vector-virus pathosystem, the submodels for crop growth and soil water-balance are used in three ways in the model EPOVIR: to calculate yield, to describe the influence of plant physiology on virus multiplication and translocation in plants and to describe the effect of plant phenology on vector behaviour.

Plants

Forecasting tuber yield and tuber size are, of course, highly important in the decision-support system. The crop submodel calculates total tuber yield and the fraction being between arbitrarily chosen size limits (see example in Figure 4 below).

Plant-virus interaction

Potato tubers are less readily infected by viruses as plants grow older (Beemster 1987). This phenomenon is called 'age resistance' or 'mature-plant resistance'. Its causes are unknown, but it seems to be related to the physiological activity of the leaves (Venekamp et al. 1980). The submodel keeps track of the leaf mass produced per day as well as its physiological age (Johnson et al. 1987; Roth et al. 1995). We use the fraction l_y of leaf mass younger than a physiological age threshold ($0 \leq l_y \leq 1$) to provide an estimate of the age dependent susceptibility to virus infections *AS*. The susceptibility to virus infections depends also on drought: Wislocka (1982) demonstrated that drought (suboptimal water supply) can favour tuber infection, apparently by partially breaking the age resistance. The effect of water stress *wStress* ($0 \leq wStress \leq 1$, 1 = optimal growing conditions, 0 = no growth, see Roth et al. 1995) is also included in the equation for *AS*:

$$AS = \begin{cases} 1-(1-ly)(1-0.5(1-wStress)) & \text{if drought} \\ l_y & \text{otherwise} \end{cases}$$

The constant 0.5 is a weighting factor that describes the effect of drought (soil water content below critical soil moisture, Roth et al. 1995) on the susceptibility. Figure 2 shows a typical evolution of *AS*. In young plants, *AS* equals 1, irrespective of drought. As the plants grow older, l_y decreases gradually to 0 and consequently *AS* decreases also. Drought can increase the susceptibility at

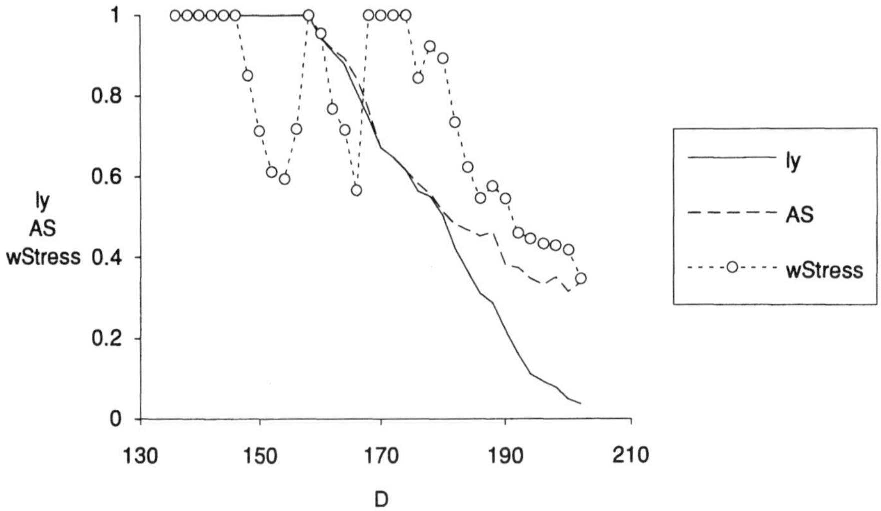

Figure 2. Age dependent susceptibility to virus infections *AS*, which is a function of the fraction l_y of leaf mass younger than a physiological age threshold and the water stress factor *wStress*. *D* = day of the year.

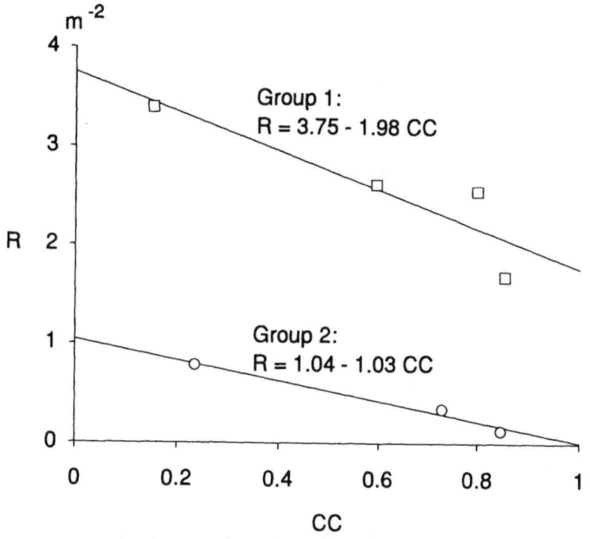

Figure 3. Relative landing rate of aphids: R = ratio between landing rate [# m^{-2} d^{-1}] in a potato field and catches in a suction trap [#/d] as functions of the canopy cover CC (= fraction of soil area covered by the canopy). The following species are contained in group 1: *Aphis spp.*, *Myzus persicae* Sulz. and *Macrosiphum euphorbiae* Thomas; group 2: *Acyrthosiphon pisum* Harris, *Brachycaudus helichrysi* Kalt., *Phorodon humuli* Schrk. and *Rhopalosiphum padi* L.

this stage, i.e. the plants become more susceptible to virus infection compared with a situation, where water supply is optimal. Without drought, AS would be equal to l_y.

Plant-vector interaction

The plants influence not only virus infection, multiplication and translocation, but also the development and behaviour of aphid populations, and therefore indirectly virus spread (Irwin and Kampmeier 1989; Nemecek 1993). The host suitability strongly influences vector behaviour and virus spread. Nemecek (1993) has shown that vector species that do not colonize potato plants, more often show sequences of behaviour that increase their efficiency as vectors. Simulation studies revealed that this attribute will lead to about twice as efficient transmission of nonpersistent viruses, compared to species that colonize potato plants.

Plant phenology influences the behaviour of vector populations. The relative landing rate R (number of aphids landing per m^2 for each aphid captured in the suction trap (Taylor and Palmer 1972) in the same time interval) decreases with increasing canopy cover (Figure 3). The landing rate was calculated indirectly from catches in a fisherline trap (Derron and Goy 1993) and their relation to the landing rates on potato plants (Derron et al. 1989). The canopy cover is simulated by the crop submodel. Regressions between the average R and canopy

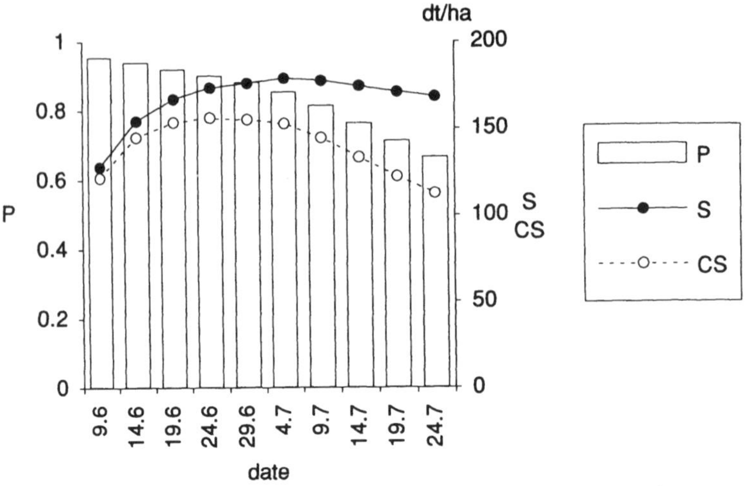

Figure 4. Development of the probability *P* that seed will be certified, the expected yield in seed grade *S* [dt/ha] and the expected average certified yield in seed grade *CS* [dt/ha]. The results are actually averages of 30 simulation runs: three situations (favourable, mean and unfavourable) were combined with the input data of the last 10 years after 14 June (up to this date the data of the current year have been used).

cover were calculated for 3, 4, 5 and 10 groups with approximately the same number of data points. The regression with the highest r^2, i.e. the best linear relation, was used in the model (Figure 3). Since the canopy cover is a variable calculated by the model, it was easy to include this relation. The landing rate and consequently virus transmissions will vary with canopy development. Further we can distinguish two groups of vector species: group 1 had a relative landing rate that was on average 8.6 times higher than that of group 2.

Model application

We show two examples of model application (see Nemecek et al. 1994 for further examples). The principal goal of 'TuberPro' is to forecast the following variables for decision making:

- the expected tuber infection by PVY or PLRV and its variance;
- the expected tuber yield of a certain size grade and its variance.

Logit transformed forecasts of virus infection are approximately normally distributed. The mean and variance describe the probability distribution of the forecasts. They include the uncertainty caused by unknown future development of input time series (weather variables and aphid abundance), as well as the error in the estimates of certain parameters. Unknown future values of inputs are replaced by data from recent years. The scope of the simulations is always one field. However, it is also possible to define average situations that are representative for regional or national production. Such an example is given in

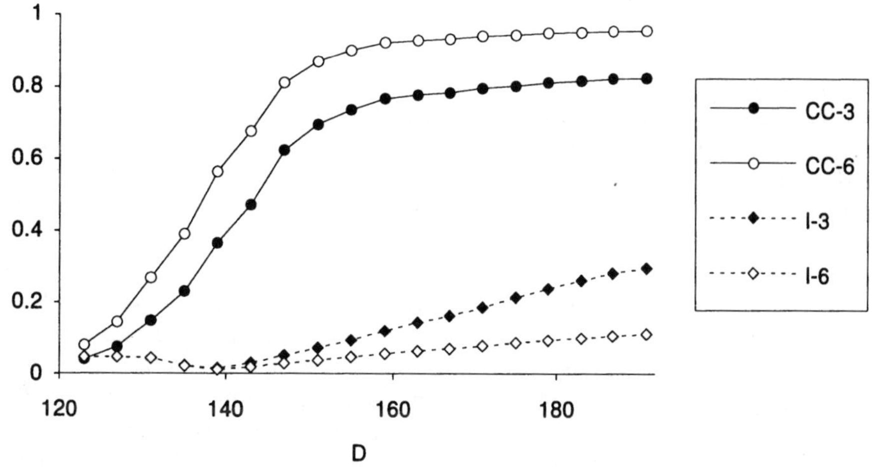

Figure 5. Effect of planting density on virus spread: I = fraction of tubers infected by PVY, CC = fraction of soil covered by the canopy, D = day of the year. The numbers 3 and 6 refer to the planting density of 3 respectively 6 seed pieces/m^2.

Figure 4. In Switzerland, a seed lot with an infection with up to 10% by PVY or PLRV can be certified as class A seed. The probability, *P*, that a lot will be certified is therefore the probability that the virus infection of tubers is ≤ 10%, which is also the fraction of lots that are expected to be certified. The maximum expected yield in seed grade, *S*, is reached on 4 July. *P* and *S* allow calculation of the expected average certified yield in the seed grade, *CS*, which is simply the product of the two variables ($CS = S \times P$). The maximum of *CS* is reached as early as on 24 June. So, 24 June would be the optimal haulm destruction date for production of a maximum of certified seed.

Another strategy would be to maximize the farmer's income. The farmer is also interested in the larger tubers, since they can often be sold for human consumption. For this reason maximum expected income will usually be reached later than maximum expected certified seed. The user of TuberPro must define the strategy to achieve his objectives.

The model can also be used to evaluate cultural practices. Simulation studies show that virus infection can be substantially reduced by increasing planting density (Figure 5). This is due to increased canopy cover (see Figure 3) and a dilution effect (Power 1990; 1992). Both factors lead to a lower number of vectors per plant, which will lower the proportion of plants infected. The high density planting reduced the average number of vectors per plant by 58% and the virus infection by 62% compared with the low density planting, i.e. virus infection was reduced even more than vector density. This difference is due to the fact that vector abundance was decreased mainly early in the growing season, when virus transmission is most dangerous (Nemecek 1993).

Discussion

The crop reacts to influences exerted by the natural environment and by man. As the multiplication of potato viruses depends completely on the host plant metabolism, changes in it will also affect the epidemic process. For example if leaf growth continues longer, this will most likely favour virus spread by delaying age resistance. The vectors have a very close relationship to the crop as well. Such indirect influences on the epidemic process are best described by a virus epidemic model coupled to a crop-growth model.

Since well-validated models are now available for many crops, we think that crop models should be more widely used in modelling virus epidemics. The use of a crop submodel makes the entire model more flexible and adaptable to other conditions, e.g. in other countries with different production schemes. By including the plant-virus and plant-vector relations, the model also becomes more explicit and offers better possibilities for testing hypotheses and evaluating cultural measures (Nemecek 1993). A later unpublished version of the model of Ruesink and Irwin (1986) also considers the effect of the canopy cover on the landing rate. However, the authors use measured time series as inputs and so, new experiments will be required each time growth parameters are changed. That is not necessary in our approach. Our last argument for the use of crop submodels is perhaps the most obvious one: the crop submodel calculates tuber yield and so allows seed potato production to be optimized according to different criteria (see Figure 4). It allows also evaluation of the effect of virus control measures on yield.

Coupling a crop submodel to a virus epidemic model has also some drawbacks. The complexity of the whole model increases considerably and a lot more parameters are needed. As these parameters will not be available in all cases, default or average values must be used instead. By doing so, one loses some of the advantages of the crop submodel. Finally, the development and coupling of models is laborious and time consuming, especially where no suitable crop model is already available. The decision to use a crop submodel will depend on the objectives of the research, the availability of a suitable model and the available resources.

TuberPro is a powerful system management tool for seed potato production. It allows production to be optimized according to various strategies which are chosen by the decision maker. A farmer will probably try to assure a certain income. A seed growers' organization will perhaps try to increase its profits and the national service is mainly preoccupied with maintaining a high quality. TuberPro calculates the variables necessary to find optimal solutions at the different levels, especially to determine the haulm destruction dates according to the specified goal. Moreover, it has proved to be useful for the instruction of farmers and experts and for the evaluation of cultural and virus control measures (Nemecek et al. 1994). TuberPro is now being tested in practice in Switzerland. As it is very flexible, it should be possible to adapt it to other conditions with a reasonable amount of effort.

Acknowledgements

We thank R. Schwärzel for his valuable contributions. We are also grateful to the Swiss Alcohol Board and the Swiss National Science Foundation (grant no. 31–8766.86) for their financial support.

Prescriptive crop and pest management software for farming systems involving potatoes

W.R. STEVENSON[1], J.A. WYMAN[2], K.A. KELLING[3], L.K. BINNING[4], D. CURWEN[4], T.R. CONNELL[4] and D.J. HEIDER[4]

Departments of [1] Plant Pathology, [2] Entomology, [3] Soil Science, [4] Horticulture, University of Wisconsin, Madison, WI 53706, U S A

Abstract. Societal concerns about pesticide and fertilizer use and the intensive management of cropping systems involving potatoes have focused attention on the choice, timing and application rates of agricultural inputs An integrated pest management (IPM) program was developed for the Wisconsin potato industry because 1) growers have traditionally been progressive and interested in IPM tactics, 2) the crop is intensively managed with multiple inputs of pesticide, fertilizer and irrigation, 3) potato is a high value/high risk crop, 4) the crop is often grown in environmentally sensitive areas, and 5) a strong research base existed to support the IPM effort An important aspect of the Wisconsin IPM program was the development of specialized computer software to assist growers in management decisions Computer software has provided an effective on-farm tool for analyzing complex environmental and crop information and providing specific management recommendations The Potato Crop Management (PCM) software was developed at the University of Wisconsin through a multidisciplinary team effort The program contains modules for predicting crop emergence, scheduling irrigation, managing pests (diseases, insects and weeds) and assessing storage ventilation needs Since its release to growers and IPM consultants in 1989, use of the PCM software has continued to increase Growers and consultants now use the software on approximately 28,300 ha of potato production Savings related to reduced use of pesticides and irrigation are estimated to exceed US $ 5,890,000 annually when compared with inputs prior to use of this software The current software, however, has application to only a few of the many decisions confronting growers Software enhancements are underway that focus on potato canopy development, crop nutrition, seedpiece decay and farm record-keeping The PCM program is also being converted from an MS-DOS® environment to a Microsoft Windows™ platform The new software entitled WISDOM will allow easier data input and exchange, graphical presentation of data and comparison of environmental and crop/pest data between fields and years The WISDOM software will expand opportunities for managing the farming enterprise including crops often grown in rotation with potato Modules are being developed for management of snap beans (irrigation scheduling, insect management, and risk assessment for white mold) and sweet corn (irrigation scheduling and insect management) Program modularity will facilitate the future addition of new potato and rotational crop modules as new information becomes available from field and laboratory research

Introduction

Potato production for processing, fresh market and seed is an important asset to Wisconsin's economy. Grown on approximately 27,500 ha, the crop is valued at $ 120,000,000. Wisconsin is ranked fourth in the United States in total potato

A J Haverkort and D K L MacKerron (eds) Potato Ecology and Modelling of Crops under Conditions Limiting Growth 291–304
© 1995 *Kluwer Academic Publishers*

production (Pratt 1993). The total value to Wisconsin in jobs, processing and allied agribusiness approaches $ 350,000,000 per year. Production costs for irrigated potatoes are typical of national costs and currently average $ 4,199 per ha for long season russet potatoes while gross returns often exceed $ 4,900 per ha.

Midwestern US weather conditions not only favour the production of cool weather crops such as potato, but they also contribute to the development of a broad range of important pest problems. Production of potatoes in Wisconsin and throughout the North Central region typically includes the intensive use of pesticides to manage diseases, insects and weeds; relatively high rates of fertilizer, particularly on irrigated sandy soils; and irrigation water for optimal production. Fifteen years ago, typical Wisconsin crops of Russet Burbank on a loamy sand soil were treated with fungicide dust on the seedpieces, up to 12 applications of foliar fungicides for control of early and late blight, a systemic insecticide at planting and up to four sprays of foliar insecticides, one to three herbicide sprays for control of grass and broadleaf weeds, one to two applications of vine desiccant, irrigation consisting of three applications per week totaling 5 cm per week and fertilizer treatments totaling 280–392, 134–179 and 336–448 kg/ha of N, P_2O_5 and K_2O, respectively.

During the early 1980s, these intensive inputs began to induce concerns in citizens and growers. Their concerns focused on environmental risks, grower and consumer safety and the potential for developing pest resistance to commonly used pesticides. The reliance on intensive use of pesticides and other inputs as well as environmental concerns associated with potato production were the driving forces behind the development of a comprehensive integrated crop and pest management program for Wisconsin growers. This integrated program was designed in partnership with growers, processors and allied industries to maintain a competitive potato industry that optimizes inputs and attempts to minimize or eliminate adverse environmental inputs associated with potato production.

The foundation of this crop and pest management program was a long history of research in areas such as disease management (judicious use of fungicides in conjunction with disease forecasting), insect management (targeting vulnerable stages of pest development and economic thresholds), weed management (prediction of crop emergence, plant canopy development and the timing of herbicide applications), fertility practices (rates, placement, source and timing of required nutrients) and irrigation management (matching applied irrigation with crop needs). Research in these and other critical areas created an extensive database that proved useful in making recommendations for solving specific production problems. During the mid-1980s, however, it became apparent to both the potato industry and University of Wisconsin researchers that to be most useful, this information must take on a more integrated and accessible format. The computer, in its early evolution for on-farm use, turned out to be a convenient vehicle for this integration and improved accessibility.

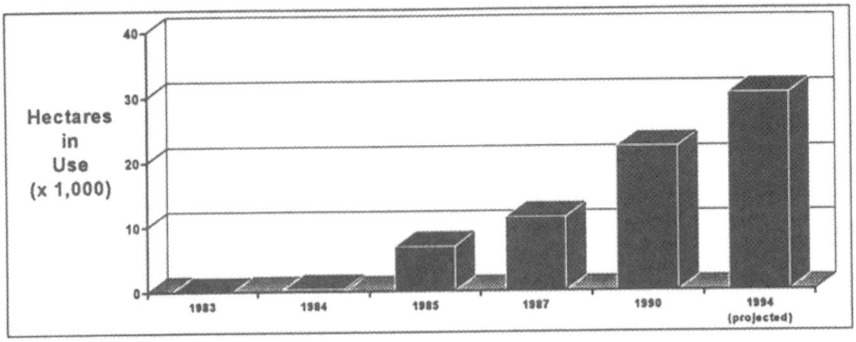

Figure 1. Adoption of the PDM/PCM software by the potato industry since 1983.

Initial development of computer aplications

Initial development of computer software included the Potato Disease Management (PDM) and Wisconsin Irrigation Scheduling (WISP) programs as stand-alone software. The PDM program addressed the prediction and control of early blight (an annual problem in Wisconsin) (Pscheidt and Stevenson 1986) and late blight (a sporadic problem occurring as a function of inoculum and weather) using a modified BLITECAST approach (Krause et al. 1975; Stevenson 1983). Growers experiencing losses to these diseases in the early 80s were receptive to computerized information processing that enabled them to predict the appearance of these diseases and then to improve disease control through optimized fungicide application scheduling and selection of fungicide application rate. The PDM program provided growers with prescriptive pest management software using weather and crop information collected in individual fields. Because of the success achieved in disease control, PDM served as a successful cornerstone for future development of more comprehensive computer software. Subsequent to the introduction and adoption of the PDM program, improved computer technology enabled our IPM computer programmers to combine the PDM and WISP programs into a more comprehensive Potato Crop Management (PCM) program. This software included modules for predicting and controlling disease, scheduling irrigation (Curwen and Massie 1984), predicting crop emergence, predicting development and managing harmful insects through improved targeting of insecticides and use of economic thresholds (Walgenbach and Wyman 1984a,b) and calculating the ventilation needs for storing potatoes. Each of these areas was requested by growers and represented areas of intensive research. The PCM program facilitated integration between modules and allowed growers to use data collected in their fields for multiple management decisions.

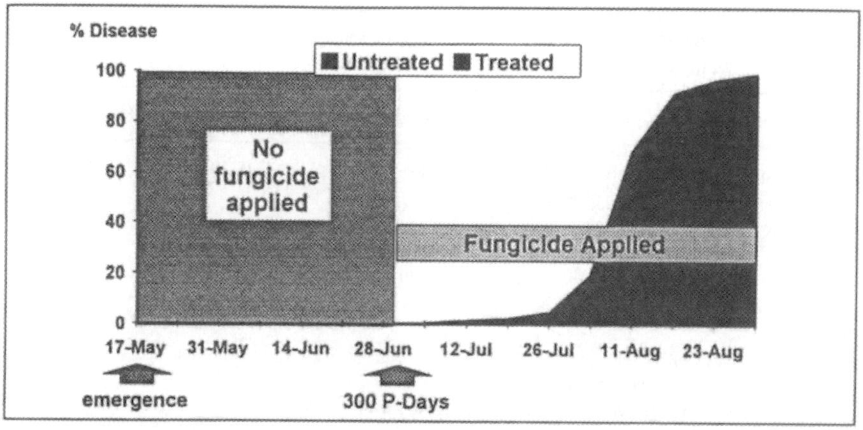

Figure 2. Delayed treatment with fungicide for early blight control using a treatment threshold of 300 P-Days. After 300 P-Days, fungicide is applied at 5- to 10-day intervals, depending on environmental conditions, for the duration of the growing season. Fungicide treatment slows disease progress.

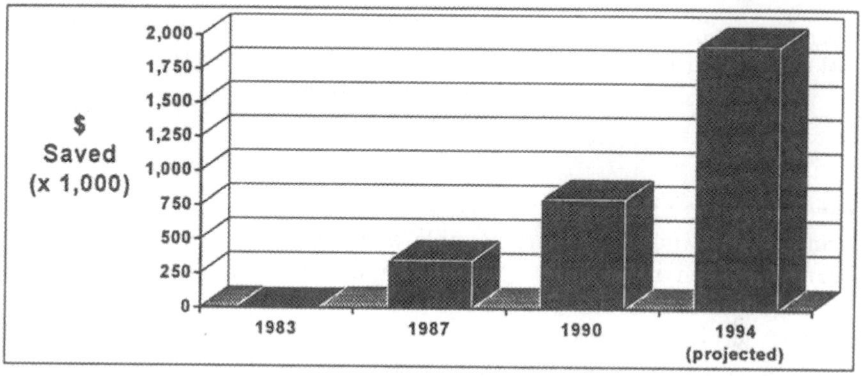

Figure 3. Estimated savings from reduced fungicide inputs related to use of the PDM/PCM software.

Value related to software use

Growers and IPM consultants report use of the PCM software on approximately 16,200 of Wisconsin's 27,500 ha of potatoes. The program is also used in neighbouring states where growers are using the software on an additional 12,100 ha (Figure 1). Use of the software saves growers an estimated $ 49 per ha (Connell et al. 1991) by eliminating 2–3 fungicide sprays early in the growing season before the threshold accumulation of 18 severity values (Krause et al. 1975) and 300 P-Days (Sands et al. 1979; Pscheidt and Stevenson 1986) (Figure 2). The estimated savings related to the delayed initiation of early blight and late blight controls for Midwestern potato growers exceed $ 1.4 million per

Table 1. Origins of reduced inputs for production of long season 'Russet Burbank' potatoes using the Potato Crop Management software and associated technology

Description of Reduced Inputs	Input Reduction/Hectare	Input Value/Hectare
Elimination of two sprays using EBDC fungicide each 1.7 kg ai/ha plus application costs ($ 12.35/ha)	3.4 kg ai	$ 49.40
Reduction in fungicide rate for remaining sprays – e.g. EBDC – 18% reduction from historical use pattern of 13.4 kg/ha	2.52 kg ai	$ 18.53
Herbicide use reduction (50%) – Eliminate one spray post emergence; Reduce rates of preemergence spray	metolachlor – 1.7 kg ai; metribuzin – 0.56 kg ai	$ 56.81
Insecticide use reduction (40%) – Eliminate systemic at planting and 1–2 foliar sprays	2.8 kg ai	$ 49.40
Reduction in irrigation and energy costs – application of water to match crop needs	13.2 cm	$ 11.12
Nitrogen fertilizer reduction (20%)	56 kg	$ 22.72
Total Input Reduction	10.9 kg ai pesticide, 56 kg N, 13.2 cm water	$ 207.97

year (Figure 3). Further reductions (15–20%) in the amount of fungicide applied during the growing season are possible when growers adjust fungicide rates according to environmental conditions and the accumulation of P-Days. Additional benefits from using PCM and associated technology include: (i) reducing herbicide rates (up to 50%) and improving the timing of herbicide applications; (ii) reducing insecticide treatment(40–50% reduction) by using economic thresholds and by targeting vulnerable life stages of economically damaging insects; (iii) reducing irrigation inputs 10–15% by matching irrigation inputs to crop demand; and (iv) reducing nitrogen inputs by 20% (Table 1 and Figure 4). Potential total savings from using the PCM software and associated technology are approximately $ 208/ha. Growers and consultants who have used the software for several years appear to achieve the greatest benefits. It should be noted, however, that these savings are partially offset by the costs of soil and plant tissue analysis, scouting services and environmental monitoring. These costs appear to range from $ 37–$ 49/ha in Wisconsin, but can be higher in other regions of the US. Wisconsin growers and allied industry are currently being surveyed to determine the full extent, value and costs of using integrated pest and crop management technology. The survey will also assist in identifying barriers to further adoption of management technology.

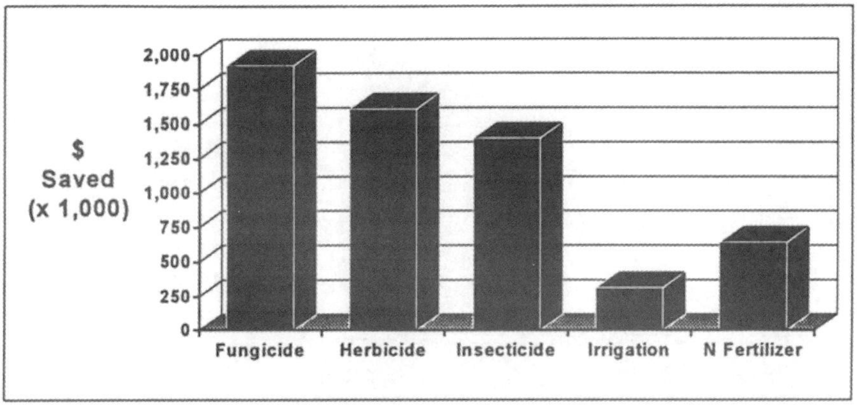

Figure 4. Projected savings from reduced inputs related to use of PCM software and associated technology on 28,300 ha in 1994.

Opportunities for enhanced use of prescriptive pest management software

Disease management

Early and late blight control. Foliar application of fungicides is directed at the control of both early (*Alternaria solani*) and late (*Phytophthora infestans*) blights. The appearance of a second mating type of *P. infestans* (A2) in many states in the US and the appearance of several strains insensitive to metalaxyl fungicide has placed added emphasis on multiple applications of protectant fungicide when weather conditions favor disease development. Potato cultivars currently grown in Wisconsin and other potato production areas of the US and Canada are susceptible to both early blight and late blight. Field evaluation of a broad spectrum of potato cultivars and breeding lines indicate potential for reducing fungicide inputs for early blight control on some cultivars and breeding lines, but the presently available cultivars do not have worthwhile levels of resistance to late blight. Field trials in Wisconsin have demonstrated the benefits of cultivar resistance to early blight for reducing disease progress in both sprayed and unsprayed blocks (Figure 5). Combining disease forecasting technology with information on cultivar susceptibility to foliar diseases would enable the grower to make informed decisions on the timing and rate of fungicide application.

Seedpiece handling. Seedpiece decay, stand loss, and loss of plant vigor pose major production risks to Wisconsin growers. The seed handling and planting aspects of potato production are among the most critical decisions faced by growers. Failure to recognize potential problems related to seedpiece decay and failure to take corrective action can result in reduced or delayed emergence. This in turn may require replanting or result in retarded growth which then affects

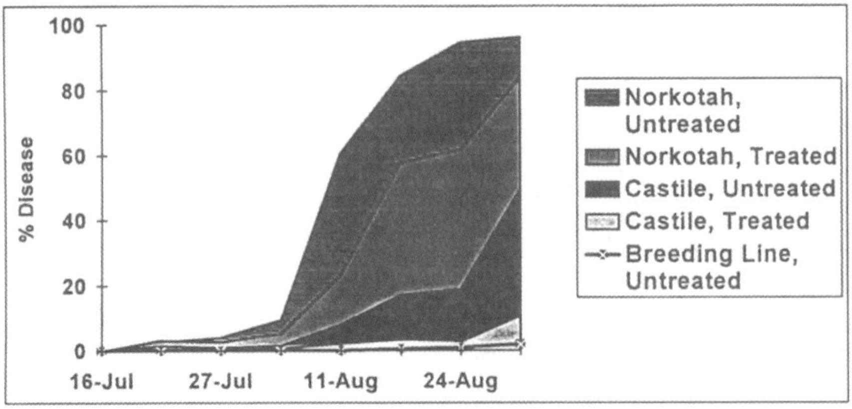

Figure 5. Changing the progress of early blight with combinations of field resistance and fungicide.

harvest scheduling, yield, marketability and crop profitability. To offset the risk of seedpiece decay, growers have traditionally coated cut-seedpieces with a fungicide dust before planting. Planting 2 metric tonnes of seedtubers/ha and applying a dust treatment at the rate of 1 kg/100 kg of seed results in each planted hectare receiving 20 kg of fungicide dust. Research at the University of Wisconsin has shown that seedpiece decay in Wisconsin is generally caused by bacteria such as *Erwinia carotovora atroseptica* and a fungicide dust treatment does little to protect the seedpiece from decay. Research has also shown that seedpiece decay is directly related to how the seedpiece is handled before and during planting and to the temperature and moisture at the time of planting and during the week after planting. Seedpiece decay is, therefore, not a random event, but rather a situation that is under the grower's control. Field and laboratory data provide the basis for development of an expert system designed to analyze information on seed handling and weather conditions and to provide site-specific recommendations for reducing the potential for seedpiece decay with minimized fungicide inputs. The addition of this expert system to the PCM program will provide an additional avenue for adoption of IPM technology and for further reduction in fungicide inputs.

Insect management. The increased interactive capacity provided by the new PCM software presents opportunities for growers to improve decisions on insect management by using real-time scouting data cross referenced with other fields and historical data. Differences in cultivar tolerance to key insect pests could be accounted for in control decisions by adjusting existing treatment thresholds to reflect higher or lower tolerance. Temporal changes in susceptibility that occur through the growth cycle could also be addressed in a similar fashion. The refinement of treatment thresholds to reflect known cultivar differences would reduce pesticide use on resistant varieties. As insecticidal inputs are decreased, the potential for integration of biological control into the general insect management program will be enhanced. Graphical presentation of real-time information, generated by field scouts, on populations of pests and beneficial insects would allow growers to assess the potential for biological regulation in individual fields and significantly increase the possibility of success. Insect resistance to chemicals continues to be a major threat to the effective management of insect pests such as the Colorado potato beetle and the green peach aphid. On-line availability of historical data on pesticide use will allow growers to follow resistance management strategies more effectively for individual species while ensuring that control decisions for other pests do not compromise such strategies.

Weed management. Weed control efforts in most years consist of pre-emergence herbicide sprays, cultivation and occasionally, post emergence sprays. The reliance on chemical controls for weed management in potatoes has directed research efforts to methods that reduce the potential environmental impacts of these applications. Recent research has focused on modelling the growth of the potato crop and evaluating the effect of crop shading on weed growth (Connell and Binning 1991; Raby and Binning 1986). A potato cultivar such as 'Russet Burbank' forms a dense canopy and provides approximately 95% shading of the soil surface seven to nine weeks after emergence. Using this knowledge, we have been able to adjust the timing and reduce the rate of herbicide applications such that only the required seven to nine weeks of control are provided from a single herbicide application. Other experiments have determined that, at these reduced rates, the potential for herbicide movement through the soil and into the groundwater is greatly reduced or is eliminated. To make this information useful for a wide range of geographical areas and different cultivars (which vary in the amount of competition they provide), we are focusing on modelling the weed growth under a variety of shading conditions. By evaluating the growth of different weed species under differing light conditions, we will be able to predict weed problems before they occur, based on historical weed geography and the amount of shading that different cultivars provide. This information can then be combined into a total weed management program to guide the grower in making environmentally and economically sound decisions.

Irrigation management. Improved management of irrigation has the potential to reduce the impact of some plant diseases, improve tuber quality and reduce the potential for leaching of pesticides and nitrogen. Over irrigation from emergence to tuberization appears to increase root infection by *Verticillium dahliae* but avoiding water stress during tuber bulking may decrease disease severity (Powelson et al. 1993). Since the activity of the common scab pathogen (*Streptomyces scabies*) is inhibited in moist soil, disease severity can be limited by maintaining 80–90% of available soil moisture during tuber initiation and bulking (Powelson et al. 1993). Soil moisture levels that deviate from the optimum during plant and tuber growth also affect the development of hollow heart, sugar ends and tuber malformation. The fine-tuning of irrigation by carefully adjusting water inputs during important phases of crop growth offers opportunities for reducing water inputs and perhaps more importantly, for improving the quality and marketability of the harvested tubers.

Nitrogen fertilizer management. Historically nitrogen fertilizer has been applied at rates totaling 280–392 kg N/ha. In most cases, yield and quality have been optimized with rates closer to 224 kg N/ha (Fixen and Kelling 1981). Recent research has shown that, through the use of a calibrated petiole nitrate-N test, the amount of early-season applied N can be reduced and later-season applications made only where the need exists. This approach results in fewer instances of excessive N being applied that can then leach to groundwater. The calibrations have been completed for several potato cultivars (Kelling and Wolkowski 1992). Subsequent research has also shown that these tests help growers to accurately account for on-farm N sources when potatoes are preceded by a forage legume (Kelling et al. 1993). This information is being incorporated into a nitrogen fertilizer management module which will help the grower to select the best rate and timing of N application for the cultivar being grown and the management options available for a given field. It will also help growers interpret in-season petiole tests for selected cultivars and, where appropriate, will recommend rates of N to apply.

Pest and crop interactions. There are often complex interactions between the crop, pests and cropping inputs which require integrated solutions. For example, plants deficient in nitrogen are more susceptible to early blight than well nourished plants. Early blight progresses rapidly on nitrogen deficient plants and prematurely defoliates infected plants even with intensive treatment with multiple fungicide sprays. Excessive irrigation promotes the leaching of nitrogen below the root zone of the potato and leads to nitrogen deficiency in the plant. Frequent and excessive irrigation also keeps the foliage wet for extended periods and thus favors spread of disease. As the nitrogen deficient plants begin to senesce prematurely, foliage is lost to early blight infection and the canopy density and shading is reduced. Light penetration stimulates weed growth which leads to weed competition for light, water and nutrients and may cause harvesting difficulties. Insects may also be attracted to stressed potato plants leading to additional defoliation. The use of prescriptive software provides opportunities to improve the management of complex problems with timely and effective remedies.

Enhancement of computer applications

The Wisconsin potato industry has requested additional program functions such as modules for soil fertility and farm record keeping that require expanded software capabilities not possible within the current PCM program in an MS-

DOS® environment. Currently, the PCM program is being converted to a Microsoft Windows™ application entitled WISDOM. This conversion allows greater flexibility in programming and it also provides many new features to the program user including:

1. An improved spreadsheet format for easier data entry and exchange of data between files: Environmental and crop data are manually entered into a spreadsheet format on the screen. Moving between days, weeks and months of the growing season is achieved by touching left or right arrows. Some growers report that they use data collected at a single monitoring station for management decisions in several nearby fields planted with different cultivars, planting dates, growth rates, etc. With the new program, data can be copied easily from one file into other files. The copy routine can be set to automatically copy newly entered data into several designated files, thus reducing data entry time for the user.
2. Presentation of data summaries in a graphical format: Information needed for grower decisions is presented as line graphs to show how various parameters (e.g. P-Days, Severity Values, Degree Days, Rainfall/Irrigation, days after emergence) change over time. Program users can be alerted when these units exceed specific thresholds so that key management decisions can proceed in a timely manner.
3. Graphical comparison of data between years: Historical data for a specific field or region can be stored in a separate file and used for comparison of production parameters with the current growing season. By using multiple comparisons per graph, program users will be able to further refine their crop and pest management activities.
4. Crop and pest record keeping: Records of field activities such as pesticide and fertilizer applications and pest scouting information can be stored for future reference. These records will play a greater significance in the future as they are linked to various program modules for tailoring specific recommendations to the occurrence of specific pests and implementation of specific management decisions. Records of pesticide application will also play useful roles in implementing strategies for managing or preventing problems of pest resistance.
5. A pictorial database of pest and crop problems: Scanned slides of insects, diseases, weeds, deficiencies, toxicities and physiological problems, stored on disk or CD-ROM, are being linked with existing software. The stored photographs, complete with descriptive information and recommendations for control can then be selected by the software user for review. This feature will help in the diagnosis of problems and in the application of the correct cultural, biological or chemical remedies.

Expanding the horizons of PCM to include the entire farming enterprise

The PCM program is comprehensive, but by design, is targeted at only a portion of the total farming enterprise, the potato crop. A need still exists to incorporate IPM information for the crops commonly used in rotation with potatoes so that growers have a guide for the entire farming system. Potato production in Wisconsin usually includes two- to three-year rotational sequences involving marketable commodities (snap beans, sweet corn, field corn, peas) and non-harvested cover crops (alfalfa, red clover, sorghum-sudan, rye). Both types of rotational crops have advantages to the whole farming enterprise. Cover crops may serve to hold the soil and nutrients in place, act as trap crops for insect pests or reservoirs for natural enemies, and have allelopathic effects on weeds and diseases. Many cover crops do not provide short-term economic returns to growers, but may promote long-term benefits to the overall enterprise which may be of equal or greater value. In contrast, marketable rotational crops provide short-term income to the grower and may also provide some additional long-term benefits. Some rotational crops, however, may contribute to the survival and increase of important pest problems and so would represent risks to the potato component of the farming enterprise.

Table 2. Six specific two- and three-year rotations involving potatoes as the crop of primary focus

Rotation	Year of Rotation					
No.	1	2	3	4	5	6
1	Potatoes	Snap Beans	Potatoes	Snap Beans	Potatoes	Snap Beans
2	Potatoes	Sorghum-Sudan	Potatoes	Sorghum-Sudan	Potatoes	Sorghum-Sudan
3	Potatoes	Sweet Corn	Potatoes	Sweet Corn	Potatoes	Sweet Corn
4	Potatoes	Snap Beans	Sorghum-Sudan	Potatoes	Snap Beans	Sorghum-Sudan
5	Potatoes	Snap Beans	Sweet Corn	Potatoes	Snap Beans	Sweet Corn
6	Potatoes	Snap Beans	Red Clover	Potatoes	Snap Beans	Red Clover

An integrated research project was initiated in 1991 to investigate the short- and long-term benefits and risks of six specific rotational sequences involving marketed and non-marketed crops grown in rotation with potatoes. The project was designed as a long-term (six-year) experiment to investigate six specific two- and three-year rotations involving potatoes (Table 2). Factors are being identified that contribute to or interfere with the preventive management of pests in the potato crop. Information developed from this research will be integrated with existing preventive and therapeutic strategies currently used in the PCM software. Expansion of the PCM software which will include modules for the management of both snap beans and sweet corn is underway. Management modules for snap beans include irrigation scheduling, a degree day calculator for predicting insect development and crop maturity and assessment of the potential the development of white mold, a disease affecting both potatoes and snap beans. Modules for sweet corn include irrigation scheduling and insect management. All of these modules will be included in an enterprise management program that addresses the improved management of potato and the crops grown in rotation with potato.

Summary

Growers and associated industry have been quick to adopt new IPM practices that enhance productivity, reduce crop inputs, and maintain their competitiveness with other major production areas. Computer technology has provided a convenient vehicle for delivering new IPM technology to the farm for implementation. Feedback from growers indicates many needs and opportunities for further adoption of IPM technology. This feedback has spawned many component and integrated research projects that will lead to additional solutions to current problems. The complexity of the PCM program has increased greatly since its first release in 1989. As we move from a single crop focus to the broader perspective of an entire farming enterprise, we are keenly aware of the risks and opportunities of this endeavor.

Acknowledgements

The authors gratefully acknowledge the support of the many undergraduate and graduate students, postdoctoral trainees and research specialists who worked with program development, growers who contributed time and facilities for program testing, staff of the UW Agricultural Research Stations where much of the component research was conducted, County Extension agents in key areas of Wisconsin potato production and independent crop consultants who continue to provide input. The authors are especially grateful to George Rice and Paul Kaarakka and Roger Schmidt, computer programming specialists, who have been instrumental in expanding the capabilities of the PCM software and who are rewriting this software in a Windows™ format.